VCE Units 3 & 4
BIOLOGY

GEORGINA BONNINGTON | KATHERINE NICHOLLS

T0360369

A+

2022-2026 STUDY DESIGN • 2022-2026 STUDY DESIGN •

PRACTICE EXAMS

+ 14 topic tests
+ two complete practice exams
+ detailed, annotated solutions

A+ Biology Practice Exams VCE Units 3 & 4
1st Edition
Georgina Bonnington
Katherine Nicholls
ISBN 9780170479448

Publisher: Alice Wilson
Project editor: Felicity Clissold
Editor: Catherine Greenwood
Proofreader: Marcia Bascombe
Cover design: Nikita Bansal
Text design: Alba Design
Project designer: Nikita Bansal
Permissions researcher: Liz McShane
Production controllers: Renee Tome, Karen Young
Typeset by: SPi Global
Reviewers: Carol-Anne Glynne, Dawn Duncan, Cathy Jackson

Any URLs contained in this publication were checked for currency during the production process. Note, however, that the publisher cannot vouch for the ongoing currency of URLs.

Acknowledgements

Selected VCE Examination questions and extracts from the VCE Study Designs are copyright Victorian Curriculum and Assessment Authority (VCAA), reproduced by permission. VCE® is a registered trademark of the VCAA. The VCAA does not endorse this product and makes no warranties regarding the correctness or accuracy of this study resource. To the extent permitted by law, the VCAA excludes all liability for any loss or damage suffered or incurred as a result of accessing, using or relying on the content. Current VCE Study Designs, past VCE exams and related content can be accessed directly at www.vcaa.vic.edu.au

For product information and technology assistance,
in Australia call **1300 790 853**;
in New Zealand call **0800 449 725**

For permission to use material from this text or product, please email **aust.permissions@cengage.com**

ISBN 978 0 17 047944 8

Cengage Learning Australia
Level 7, 80 Dorcas Street
South Melbourne, Victoria Australia 3205

Cengage Learning New Zealand
Unit 4B Rosedale Office Park
331 Rosedale Road, Albany, North Shore 0632, NZ

For learning solutions, visit **cengage.com.au**

Printed in Australia by Ligare Pty Limited.
1 2 3 4 5 6 7 25 24 23

CONTENTS

UNIT 3

UNIT 4

9780170479448

HOW TO USE THIS BOOK

The A+ Biology resources are designed to be used year-round to prepare you for your VCE Biology exam. *A+ Biology Practice Exams* includes 14 topic tests and two practice exams, plus detailed solutions for all questions in this resource. This section gives you a brief overview of the features included in this resource.

Topic tests

Each topic test is on one key knowledge area of the study design. The tests follow the same sequence of the study design, starting with the first key knowledge area of unit 3, 'The relationship between nucleic acids and proteins', and ends with the final key knowledge area of unit 4, 'Experimental design'. Each topic test includes multiple-choice and short answer questions.

Practice exam section

Both practice exams cover all of units 3 and 4 of the *VCE Biology 2022–2026 Study Design*. The practice exams have perforated pages so that you can remove them from the book and practice under exam-style conditions.

Solutions

Solutions to topic tests and practice exams are supplied at the back of the book. They have been written to reflect a high-scoring response and include explanations of what makes an effective answer.

Explanations

The solutions section includes explanation of each multiple-choice option, both correct and incorrect, and explanations to written response items explain what a high-scoring response looks like and signposts potential mistakes.

Multiple-choice solutions

Question 1   A+ Study Notes p.11

D 21%

Adenine (29%) pairs with thymine (29%) and the remaining 42% of DNA consists of guanine (21%) pairing with cytosine (21%). **A** is incorrect because 29% would be thymine as it is complementary to adenine. **B** is incorrect because 42% is the total remaining DNA % required for both guanine and cytosine. **C** is incorrect because 29% × 2 and 42% × 2 do not add up to 100%.

Short-answer solutions

Question 1   A+ Study Notes p.11

a DNA (deoxyribonucleic acid)

This is because there are two strands bonded in an antiparallel manner and thymine is present.

Icons

You will notice the below icons occurring in the topic tests and solutions section.

This icon appears next to official past VCAA questions.

This icon signifies the difficulty of each question in the topic tests. One of these icons appears next to all questions to indicate whether the question is easy, medium or hard.

This icon appears in the solutions and signifies which topic the topic test is covering.

This icon appears in the solutions section and shows you what page of *A+ Biology Study Notes* you can find revision notes for each question.

About A+ Study Notes

A+ Biology Practice exams can be used independently, or alongside the accompanying resource *A+ Biology Study Notes*. *A+ Biology Study Notes* includes topic summaries and exam practice for all key knowledge in the *VCE Biology Study Design 2022–2026* that you will be assessed on during the exam, as well as detailed revision and exam preparation advice to help get you ready for the exam.

 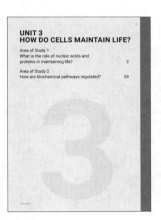 

A+ DIGITAL FLASHCARDS

Revise key terms and concepts online with the A+ Flashcards. Each topic for this subject has a deck of digital flashcards you can use to test you understanding and recell. Just scan the QR code or type the URL into your browser to access them.

https://www.get.ga/
a-biology-vce-u34

Note: You will need to create a free NelsonNet account.

ABOUT THE AUTHORS

Katherine Nicholls

Kate Nicholls completed her undergraduate degree in Science at the University of Melbourne, majoring in Pathology before a short period in research studying colorectal cancer. After this research she returned to University of Melbourne to undertake her Masters of Teaching. Since completing this qualification, she has taught Junior Science and VCE Biology for the last four years at University High School in Melbourne. Kate was an assessor of the 2020 VCAA VCE Biology exam.

Georgina Bonnington

Georgina began her teaching career in the UK, where she was head of Biology for five years at a prestigious girls school in Surrey. Since moving to Melbourne in 2013 she has taught Biology at Wesley College and Camberwell Girls' Grammar School, as well as assessing the VCAA VCE Biology exam.

Test 1: The relationship between nucleic acids and proteins

Section A: 20 marks. Section B: 67 marks. Total marks: 87 marks.
Suggested time: 45 minutes

Section A: Multiple-choice questions

Instructions to students
- For each question, circle the multiple-choice letter to indicate your answer.

Question 1

The DNA of the common mouse *Mus musculus* was determined as containing 29% adenine bases. What percentage of the mouse's bases would be guanine?

A 29%

B 58%

C 42%

D 21%

Question 2

There are structural differences between the nucleic acid in cells. Which of the following correctly shows these differences?

	DNA	RNA
A	Double-stranded molecule with a ribose sugar in the backbone	Single-stranded molecule with a deoxyribose sugar in the backbone
B	Contains four nitrogenous bases	Contains five nitrogenous bases
C	Two nucleotide polymers joined together by hydrogen bonds	Polymer of nucleotides joined by phosphodiester bonds in the backbone
D	A polymer of nitrogenous bases formed through condensation reaction	A monomer of nitrogenous bases formed through hydrolysis

Question 3

Some of the events involved in the production of a protein are listed below.

V: Production of mRNA from a DNA template

W: Attachment of mRNA to a ribosome

X: Transfer of mRNA to the cytoplasm

Y: Formation of peptide bonds between amino acids

Z: Loaded tRNA enters with a complementary anticodon to the codon

The correct sequence of events is

A V → Z → Y → X → W

C V → X → W → Z → Y

B Z → W → X → Y → V

D V → X → Y → W → Z

Question 4 ○●●

The diagram at the right shows a nucleic acid monomer.

What happens when this type of molecule forms a polymer?

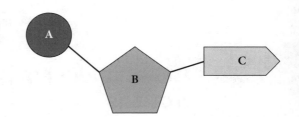

A Bonds are formed between chemical A of one monomer and chemical C of the next monomer.

B Bonds are formed between chemical A of one monomer and chemical B of the next monomer.

C Bonds are formed between chemical C of one monomer and chemical B of the next monomer.

D Bonds are formed between chemical B of one monomer and chemical B of the next monomer.

Question 5 ●●●

The graph shows the folding time of proteins with different numbers of amino acids.

Which of the following statements is correct?

A The folding time decreases as the number of amino acids decreases.

B The folding time increases as the number of amino acids decreases.

C The folding time depends on the type of amino acids present.

D There is no relationship between the folding time and the number of amino acids.

Folding time of proteins

(graph: y-axis "Time (microseconds)" marked 200, 400, 600, 800, 1000, 1200, 1400, 1600, 1800; x-axis "Number of amino acids" marked 500, 1000, 1500, 2000, 2500, 3000)

Question 6 ©VCAA VCAA 2009 (2) SA Q20 ●●●

Regulatory and structural genes differ in their arrangement in the genomes of prokaryotic and eukaryotic cells. In prokaryotic cells, regulatory genes are arranged side by side. This arrangement is known as an operon. In eukaryotic cells, the regulatory genes and structural genes may be located on different chromosomes. Therefore, it would be reasonable to say that

A eukaryotic stem cells have all genes switched on.

B the environment has no impact on whether a gene is switched on or off.

C all bacterial operons are located on a large circular chromosome within the cell.

D mutations in distant regulatory genes will have no effect on their related structural genes in eukaryotic cells.

Question 7 ©VCAA VCAA 2018 SA Q4 ○●●

The genetic code is described as a degenerate code. This means that

A in almost all organisms the same DNA triplet is translated to the same amino acid.

B some amino acids may be encoded by more than one codon.

C a single nucleotide cannot be part of two adjacent codons.

D three codons are needed to specify one amino acid.

Question 8 ⬤⬤⬤

The genome of a prokaryote cell, such as *E. coli*, contains all the genetic information needed by the bacteria to metabolise, grow and reproduce. Many of the genes are always turned on; however, others are active only when their products are needed by the cell.

If the amino acid tryptophan is added to a bacterial culture, the bacteria soon stop producing the five enzymes needed to synthesise tryptophan from intermediates. In this case, which of the following statements is true?

A Tryptophan promotes transcription of the enzymes involved.

B Translation of the enzymes is not affected by tryptophan.

C Tryptophan is not needed by the *E. coli* cell.

D The presence of tryptophan represses enzyme synthesis.

Question 9 ⬤⬤⬜

The *trp* operon is classified as

A a repressible system that is active if *trp* is not present.

B an inducible system that is active if *trp* is present.

C an inducible system if only lactose is present for the cells' energy source.

D a repressible system if other operons are being expressed in the bacteria.

Question 10 ⬤⬤⬤

The ACU codon in the leader section of the *trp* operon is transcribed in the mRNA as UGA. This is important in successful attenuation because

A UGA codes for a stop codon causing the ribosome to stop and overlap three sections of the mRNA preventing the formation of hairpin loops.

B UGA codes for a stop codon which stops the ribosome and enables a hairpin loop to form which pulls the mRNA away from the DNA thereby terminating transcription.

C the codon codes for the amino acid tryptophan and the ribosome pauses and waits until tRNA brings tryptophan to join to the growing polypeptide chain.

D the codon codes for the amino acid tryptophan and as there is no tryptophan in the cell the ribosome will stop thereby terminating translation.

Question 11 ⬤⬤⬜

The one gene, one polypeptide theory has been superseded by the one gene, many polypeptides theory of gene action in eukaryotes. According to one gene, many polypeptides theory, in order to produce all of the proteins found in a cell, it is necessary for

A different enzymes to read the mRNA.

B the mRNA to be read in either direction.

C there to be different editing of the pre-mRNA.

D the mRNA to mutate to produce all the types needed.

Question 12 ⬤⬜⬜

While examining a sample of a nucleic acid taken from a cell, the following sequence of bases was established:

U C A C C A U G U C A C U C C A G U

This nucleic acid sample could have been

A a DNA sample from a prokaryotic cell.

B part of a DNA molecule that codes for a protein.

C a segment of RNA that codes for a phospholipid.

D an RNA molecule found in the cytoplasm of a eukaryotic cell.

Question 13 ⬤⬤⬜

Regulator genes are genes that

A are responsible for the production of all proteins.

B produce chemicals that control the action of other genes.

C regulate the rate of lipid synthesis.

D are involved in the splicing of introns from a gene.

Question 14 ⬤⬜⬜

The three-dimensional structure of a protein

A is determined by its sequence of amino acids.

B varies depending on the interaction of the protein molecule with other molecules.

C is irrelevant to the function of the protein molecule.

D is determined by the active site.

Question 15 ⬤⬤⬜

One section of a polypeptide has the amino acid sequence

Ala–Cys–Lys–Ile–Asn

The codons for these amino acids are

Ala	GCA, GCC, GCG or GCU
Cys	UGC or UGU
Lys	AAA or AAG
Ile	AUU, AUC or AUA
Asn	AAC or AAU

The sequence of DNA coding for this section of the polypeptide could be

A CGTACGTTTTATTTG

B CGTTCGTTTTATTTG

C CGTACTTTTTACTTG

D CGAACATTCTATTTT

Question 16 ⚫⚫⚫

The diagram at the right shows a section of the process involved in protein production within in a prokaryotic cell.

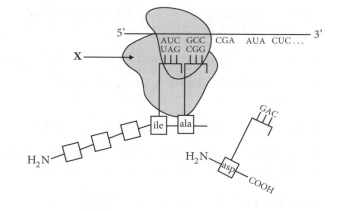

From the information above, it is reasonable to conclude that the

A amino acids brought to molecule X are connected to tRNA. The specific tRNA contains a codon that aligns with the mRNA strand.

B polymer being formed at molecule X is a polypeptide chain formed through an anabolic reaction.

C reaction between Ile and Ala is a hydrolysis polymerisation reaction.

D anticodons on the 5'–3' sequence at molecule X code for specific amino acids.

Question 17 ©VCAA VCAA 2007 (2) SA Q11 ⚫⚫⚫

A haemoglobin molecule is composed of four protein (globin) chains each attached to an iron-containing haem group. Two are identical alpha chains and two are identical beta chains. The following diagram is a stylised representation of a haemoglobin molecule.

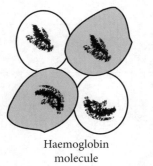

Haemoglobin
molecule

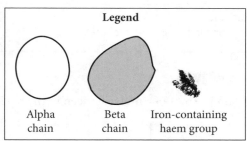

A haemoglobin molecule would be classified as having a

A primary protein structure.

B secondary protein structure.

C tertiary protein structure.

D quaternary protein structure.

Question 18 ⚫⚫⚫

Calcitonin is a protein hormone that regulates blood calcium levels. It is secreted by the thyroid gland. Which of the following correctly represents the pathway synthesised molecules of calcitonin take before they leave a thyroid cell?

A nucleus → mitochondrion → Golgi body → vesicle

B ribosome → Golgi body → endoplasmic reticulum → membrane

C ribosome → endoplasmic reticulum → Golgi body → vesicle

D nucleus → ribosome → endoplasmic reticulum → Golgi body

Question 19 ●●●

A human body is made up of many different types of cells. Would a skin cell have the identical proteome as a white blood cell?

A Yes, all the genes in every cell are expressed but only some of the proteins become functional.

B Yes, a skin cell and a white blood cell have similar functions and therefore would require similar proteins.

C No, different genes are expressed in different cells depending on the function of the cell.

D No, there would be a different set of chromosomes in a skin cell compared to a white blood cell and therefore different proteins would be produced.

Question 20 ©VCAA VCAA 2017 SA Q1 ●●●

Consider the structure and functional importance of proteins. Which one of the following statements about proteins is correct?

A A change in the tertiary structure of a protein may result in the protein becoming biologically inactive.

B Proteins with a quaternary structure will be more active than proteins without a quaternary structure.

C Two different proteins with the same number of amino acids will have identical functions.

D Denaturation will alter the primary structure of a protein.

Section B: Short-answer questions

Instructions to students
• Answer all questions in the spaces provided.

Question 1 (6 marks) ●●●

The following diagram is a part of a molecule seen in the nucleus of cells.

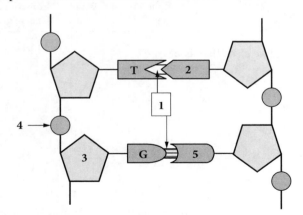

a What name is given to this molecule?   1 mark

_____

b Write the names of parts 1–5.   2 marks

_____

_____

_____

c The diagram is composed of many monomers to produce a polymer. Circle a monomer.   1 mark

d Explain the process and the bonds involved in forming the polymer in the diagram.   2 marks

_____

_____

_____

Question 2 (5 marks)

Part of a nucleic acid molecule that codes for the protein insulin was extracted from a human pancreatic cell and examined in a laboratory. The sequence of bases in the molecule was determined as:

TAC ACA GGC GAT AGG CCG

a Which type of nucleic acid is the sequence above? Justify your response.   1 mark

_____

_____

b Where in the cell would the above molecule be found?   1 mark

_____

c What would be the sequence of anticodons on the tRNA and the amino acid sequence if the strand above were the template strand?   2 marks

_____

_____

_____

d The lab had previously analysed genetic material from bacteria. The sequence of bases TAC in human DNA codes for the amino acid methionine. Explain whether the sequence of bases TAC would code for methionine in the bacterial cells.   1 mark

_____

_____

_____

_____

Question 3 (9 marks) ©VCAA VCAA 2017 SA Q8–10 (adapted) ●●●

The following diagram outlines the production of protein in a cell when DNA is inactivated.

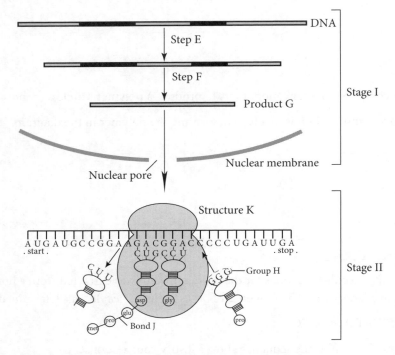

a Name and explain stage II. Refer to bond J, group H and structure K in your answer.   4 marks

_____

_____

_____

_____

_____

b i Explain **two** differences in the composition of the DNA strand compared to product G.   2 marks

_____

_____

_____

ii Outline **one** way in which RNA processing can affect protein expression.   2 marks

_____

_____

_____

c Would this process be in a eukaryotic or in a prokaryotic cell? Justify your response.   1 mark

_____

_____

Question 4 (7 marks)

Use the mRNA codon chart to analyse the sequence below.

- Series 1 represents four mRNA codons.

- Series 2 represents the same sequence of codons transcribed from a DNA molecule that has undergone a mutation.

Series 1: GGA GCG GUC CCU

Series 2: GCA GCG GUC CCU

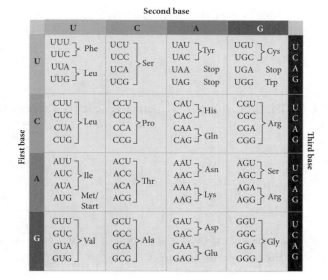

a i For series 1, which amino acids would be incorporated into the polypeptide chain?   1 mark

ii What is the sequence of nucleotides in the mutated DNA that acted as the template molecule for the mRNA molecule shown as series 2?   1 mark

b i What type of mutation has occurred and how has this altered the polypeptide chain?   2 marks

ii Would this have an effect on the protein function?   1 mark

c Explain whether there are other base sequences in DNA that code for proline; if so, provide an example. What are the consequences for the DNA sequence?   2 marks

Question 5 (11 marks)

a Explain what is meant by a 'structural gene'.   1 mark

b Define 'operon'.

1 mark

_____

_____

c What is the role of a regulatory gene in the expression of genes?

2 marks

_____

_____

_____

d Bacteria are exposed to tryptophan in their environment. Explain how the expression of the _trp_ operon is repressed.

2 marks

_____

_____

_____

_____

_____

e A mutation occurred in the repressor gene of the bacterial cell. Explain how this will affect the _trp_ operon's functioning and provide a potential long-term side effect for the bacteria.

2 marks

_____

_____

_____

_____

f The bacteria obtained a plasmid with a functional repressor gene embedded within. Justify whether this would be effective in repressing the _trp_ operon in the presence of _trp_.

1 mark

_____

_____

_____

_____

g Contrast the processes of repression and attenuation in the *trp* operon when there is a high level of tryptophan present in the cell.

2 marks

_____

_____

_____

_____

Question 6 (5 marks) ©VCAA VCAA 2013 SB Q2 ●●

Microtubules are hollow structures composed of the protein tubulin, which has two forms: alpha-tubulin and beta-tubulin. A tubulin dimer is formed when one alpha-tubulin molecule and one beta-tubulin molecule join. Tubulin dimers polymerise into long chains to form protofilaments. A microtubule can be formed when 13 protofilaments align side by side.

a With respect to the structure of a protofilament, explain what is meant by the term 'polymerise'.

1 mark

_____

_____

b Consider an alpha-tubulin molecule. Explain the difference between its primary structure and secondary structure.

2 marks

_____

_____

c Describe what is meant by 'tertiary' and 'quaternary' protein structures.

2 marks

_____

_____

_____

Question 7 (6 marks) ●●

Sickle cell anaemia is an inherited disorder that affects the ability of red blood cells to transport oxygen to cells in the body. Instead of being concave shaped and flexible, the red blood cells are shaped like crescent moons. These rigid structures slow the movement of blood and can become lodged within the capillaries where gas exchange occurs, resulting in blood clots and loss of oxygen to certain parts of the body.

Chemical studies have shown that the abnormal haemoglobin molecule differs from a normal haemoglobin molecule by a single amino acid. There are 147 amino acids in the polypeptide chain of both the normal and mutant haemoglobin.

a   Which of the level(s) of protein hierarchy is affected by this change? Justify your response.   2 marks

_____

_____

_____

A section of the coding sequence of DNA is: ATG GTG CAC CTG ACT CCT GAG GAG.

b   What is the mRNA sequence and amino acid sequence for the functional protein?   2 marks

	Bonds and interactions
mRNA	
Amino acid sequence	

c   The mutation in haemoglobin is at the third codon in the section shown, where cytosine is replaced with guanine. What is the outcome for the protein and how will this affect the protein structure, if at all?   2 marks

_____

_____

_____

_____

Question 8 (8 marks) ◼◼◻

The pathway that leads to the release of protein from a secretory cell is shown in the following diagram.

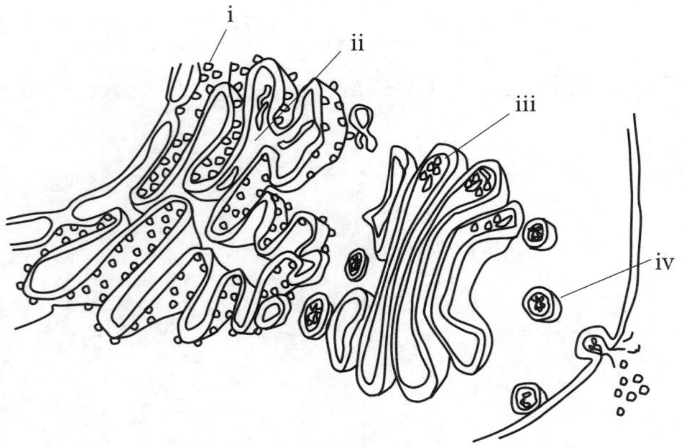

a   Name and explain the process by which the protein is released from the cell.   2 marks

_____

_____

_____

b Name the organelles and explain their role in protein production.   4 marks

	Name	Function
i		
ii		
iii		
iv		

c The organelles in the diagram have been isolated from a pancreatic cell that produces large amounts of the protein insulin, which is released into the bloodstream to lower blood glucose levels after an individual has eaten. Name **one** other organelle required in large quantities for this process and explain why this organelle would be required.   2 marks

_____

_____

_____

_____

Question 9 (4 marks) ⬤⬤⬤

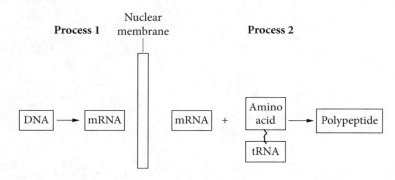

a State the name of process 1 in the diagram above.   1 mark

_____

b Explain the role of transcriptional factors in the expression of a gene.   1 mark

_____

_____

_____

c Explain **two** ways DNA differs in structure from mRNA.   2 marks

_____

_____

_____

_____

Question 10 (6 marks) ●●●

Promoter	Intron 1	Exon 1	Intron 2	Exon 2	Intron 3	Exon 3

a What is the structure and function of a promoter region in a eukaryotic organism?   1 mark

_____

_____

b The proteome in a eukaryotic organism is greater than the genome. Using the information in the image above, explain how this occurs. Use a diagram to support your answer.   2 marks

_____

_____

_____

_____

c Explain the process that forms pre mRNA.   3 marks

_____

_____

_____

_____

_____

Test 2: DNA manipulation techniques and applications

Section A: 20 marks. Section B: 66 marks. Total marks: 86 marks.
Suggested time: 45 minutes

Section A: Multiple-choice questions

Instructions to students
- For each question, circle the multiple-choice letter to indicate your answer.

Question 1

To build up resistance to disease in agriculture, scientists have isolated genes that code for defence mechanisms from particular species and inserted them into crop species, forming transgenic organisms. Which of the following is an example of a transgenic crop that would increase resistance to disease?

A Golden rice with a daffodil gene insertion to provide increased vitamin A

B Gene removal in tomatoes to increase yield and prevent softening of the fruit

C Corn modified with an insecticide gene from bacteria to produce a toxin to prevent infection

D Gene removal in cotton to prevent growth of a fungus in the roots of the plant

Question 2

PCR is a process of DNA amplification. A range of temperatures is used. The correct sequence of processes and temperatures is

A annealing 72°C, denaturing 95°C, extension 55°C.

B annealing 55°C, extension 72°C, denaturing 92°C.

C denaturing 95°C, annealing 55°C, extension 72°C.

D denaturing 72°C, annealing 91°C, extension 55°C.

The following information relates to Questions 3 and 4.

Scientists use plasmids, which are naturally occurring in bacteria, as vectors for genetic engineering. Below is a plasmid map indicating the restriction sites of four different restriction enzymes on a particular plasmid.

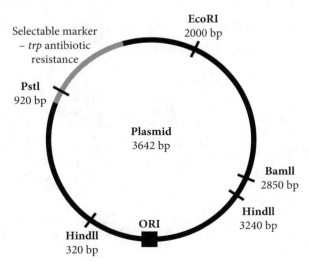

Question 3 ●●●

Which of the following restriction enzymes would **not** be useful to genetic engineering?

A HindII and BamII

B EcoRI and PstI

C HindII and PstI

D PstI and BamII

Question 4 ●●●

A scientist places the plasmid in a test tube with the restriction enzyme BamII and EcoRI. The fragment lengths that form would be

A 2000 bp and 2850 bp.

B 850 bp and 2792 bp.

C 2000 bp, 850 bp and 792 bp.

D 2000 bp, 920 bp and 1850 bp.

Question 5 ●●

DNA profiling can be used to determine parentage. Part of the DNA profiles of a mother (M) and her child (C) are shown below. The other two samples (S1 and S2) are from two men, either of whom may be the child's father.

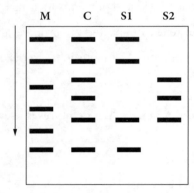

On the basis of these DNA profiles, what is a reasonable conclusion?

A Either man could be the child's father.

B S1 could be the father but S2 could not.

C S2 could be the father but S1 could not.

D Neither of the men is the child's father.

Use the following information to answer Questions 6 and 7.

Genetic engineers use restriction enzymes to cut DNA into smaller lengths. The recognition sequences of several restriction enzymes are shown in the table below. The symbol * denotes the restriction site (position of the cut).

Restriction enzyme	Recognition sequence (read in 5' to 3' direction)				
EcoRI	G*	A	A	T	T	C
	C	T	T	A	A	*G
HindIII	A*	A	G	C	T	T
	T	T	C	G	A	*A
AluI		A	G*	C	T	
		T	C*	G	A	
HaeIII		G	G*	C	C	
		C	C*	G	G	

Question 6 ©VCAA VCAA 2013 SA Q29 ●●●

Consider a length of double-stranded DNA with the sequence

5' T T A A G G A A T T C A A 3'

3' A A T T C C T T A A G T T 5'

Adding EcoRI to a solution containing one copy of this double-stranded DNA produces

A two fragments of double-stranded DNA, each with a sticky end.

B four fragments of single-stranded DNA, each with a sticky end.

C two fragments of double-stranded DNA, each with blunt ends.

D four fragments of single-stranded DNA, each with blunt ends.

Question 7 ©VCAA VCAA 2013 SA Q30 ○●●

Now consider a different length of double-stranded DNA with the sequence

5' C T T A A G C T T C C A A A T T A C C G A 3'

3' G A A T T C G A A G G T T T A A T G G C T 5'

Which enzyme(s) will cut this piece of DNA?

A EcoRI only

B HindIII only

C AluI and HindIII only

D AluI, HindIII and HaeIII only

Use the following information to answer questions 8 and 9.

Question 8 ●●●

Bacteriophages are viruses that infect bacteria. Over thousands of years, bacteria have developed a protective mechanism to prevent future infections by the virus. After a virus injects its DNA into a bacterial cell, the bacteria fragment the DNA into small sections and embed them into their genome for future exposure to the strain – the bacteria's genetic memory of the infection. The genetic material is inserted into their

A CAS gene segment.

C spacer sequences.

B repeat sequences.

D promotor regions.

Question 9 ●●

If the bacterium is reinfected by the same bacteriophage at a later time it calls on its genetic memory to quickly destroy the invading bacteriophage DNA. How does the bacterium differentiate between the bacteriophage DNA stored in its genetic memory and the same DNA from the invading bacteriophage?

A The genetic memory of the bacteriophage DNA is kept into a separate compartment in the cell and so is isolated from the bacterial DNA.

B Invading bacteriophage DNA contains small segments called protospacer adjacent motif which identify it as self.

C Bacterial DNA contains small segments called protospacer adjacent motif which identify it as self.

D Invading bacteriophage DNA contains small segments called protospacer adjacent motif which identify it as non-self.

Question 10 ○○○

CRISPR-Cas9 technology has been harnessed by scientists to modify genes in species such as fruit flies, mice and plants to remove defective genes at a specific location and replace them with a functional gene. The system involves the restriction enzyme Cas9 and single guide RNA. The function of the single guide RNA is to

A replace the gene sequence with RNA within the genome.

B cut the DNA at a specific recognition sequence.

C silence the non-functional gene and insert additional DNA into the nucleus.

D direct the Cas9 to the specific gene sequence.

Question 11 ○○

Before a region of a DNA molecule can be amplified by PCR, which of the following must be known?

A Its full DNA sequence

B The number of adenine–thymine base pairs

C The identity of the individual that the sample came from

D A portion of its DNA sequence

Question 12 ○○

When a piece of DNA is introduced into a foreign species to correct an inborn error of metabolism or to produce a large quantity of a protein, the piece of DNA must contain the

A gene only.

B gene and the upstream promoter.

C gene and the downstream repeats.

D gene and an enzyme to integrate it.

Question 13 ©VCAA VCAA 2007 (2) SA Q25 ○○○

A soil bacterium (*Agrobacterium tumefaciens*) infects roses and fruit trees, stunting their growth. A similar bacterium (*A. radiobacter*) was genetically modified to include a plasmid gene coding for an antibiotic lethal to *A. tumefaciens*. *A. radiobacter* has a gene giving resistance to this antibiotic. A transfer gene located on the plasmid enables insertion of the modified gene into *A. radiobacter*. Suspensions of genetically modified bacteria are applied to the soil around plants. For this treatment to be successful, genetic modification of *A. radiobacter* would need to include

A removal of all plasmids.

B removal of the antibiotic gene.

C destruction of the antibiotic resistance gene.

D removal of the plasmid transfer gene.

The following diagram of a DNA digest that is being run on a gel relates to Questions 14–16.

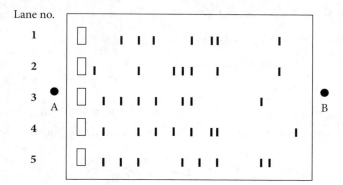

Question 14 ⬤◯◯

The terminals on the gel labelled 'A' and 'B' should show which of the following charges?

A A should be positive and B should be negative.

B A should be negative and B should be positive.

C A should be positive and B should be neutral.

D A should be negative and B should be neutral.

Question 15 ⬤◯◯

The lightest fragment of DNA is found in

A lanes 3 and 4.

B lanes 1 and 2.

C lane 2 only.

D lane 4 only.

Question 16 ⬤⬤⬤

The DNA in lane 3 is from the natural child of parents in lanes

A 1 and 2.

C 4 and 5.

B 1 and 5.

D 2 and 5.

Question 17 ©VCAA VCAA 2014 SA Q25 ⬤⬤◯

Plasmids of bacteria are used to transfer selected genes from one species to another.

The process is represented below.

Bacterial plasmid ⟶ foreign gene and plasmid mixed ⟶ plasmid with inserted foreign gene

Enzymes are used to facilitate these steps. Which of the following shows the enzymes required for the first and last steps of the process?

	Cut plasmid	Inserts genes
A	Restriction enzyme	DNA ligase
B	Restriction enzyme	DNA polymerase
C	DNA ligase	DNA polymerase
D	DNA polymerase	DNA ligase

The following information relates to Questions 18–20.

To clone a gene of interest, the following four steps are performed.

1 A plasmid is cut with a specific restriction enzyme.

2 The gene of interest is ligated into the plasmid.

3 Plasmids are transferred to bacteria.

4 Bacteria are grown on four nutrient agar plates (labelled W, X, Y and Z) that are coated with or without ampicillin and arabinose. An example of a plasmid used in cloning is shown below.

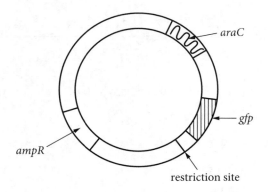

This plasmid contains a restriction site and the following three genes:

- *ampR* – confers resistance to the antibacterial agent ampicillin

- *gfp* – encodes the green fluorescent protein (GFP), which fluoresces under UV light

- *araC* – encodes a protein required to promote the expression of *gfp* when arabinose is present.

The results from a bacterial transformation experiment are shown in the table below.

Plate	W untransformed bacteria only	X untransformed bacteria only	Y transformed bacteria	Z transformed bacteria
Diagram of plate				
Added to plate	nutrient agar only	nutrient agar and ampicillin	nutrient agar, ampicillin and arabinose	nutrient agar and ampicillin
Description of result	lawn of bacteria	no growth	bacterial colonies present	bacterial colonies present

Question 18 ©VCAA VCAA 2015 SA Q25 ●●●

Bacteria are used in gene cloning because they

A contain restriction enzymes that randomly cut chromosomes into fragments of varying size.

B can replicate non-bacterial sequences of DNA in a short time.

C replicate exponentially by undergoing mitotic divisions.

D allow the entry of foreign DNA into their nuclei.

Question 19 ©VCAA VCAA 2015 SA Q26 ●●

Which plate would contain bacteria that fluoresce under UV light?

A Plate W

B Plate X

C Plate Y

D Plate Z

Question 20 ©VCAA VCAA 2015 SA Q27 ●●

Which one of the following statements is an accurate description for the purpose of plate W or X?

A Plate W shows that the plasmid was cut with a restriction enzyme.

B Plate W shows that the percentage of transformed bacteria was high.

C Plate X shows that the nutrient agar promoted the growth of viable bacteria.

D Plate X shows that ampicillin was effective in killing the untransformed bacteria.

Section B: Short-answer questions

> **Instructions to students**
> - Answer all questions in the spaces provided.

Question 1 (6 marks) ●●

A crime lab isolated two samples of blood from a crime scene. To rule out the victim's blood, the lab ran gel electrophoresis on the two samples (lanes 2 and 3) and the victim's sample (lane 4) after they performed PCR on the samples. Lane 1 contained a series of 10 fragments ranging in size from 500 to 1000 bp.

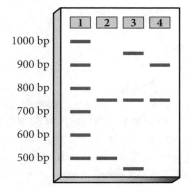

a   Explain the function of the fragments in lane 1.   1 mark

_____

_____

_____

_____

b   Explain **two** factors that would affect the rate of movement of the DNA, causing it to
migrate in gel electrophoresis.   2 marks

_____

_____

_____

c   The investigators left the gel running for 4 hours before returning and viewing it under a
UV light. They determined that the results were conclusive and could be used to determine
if either sample belonged to the victim. Is this correct?   2 marks

_____

_____

_____

_____

d   In another lab, an investigator ran a gel with one thick band at the top of the gel, but no other
bands. What did the investigator forget to do before inserting the DNA into the wells?   1 mark

_____

_____

_____

Question 2 (4 marks) ◼◼◻

Breast cancer is one of the top five cancers affecting women in Australia. Two genes – the *BRCA1* and
BRCA2 genes – when mutated are known to increase a woman's risk of developing cancer. When these
genes are functioning, they produce proteins that repair damaged DNA during a cell cycle. A potential
new drug treatment for the *BRCA1* and *BRCA2* gene mutations involves inserting the functional gene into
the at-risk patient using a plasmid vector.

a Define 'vector' in the given context.   1 mark

_____

_____

b What are the steps involved in isolating and inserting the gene of interest into the plasmid?   2 marks

_____

_____

_____

_____

_____

_____

c What is **one** advantage of using a plasmid for inserting a gene into an individual rather than a linear piece of DNA?   1 mark

_____

_____

_____

Question 3 (7 marks)

As temperatures increase around the world, farmers are experiencing pressure to produce more yield on the same amount of land with less water available. Due to the continued growth of the population, many people living on the poverty line cannot afford fresh food and many countries are experiencing an increase in the number of preventable deaths. A biotechnological company produced a genetically modified rice that was higher in nutritional value, accessible to farmers and resistant to pests to ensure a high yield.

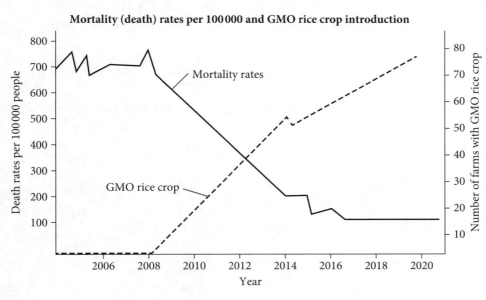

Mortality (death) rates per 100 000 and GMO rice crop introduction

a   Explain the relationship between mortality rates and the introduction of the GMO
    rice crop into the area.   2 marks

_____

_____

_____

_____

b   Suggest a reason for the peak in the death rates in 2008.   1 mark

_____

The biotechnology company inserted two genes into the rice – a daffodil gene and a soil bacterium gene –
to increase the levels of beta carotene.

| The two genes are inserted into a plasmid with promoters and taken up by bacteria. | → | Transformed bacteria are isolated and cultured. | → | Bacteria are ruptured and recombinant plasmids are isolated. | → | Plasmids are added to the petri dish with the embryos of the rice. | → | The transgenic rice embryo is bred with the local strain. |

c   What type of organism would this rice be classified as? Justify your response.   1 mark

_____

d   Explain how the biotech company would ensure only the recombinant plasmids were used for
    uptake by the embryo.   2 marks

_____

_____

e   Explain a benefit of crossing the embryo with the plasmid with the local strain of rice.   1 mark

_____

_____

_____

Question 4 (5 marks)

Sickle-cell anaemia is the result of an inherited mutation that causes the red blood cell to have an abnormal sickle shape, which effects the haemoglobin protein and makes it less efficient at carrying oxygen around the body. A DNA test can detect the presence of the defective form of the haemoglobin gene. DNA is cut by a restriction enzyme. This enzyme cuts the normal haemoglobin gene but is unable to recognise or cut the sickle-cell gene because of its altered base sequence. Each person has two forms of every gene. If an individual has two of the abnormal genes, then they cannot form normal red blood cells. A carrier of the abnormal gene has one functional gene and one non-functional gene, and a person with two normal genes has fully functioning red blood cells. When the fragments of DNA have been sorted by gel electrophoresis, they are treated with a radioactive DNA probe, which binds to all fragments of the beta-globin gene. The diagram below shows DNA samples taken from two people (A and B).

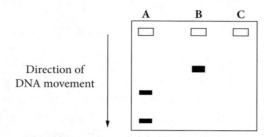

Direction of
DNA movement

a   **i**   Name the process that is used in the lab to amplify the DNA sample required for the testing.   1 mark

_____

  ii   Using the diagram, explain the process.   2 marks

_____

_____

_____

b   Using the results of the gel run, explain which individual, A or B, does not have
sickle-cell anaemia.   1 mark

_____

_____

c   In the column labelled C in the diagram, draw the banding pattern you would expect
to find for the DNA of a person who is a carrier of the sickle-cell condition.   1 mark

Question 5 (5 marks)

Non-scientists often confuse the terms 'transgenic organisms' and 'genetically modified organisms' and use the terms incorrectly.

a   Compare a genetically modified organism to a transgenic organism.   2 marks

_____

_____

_____

_____

_____

b For each of the following examples, state the organism modified and the type of modification (transgenic or genetically modified) and explain the reason for your answer.   3 marks

	Example	Organism modified	Transgenic or genetically modified	Reason
i	A rat with rabbit haemoglobin genes			
ii	A human treated with insulin produced by *E. coli* bacteria			
iii	A rice plant that has had a rice gene knocked in to produce double the amount of rice protein			

Question 6 (8 marks) ⬤⬤⬤

Cystic fibrosis (CF) is a genetically inherited disease caused by a mutation in the CF transmembrane conductance regulator gene (CFTR). The disease mainly affects the lungs by a build-up of fluid. Although there have been large developments in the management of the disease to extend the lifespan of people with CF, there is still no cure and the expected life span is 30–40 years. The possibility of curing these types of diseases with CRISPR technology could extend people's lives and reduce the pressure on the healthcare system.

a Explain how the functional gene could be inserted into a patient suffering from cystic fibrosis.   2 marks

_____

_____

_____

_____

_____

b Discuss a potential risk with this type of technology.   1 mark

_____

_____

The CRISPR system was originally isolated from bacteria and was used as a defence mechanism against bacteriophages (viruses).

Cas genes	Promoter	A	B	A	B	A

c Explain the function of the promoter upstream of the regions A and B.   1 mark

_____

d Name and explain the functions of regions A and B in the CRISPR system.   4 marks

_____

_____

_____

Question 7 (7 marks) ⬤⬤⬤

Human insulin is made up of two separate polypeptide chains: A chain and B chain. These two chains are bonded together by disulfide bonds to form a functional insulin molecule. A chain and B chain are coded for by separate genes.

a What type of protein structure is the human insulin molecule?   1 mark

_____

Bacteria can be transformed with the insulin genes and grown in culture to produce large quantities of insulin for people suffering from diabetes. Before the genes are inserted into bacterial plasmids any introns are removed.

b Explain why introns must be removed from the human insulin genes before they are
inserted into the bacterial plasmid.   2 marks

_____

_____

_____

_____

The genes encoding A chain and B chain are inserted into separate plasmids and grown in separate bacteria of the same species in separate cultures. Both genes are inserted into the plasmids next to a gene encoding β-galactosidase.

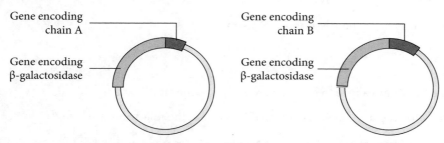

c Human insulin genes are inserted next to the *lacZ* gene that encodes for β-galactosidase.
Explain how scientists use this to separate the transformed bacterium from the bacterium
that have not taken up the insulin gene.   2 marks

_____

_____

_____

_____

d Human insulin was once obtained from the pancreas of cattle and pigs. Explain one ethical and one social implication of using insulin isolated from bacteria rather than from livestock. 2 marks

_____

_____

_____

_____

_____

Question 8 (8 marks) ◐◐◼

The restriction enzymes EcoRI, SmaI and HindIII were added to a section of DNA with a gene of interest and regions upstream and downstream of the gene. The sequence is shown below.

TTA GAA TTC CCC GGG ACA **GENE OF INTEREST** CGA ATT CGT AAG CTT AAA AAG

AAT CTT AAG GCC CCC TGT **GENE OF INTEREST** GCT TAA GCA TTC GAA TTT TTC

The restriction sites for each of the enzymes are as follows:

EcoRI G:AATTC   SmaI CCC:GGG   HindIII A: AGCTT

CTTAA:G   GGG:CCC   TTCGA:A

a State how restriction endonuclease cuts DNA strands. 1 mark

_____

b **i** Annotate the strand above to show where each enzyme cuts, if at all. 2 marks

ii How many fragments of DNA are formed for each of the following scenarios? 2 marks

Restriction enzyme combination	Number of fragments
EcoRI and HindIII	
HindIII	

c State which of the three enzymes would **not** be useful in DNA manipulation techniques. Justify your answer. 1 mark

_____

_____

d The gene of interest is being isolated and inserted into a plasmid. Which enzyme would be the most appropriate for this technique? Explain why. 2 marks

_____

_____

_____

Question 9 (10 marks) ©VCAA VCAA 2002 (2) SB Q5 (adapted) ●●

There were three suspects in an assault case. A forensic scientist found blood, other than the victim's, at the site. DNA was extracted from five blood samples:

- the victim

- the blood at the assault site (not the victim's)

- the three suspects.

Polymerase chain reaction (PCR) was used on the extracted DNA.

a A DNA polymerase enzyme is involved in the PCR process. Explain the role of the polymerase enzyme in PCR.   2 marks

One of the regions used in the forensic analysis was a short tandem repeat (STR) sequence of four bases, called HUMTHO1. This sequence, located on chromosome 11, has the sequence AATG and is repeated. It was this region of chromosome 11 that was amplified using PCR. The amplified samples were loaded onto a gel and electrophoresis was performed to separate the fragments of DNA.

b State why short tandem repeats rather than genes are used in forensic analysis.   1 mark

c The samples of the STR on chromosome 11 were compared, and the banding of suspect 3 and the banding of the blood found on the victim aligned. Is this enough evidence for the case to be solved? Justify your answer.   2 marks

d In addition to the victim's sample, the blood sample from the crime scene and the suspects' samples, what else should be included in the gel electrophoresis and why?   2 marks e   Name **two** properties of the DNA fragments that allow them to be separated from each other during gel electrophoresis.

2 marks

_____

_____

_____

_____

f   If the concentration of agarose was decreased within the gel, what effect would this have on the migration rate of the DNA?

1 mark

_____

_____

Question 10 (6 marks) ●●●

A farming corporation has edited a range of genes in their crops, including rice and soybeans, to increase productivity and decrease loss of produce due to size or damage, either through knocking out genes or introducing better functioning versions of genes. The company has seen a substantial increase in profits since these genetically modified crops were introduced into the farming sector. These genes were modified by using CRISPR technology.

a   Complete the table with **two** components involved in CRISPR and their function in the system.

2 marks

Component	Function

One such CRISPR success has been the editing of SWEET genes in rice that is grown across Asia and Africa. Scientists identified that these genes were activated by the bacterium *Xanthomonas oryzae*, which secretes chemicals that act as transcriptional factors. This leads to blight disease in the rice and loss of crops. Over time, variation within the rice has resulted in some of the population being resistant to the bacterium.

b   Explain how scientists can use CRISPR-Cas9 technology to modify the genome of these rice species to ensure the crops are no longer lost to blight.

4 marks

_____

_____

_____

_____

_____

_____

Test 3: Regulation of biochemical pathways in photosynthesis and cellular respiration

Section A: 20 marks. Section B: 72 marks. Total marks: 92 marks.
Suggested time: 45 minutes

Section A: Multiple-choice questions

Instructions to students
- For each question, circle the multiple-choice letter to indicate your answer.

Question 1

Enzymes are catalysts involved in cellular processes such as photosynthesis. These molecules are composed of

A glucose molecules.

B lipids.

C amino acids.

D nucleic acids.

The following information relates to Questions 2 and 3.

This graph shows the progress of a biological reaction in a plant cell under two different conditions.

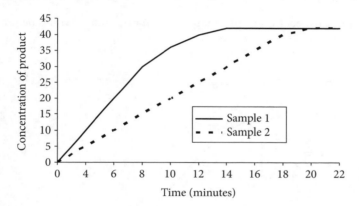

Question 2

The variable being investigated in these two samples was most likely

A amount of substrate.

B amount of reactant.

C concentration of enzyme.

D pH.

Question 3

The concentration of product stopped increasing after 14 minutes in Sample 1. What is the most likely explanation for this?

A After 14 minutes, all of the enzyme was consumed.

B After 14 minutes, all of the substrate was consumed.

C The enzyme began to break down the product.

D All of the enzyme was irreversibly bound to substrate molecules.

Question 4 ▣

Sliders are small freshwater turtles common in North America. They can remain under water for days at a time, not breathing. During a long period of being submerged, you would expect a slider to

A have high concentrations of O_2 in its blood vessels.

B have high concentrations of lactic acid in its muscle cells.

C remain very active during its time under water.

D store large quantities of ATP in its muscle cells.

Question 5 ▣

The following shows a simple biochemical pathway. Using the information shown in this pathway it would be correct to say that

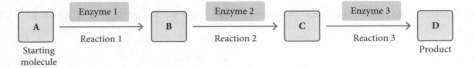

A if Enzyme 2 was denatured then B would not be produced.

B if the action of Enzyme 3 was stopped by a competitive inhibitor then C would not be produced.

C the reactant in this reaction is A.

D if all three enzymes in this reaction were denatured then no product would be produced.

Question 6 ▣

The following graph shows the change in energy content of a molecule during a biochemical reaction.

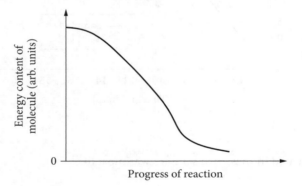

What could the molecule be?

A A molecule of glucose in photosynthesis

B A molecule of glucose in aerobic respiration

C A molecule of glucose in anaerobic fermentation

D An amino acid molecule during protein synthesis

Question 7 ●○○○

A scientist identified that enzymes denatured inside the mitochondrial matrix of a muscle cell after the cell was exposed to high temperatures for a prolonged period in the lab. An enzyme that has undergone denaturation

A has the same shape as the original functioning enzyme.

B has the same amino acid sequence as before.

C will only function if the temperature is lowered back to its optimal temperature.

D has fewer amino acids than the original functioning enzyme.

Question 8 ●○○○

In metabolism, nicotinamide adenine dinucleotide (NAD) is involved in reactions, carrying electrons from one reaction to another. The coenzyme is therefore found in two forms in cells.

- NAD^+ accepts electrons from other molecules and becomes reduced.
- This reaction forms NADH, which can then donate electrons.

  These electron transfer reactions are the main function of NAD.

NADH is used as an electron carrier in which type of cellular reactions?

A photosynthesis

B DNA synthesis

C protein synthesis

D cellular respiration

Question 9 ●○○○

There are two models of enzyme action: the lock-and-key model and the induced-fit model. These two models both propose that

A the active site is highly specific for a particular substrate.

B the shape of the active site can be modified to fit the substrate.

C the product is released from the active site.

D the enzyme is altered by the reaction and cannot be used again.

Question 10 ●●○○

The following graph shows the reaction rates for typical human enzymes and the enzymes of heat-tolerant bacteria involved in metabolism.

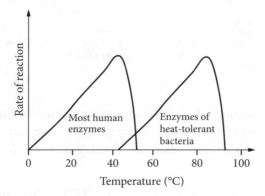

It is reasonable to conclude that

A most human enzymes are partially denatured at 20°C.

B enzymes of heat-tolerant bacteria will function well at normal human body temperature.

C a human enzyme that has been denatured at 65°C will function normally again if cooled to 35°C.

D the heat-tolerant bacteria are unlikely to survive in water at 100°C.

Question 11 ●●■

Enzyme activity in biochemical pathways such as the Krebs cycle may be inhibited by the presence of molecules that bind temporarily with the active site of the enzyme. This depends on the number of products formed in order to prevent a toxic build-up of product within the organelle. These molecules

A have a chemical structure complementary and specific to the enzyme's active site.

B have a chemical structure complementary and specific to the usual substrate molecule.

C affect the primary structure of the enzyme.

D alter the enzyme's secondary structure.

Question 12 ©VCAA VCAA 2012 (1) SA Q21 ●●■

Adenosine diphosphate (ADP) is an organic molecule found in large quantities in most cells. ADP is converted to adenosine triphosphate (ATP) by phosphorylation, as shown in the diagram below.

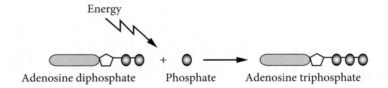

The original source of energy for this reaction is

A ADP.   **C** sunlight.

B glucose.   **D** phosphate.

The following information relates to Questions 13 and 14.

Three experiments were carried out to observe the rate of reaction. All other variables remained constant.

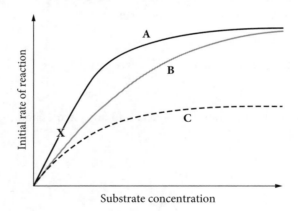

Question 13 ●●●

A scientist performed an experiment to observe the rate of a reaction with the enzyme Rubisco in photosynthesis under different conditions. The results are shown in the graph. Which of the following would account for the difference in rates between A (optimal conditions), B and C when substrate is not a limiting factor?

A A is with no inhibitor, B is with a competitive inhibitor and C is with a non-competitive inhibitor.

B A is at optimal temperature, B is when the temperature is above optimal, and C is at 0°C.

C A is a different enzyme, which is more specific and complementary for the substrate than enzymes B and C.

D A is a plant in full sunlight, B is a plant placed in the shade and D is a plant placed in the dark.

Question 14 ⬤⬤

At point X on the graph above, which is the limiting factor?

A temperature

B substrate concentration

C enzyme concentration

D product saturation

Question 15 ©VCAA VCAA 2011 (1) SA Q23 ⬤⬤

The following graphs depict two different reactions.

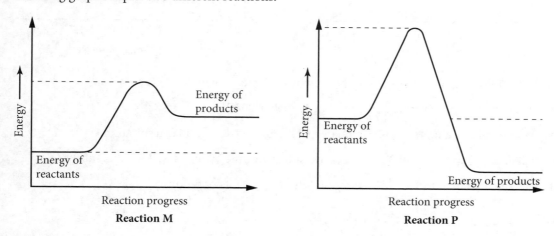

Reaction M   **Reaction P**

From the two graphs, it is reasonable to conclude that

A in reaction P, the energy level of the products is greater than that of the reactants.

B activation energy of reaction M is greater than that of reaction P.

C graphs M and P represent reactions that consume energy.

D energy is released in reaction P only.

Question 16 ©VCAA VCAA 2008 (1) SA Q19 ⬤

Activation energy in a biological reaction

A increases in the presence of an enzyme.

B increases with an increase in temperature.

C is the energy required to start the reaction.

D is involved in the formation of complex molecules only.

Question 17 ©VCAA VCAA 2014 SA Q13 (adapted)

The following graphs show the way four enzymes W, X, Y and Z change their activity in different pH and temperature situations.

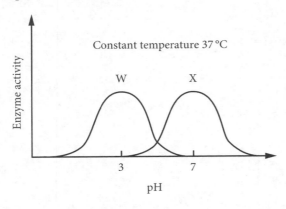 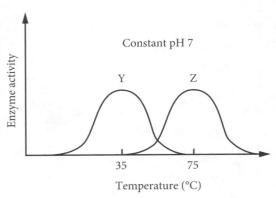

Which of the following statements about the activity of the four enzymes is true?

A At pH 7, enzyme Y is denatured at temperatures below 20°C.

B Enzyme Z could be an intracellular human enzyme involved in cellular respiration.

C At pH 3 and a temperature of 37°C, the active site of enzyme W binds well with its substrate.

D At pH 3 and a temperature of 37°C, enzyme X functions at its optimum.

Question 18 ©VCAA VCAA 2015 SA Q7 (adapted)

The production of adenosine triphosphate (ATP) is represented by the following equation.

$$ADP + P_i \longrightarrow ATP$$

The production of ATP

A requires complex molecules being broken down into simpler ones.

B requires an overall input of energy.

C only occurs in the absence of oxygen.

D occurs only in the mitochondria of a cell.

Question 19

A student was asked to identify differences between the overall process of photosynthesis and aerobic respiration in eukaryotic cells. The student prepared the table below to outline the differences. The only correct comparison listed by the student is

	Photosynthesis	**Aerobic respiration**
A	Releases energy	Requires energy
B	All stages occur within chloroplast	All stages occur within mitochondria
C	Electron transport not involved	Electron transport involved
D	Uses water as a reactant in the first stage	Forms water as a product in the final stage

Question 20 ©VCAA VCAA 2008 (1) SA Q5 ●●

The enzyme maltase catalyses the breakdown of maltose into glucose. Maltase was added to a tube containing a solution of maltose in water and incubated at 37°C. The amount of glucose produced was monitored over a period of time. No maltose remained at the end.

The graph showing the change in glucose concentration in the tube is

A.
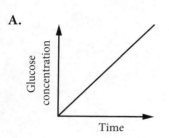

B.
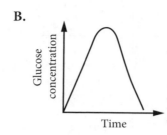

C.
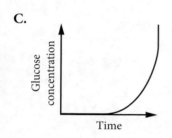

D.
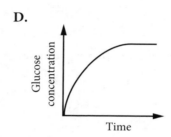

Section B: Short-answer questions

Instructions to students
- Answer all questions in the spaces provided.

Question 1 (8 marks) ●●

A group of students carried out an experiment to investigate the rate of breakdown of glucose by an enzyme extracted from yeast cells. The students set up eight test tubes, each containing the same amount of the enzyme. Each tube contained a different concentration of glucose in water (H_2O). The volume of solution in each tube was the same. All of the tubes were kept at the same temperature and pH. The results are given in the table.

Tube no.	Glucose concentration (mM)	Rate of glucose breakdown (μmol/min)
1	0	0
2	0.5	30
3	1.0	55
4	1.5	75
5	2.0	90
6	2.5	100
7	3.0	105
8	3.5	105

a   Sketch a graph of the results of this experiment below. Label both axes clearly.   2 marks

b   i   List the contents of tube 1.   1 mark

_____

ii   Explain the purpose of tube 1 in this experiment.   1 mark

_____

_____

c   The rates of breakdown of glucose in tubes 7 and 8 were the same (105 μmol/min).
    Why is this the case?   1 mark

_____

_____

In a second experiment, a new variable was investigated. In this experiment, the amount of the enzyme
and the glucose concentration in each tube were kept constant. The other variable was increased in
tubes 1–4. The results recorded by the students are shown in the table below.

Tube no.	Rate of glucose breakdown (μmol/min)
1	5
2	20
3	80
4	5

d   i   Name **one** variable that the students may have been investigating in this experiment.   1 mark

_____

ii   For the variable you suggested in part **d i**, explain how the results support your conclusion. 2 marks

_____

_____

_____

Question 2 (5 marks) ©VCAA | VCAA 2011 (1) SB Q7 (adapted) | ⚫⚫⚫

a Write the word or balanced chemical equation for aerobic cellular respiration.   1 mark

_____

b Cyanide inactivates metabolic reactions at the cristae of the mitochondria. Cyanide poisoning
often results in death. Why?   2 marks

_____

_____

c Some species are sensitive to cyanide while others appear to be unaffected by this chemical.
Why?   2 marks

_____

_____

Question 3 (10 marks) ⚫⚫⚫

a Define 'biochemical pathway'.   1 mark

_____

_____

The diagram below shows a simplified representation of the ten steps of glycolysis.

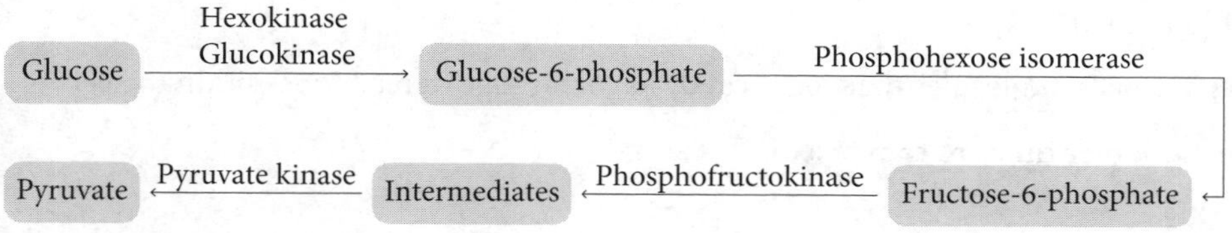

b Where does this reaction take place?   1 mark

_____

c Explain the effect a mutation in the phosphohexose isomerase gene would have on the amount of
pyruvate produced.   2 marks

_____

_____

_____

High concentrations of ATP in the cell bind to the enzyme phosphofructokinase at a site other than the
active site, reducing its affinity for the fructose-6-phosphate.

d **i** State the name of this type of inhibition.   1 mark

_____

ii Using annotated diagrams, demonstrate how ATP works to lower the enzyme activity.   2 marks

iii Other than inhibitors, what is **one** way in which this pathway could be regulated?   2 marks

e Explain the function of ATP in cells.   1 mark

Question 4 (4 marks)

Enzymes are biomacromolecules that act as catalysts for a range of reactions within the cell.

a State the monomer they are composed of.   1 mark

b Explain the function of **two** organelles in the production of extracellular enzymes.   2 marks

c An enzyme was isolated from a mouse cell and inserted into a human cell, but the enzyme underwent no reaction. State a potential reason why.   1 mark

Question 5 (7 marks) ⬤◖◗

Adenosine triphosphate is a useable energy source for the cell.

a Explain how ATP is formed.   2 marks

_____

_____

_____

b Name the stages and state the ATP output during each stage of aerobic respiration.   2 marks

_____

_____

c Explain how the amount of oxygen in a cell and the amount of ATP are directly related.   1 mark

_____

_____

d Contrast the amount of ATP required in a muscle cell with the amount required in a skin cell.
How would the organelle composition of the two cells differ?   2 marks

_____

_____

_____

Question 6 (5 marks) ©VCAA VCAA 2017 SA Q6 (adapted) ⬤⬤⬤

Hydrogen peroxide is a toxic by-product of many biochemical reactions. Cells break down hydrogen peroxide into water and oxygen gas with the help of the intracellular enzyme catalase. The optimum pH for catalase is 7.

A student measured the activity of catalase by recording the volume of oxygen gas produced from the decomposition of hydrogen peroxide when a catalase suspension was added to it. The catalase suspension was made from ground, raw potato mixed with distilled water. The student performed two tests and graphed the results.

Test 1 used 5 mL of 3% hydrogen peroxide solution and 0.5 mL of catalase suspension and was conducted at 20°C in a buffer solution of pH 7.

Test 2 was carried out under identical conditions to test 1, except for one factor that the student changed.

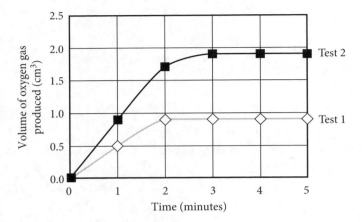

a State the variable that was changed between tests 1 and 2.

1 mark

_____

b Explain how the rate of reaction would alter if the temperature was lowered to 10°C and then raised to 50°C.

2 marks

_____

_____

_____

c In the space below, draw a graph showing the rate of reaction for the catalase enzyme at different pH values.

2 marks

Question 7 (7 marks) ●●▮

Consider the reactions involved in photosynthesis and cellular respiration.

a Complete the following table to show the levels of CO_2, O_2 and glucose under
   different conditions.   3 marks

Condition	CO_2 level (increase, decrease or constant)	O_2 level (increase, decrease or constant)	Glucose level (increase, decrease or constant)
Rate of photosynthesis is greater than rate of cellular respiration			
Rate of photosynthesis equal to rate of cellular respiration			
Plant exposed to 0 light intensity			

b Explain why the reactions of photosynthesis and cellular respiration are often referred to as
   complementary to each other. Use an example in your response.   2 marks

_____

_____

_____

c How can temperature affect both photosynthesis and cellular respiration?   2 marks

_____

_____

_____

_____

Question 8 (9 marks) ●●▮

Biological processes involve a range of molecules that facilitate and enable the reactions to occur within
cells to ensure the organism survives and homeostasis can be achieved.

The mitochondria are involved in cellular respiration, regulating calcium levels and apoptosis.

a State the function of an enzyme.   1 mark

_____

_____

b Explain **two** factors that need to be controlled to ensure the enzymes in the mitochondria
   function optimally.   2 marks

_____

_____

_____

_____

c Enzymes can be inhibited in two ways. Compare these processes.   2 marks

_____

_____

_____

d A new pesticide is being developed to inhibit the functioning of an enzyme involved in the process of cellular respiration. Develop an experiment to determine if the pesticide is acting as a competitive or non-competitive inhibitor.   4 marks

_____

_____

_____

_____

_____

_____

Question 9 (9 marks) ●●

Autotrophs can produce organic molecules through reactions using inorganic substances such as CO_2, whereas other organisms (heterotrophs) must obtain these organic molecules by consuming plants or animals to obtain their nutrients and energy.

Consider the following diagram of two biological processes that occur in some organisms and are said to be complementary.

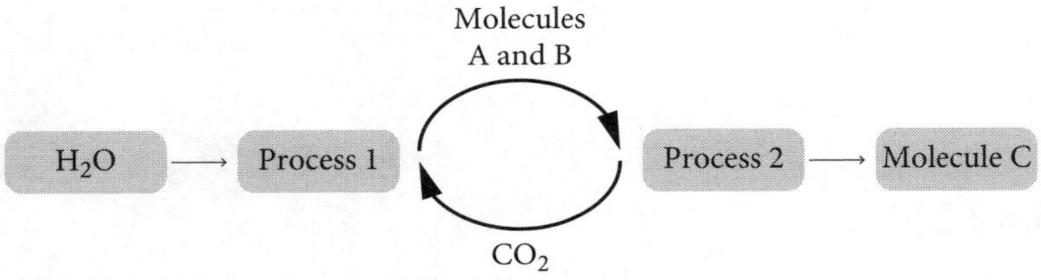

a Identify the molecules or processes in the table.   3 marks

Molecule	Name
Process 1	
Molecules A and B	
Molecule C	

b Name the coenzymes and explain their role in process 2.   2 marks

_____

_____

_____

c What is missing from the diagram to facilitate process 1 from occurring? Explain how this facilitates the reaction.   2 marks

_____

_____

d Some organisms are unable to undergo process 1 in the diagram. In this case, how does the organism obtain molecules A and B for process 2?   2 marks

_____

_____

Question 10 (8 marks) ▢▢▢

A group of students set up a range of experiments to observe the effect of different factors on the rate of photosynthesis in tomato plants in a sealed clear container.

- **Experiment 1:** Three experimental groups at different temperatures (0, 20, 40°C) and a control group at each temperature, which included the sealed container but lacked the tomato plant.

- **Experiment 2:** Four experimental groups at different soil pH (pH 3, 5, 9 and 11) and a control group at each pH, which lacked the tomato plant.

- **Experiment 3:** Three experimental groups at different light intensities (low, medium and high).

    Each experimental group contained 50 tomato plants under the conditions for 1 week. Each experiment tested the rate of oxygen release (ppm) throughout the experiment.

a Other than the controlled variables stated above, identify **two** other variables that should be controlled during all of these experiments.   1 mark

_____

_____

b Choose one of the experiments above and explain a weakness of the experiment. Explain how this experiment could be improved to produce more reliable data.   2 marks

_____

_____

A student conducting experiment 1 hypothesised that the oxygen concentration would continue to decrease from 0 to 40°C.

c Form a conclusion that would refute this hypothesis.   1 mark

_____

_____

d Using your biological knowledge, explain what you would expect to see in the results for experiment 3.   2 marks

_____

_____

_____

e Explain how the pH in experiment 2 would influence the rate of photosynthesis.   2 marks

_____

_____

_____

_____

Test 4: Photosynthesis as an example of biochemical pathways

Section A: 20 marks. Section B: 58 marks. Total marks: 78 marks.
Suggested time: 45 minutes

Section A: Multiple-choice questions

Instructions to students
- For each question, circle the multiple-choice letter to indicate your answer.

Question 1

The cycling of coenzymes occurs in living cells. During photosynthesis, and in particular the light-dependent reactions, the reaction involves the reduction of $NADP^+$ to form NADPH. A possible reaction for this is

A $NADPH + H^+ \rightarrow NADP^+ + 2e^- + 2H^+$

B $NAD^+ + 2e^- + 2H^+ \rightarrow NADH$

C $NADP^+ + 2e^- \rightarrow NADP^+ + 2H^+$

D $NADP^+ + 2e^- + 2H^+ \rightarrow NADPH + H^+$

Question 2

Which of the following factors does **not** affect the rate of glucose production?

A ambient temperature

B amount of carbon dioxide in the air

C amount of oxygen in the cell

D light intensity

Question 3

Consider the diagram shown below with areas A and B labelled. Which one of the following is a correct statement about the diagram?

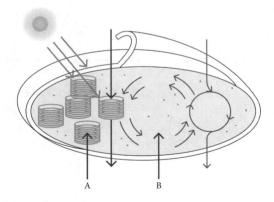

A The light-dependent stage of photosynthesis occurs at B.

B CO_2 is a product of the reaction that occurs at B.

C O_2 is a product of the reaction that occurs at A.

D ATP is required for the reaction at A to occur.

Question 4

Which molecule is produced in the light-dependent reactions of photosynthesis and then used in the light-independent reactions of photosynthesis?

A CO_2

B ATP

C H_2O

D O_2

Question 5

The Calvin cycle does not normally take place at night because

A it is too dark for the light to be absorbed for the reaction.

B it relies on the products of the light-dependent reactions.

C it is too cold.

D plants are unable to obtain water at night.

Question 6

Radioactive oxygen (^{18}O) is introduced into the environment of an experimental plant. After a short period of exposure to sunlight, the radioactive oxygen is detected in sugars in the leaf cells of the plant. The ^{18}O was most likely introduced to the plant's environment in which molecule?

A CO_2 **C** H_2O

B ATP   **D** O_3

The following information relates to Questions 7 and 8.

The following graph shows the relationship between light intensity and net oxygen uptake or output by a particular green plant.

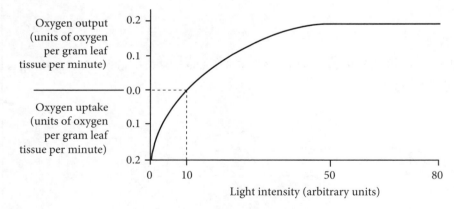

Question 7 ©VCAA VCAA 2008 (1) SA Q6

At a light intensity of 10 units

A the rate of photosynthesis is zero.

B the rate of aerobic respiration is zero.

C oxygen produced by photosynthesis is equal to the oxygen used by aerobic respiration.

D oxygen produced by photosynthesis is equal to twice the oxygen used by aerobic respiration.

Question 8 ⬤⬤

A limiting factor at 80 units could be

A glucose input.

B carbon dioxide concentration.

C light intensity.

D oxygen concentration.

Question 9 ⬤⬤

CAM plants differ from C3 plants because

A CAM plants process the gas exchange in the leaves during the middle of the day to maximise photosynthesis.

B CO_2 is absorbed at night and converted into a storage molecule for use during the day.

C CAM plants absorb the light energy in the daytime and the Krebs cycle occurs during the night.

D CAM plants fix carbon dioxide directly from the air into the 3-carbon compound during the Krebs cycle.

Question 10 ⬤⬤⬤

Oxygen in high amounts can act as a competitive inhibitor during which part of photosynthesis?

A Splitting of the water in the light-dependent reaction

B Accepting the electrons for NADP at the thylakoid membranes

C Binding to the active site of the Rubisco enzyme

D Denaturing the enzymes in the Calvin cycle

Question 11 ⬤⬤

Stomata act as pores for plants on land and in the ocean to exchange materials with their surroundings. Stomata are typically located on the underside of the leaf and are surrounded by guard cells that enlarge and contract to regulate the stomatal pores. When stomata are open, they allow the

A entry of oxygen and release of CO_2 into the atmosphere.

B loss of water and input of oxygen and CO_2 for photosynthesis.

C input of CO_2 and release of O_2 during photosynthesis.

D input of water through the pores and release of CO_2 and O_2 from the air spaces in the leaf.

Question 12 ⬤⬤⬤

A person purchased some plants from the local nursery. The plants were classified as CAM and C4 plants. These plants would be best suited to

A hot and damp conditions.

B cold and dry conditions.

C hot and dry conditions.

D humid and damp conditions.

Question 13 ⬤⬤⬤

C4 plants have adapted to warm and tropical weather through

A closing their stomata throughout the night and only up taking CO_2 in the afternoon.

B separating the reactions between the mesophyll and bundle sheath cells.

C converting carbon dioxide to a 3-carbon compound to be stored before use.

D isolating the glucose molecule through the process of photorespiration.

Question 14 ©VCAA VCAA 2002 (1) SA Q4 ⬤⬤

The amount of water lost per unit area from the leaf surface of two different plants was measured. Both plants were grown in the same conditions. The results are shown in the graph.

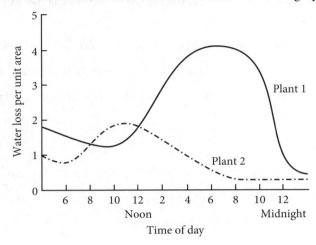

From the information given, you can conclude that

A plant 2 will be more likely to survive in a dry environment than plant 1.

B at 12.00 pm, plant 1 has a greater water loss per unit area than plant 2.

C at 10.00 am, the average stomatal aperture will be greater in plant 1 than in plant 2.

D at 5.00 pm, the rate of photosynthesis will be greater in plant 2 than in plant 1.

Question 15 ⬤⬤

The chloroplast is the site of photosynthesis in plants. In the chloroplast

A water is the reactant for the light-independent reaction, which occurs on the thylakoid membranes.

B oxygen is the reactant for the light-dependent reaction, which occurs on the thylakoid membrane.

C glucose is the product of the reaction occurring on the thylakoid membranes.

D carbon dioxide is the reactant for the light-independent reaction, which occurs in the stroma.

Question 16 ©VCAA VCAA 2002 (1) SA Q3 ⬤⬤

A plant wilted but later recovered. During the time the leaves were wilted, the rate of photosynthesis decreased because

A enzymes in the leaf cells were denatured.

B chlorophyll in the wilting leaves was broken down.

C the amount of light reaching the plant was reduced.

D the amount of carbon dioxide entering the leaf decreased.

Question 17 ●●●

The table below shows a summary of the inputs and outputs for six cycles of the light-independent stage of photosynthesis. Two numbers are missing from the table and have been replaced by the letters X and Y.

Inputs		Outputs	
Molecule	Total number	Molecule	Total number
CO_2	6	H_2O	6
NADPH	X	$NADP^+$	12
ATP	Y	ADP	12
		Pi	12
		Glucose	1

X and Y are

A 12; 12

B 12; 24

C 6; 12

D 6; 6

Question 18 ©VCAA VCAA 2014 SA Q7 ●○○○

During photosynthesis, light energy is used to split water, forming oxygen and hydrogen ions.

The splitting of water occurs

A in the stroma during the light-independent reaction.

B in the grana during the light-dependent reaction.

C on the membrane of the thylakoids during the light-independent reaction.

D on the surface of the outer chloroplast membrane during the light-dependent reaction.

Question 19 ©VCAA VCAA 2014 SA Q8 ●●○○

An increase in the atmospheric CO_2 levels increases the rate of photosynthesis. The rate of photosynthesis increases because

A the rate of the light-independent reactions on the thylakoid membranes of the chloroplasts increases.

B water loss from the leaf decreases, resulting in the availability of water for photosynthesis increasing.

C the increased CO_2 level lowers the pH inside the chloroplasts and increases the rate of enzyme-catalysed reactions.

D the rate of the light-independent reactions in the stroma increases with the increase in CO_2 level.

Question 20 ●●○○

Plants containing chlorophyll absorb

A green light for photosynthesis to then reflect and appear green.

B all wavelengths from the visible light spectrum at an equal rate to use in photosynthesis.

C other wavelengths from the visible light spectrum and reflect green wavelengths.

D no wavelengths and reflect the green light wavelengths hitting the chlorophyl pigment.

Section B: Short-answer questions

> **Instructions to students**
> · Answer all questions in the spaces provided.

Question 1 (8 marks) ⬤⬜⬜

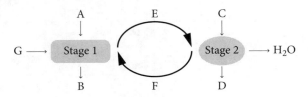

a Name and explain the roles of the following components that appear in the diagram.   3 marks

Label	Name	Role
A		
F		
G		

b Explain the process that occurs at stage 2.   3 marks

_____

_____

_____

_____

c A plant is exposed to a low temperature. Explain how this will specifically affect the rate of these reactions.   2 marks

_____

_____

_____

_____

Question 2 (7 marks) ⬤⬤⬤

a Name the enzyme involved in the cyclic set of reactions that occur in photosynthesis.   1 mark

_____

b Compare and contrast C3 and C4/CAM plants.   2 marks

_____

_____

_____

_____

c Explain how oxygen interacts with the enzyme stated in part **b**. What would an increase in CO_2 do to this interaction?   2 marks

_____

_____

_____

_____

d What advantages does a CAM plant in a hot environment have compared to a C3 plant?   1 mark

_____

_____

_____

e Compare photorespiration to photosynthesis.   1 mark

_____

_____

_____

Question 3 (5 marks) ●●■

The following graph shows the rate of oxygen exchange between a leaf and the external environment in varying light intensities.

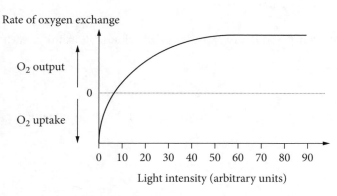

a Name the process responsible for the output of oxygen.   1 mark

_____

b Write a word equation for this process.   1 mark

_____

_____

c A student concluded that because the leaf did not release oxygen at light intensities of less than 10 units, respiration was the only process occurring in the leaf cells at these light intensities.

Use the information contained in the graph and your knowledge of the cellular processes in plants to justify whether this student's conclusion is supported. 2 marks

_____

_____

_____

d Explain how you could account for the fact that the rate of O_2 output remained constant between light intensities of 50 and 80 units. 1 mark

_____

_____

_____

Question 4 (4 marks) ●●●

The following diagram shows a section of a chloroplast.

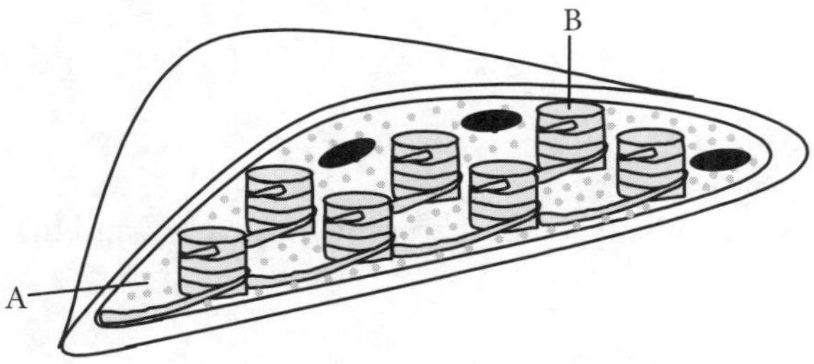

a Name the regions labelled A and B and identify the reaction that occurs at each region.

  i A _____ 1 mark

  ii B _____ 1 mark

b Photosynthesis involves two sets of reactions in the chloroplasts. Name the products of these reactions.

  i First stage _____ 1 mark

  ii Second stage _____ 1 mark

Question 5 (7 marks)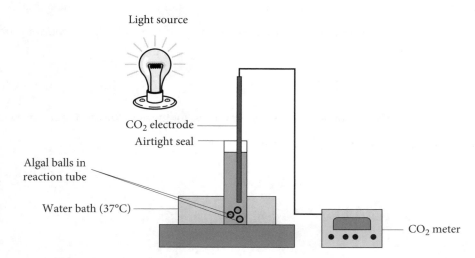

The apparatus shown below is used to study photosynthesis by measuring the levels of oxygen and carbon dioxide within a chamber. The chlorophyll pigments respond to all wavelengths of light in different proportions.

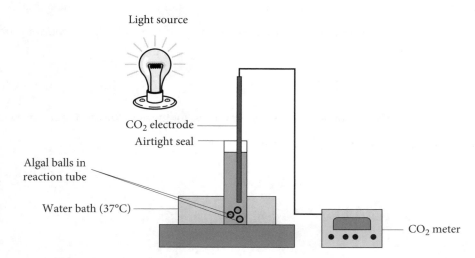

Some students performed three experiments and observed the reactions that took place in each.

- **Experiment 1:** A white light was shone on the reaction tube.

- **Experiment 2:** No light was shone on the reaction tube.

- **Experiment 3:** A green light was shone on the reaction tube.

The temperature of the tubing was carefully controlled to 37°C throughout the experiment and the pH of the solution was initially the same in all test tubes.

In each experiment, oxygen was present in the tubing at the start.

a   Using your knowledge and understanding of photosynthesis, complete the table below with your prediction of the O_2 and CO_2 concentrations (increase, decrease or no change) after 24 hours.   3 marks

Experiment number	Carbon dioxide	Oxygen
1		
2		
3		

b   From the information provided above, state if there would be a net increase, decrease or no change in glucose within the algal balls.   2 marks

Experiment number	Glucose level	Reasoning
1		
2		

c   The students did not use a pH buffer throughout the process. Explain how a large change in pH in the solution could affect the rate of photosynthesis.   2 marks

Question 6 (5 marks) ©VCAA VCAA 2010 (1) SB Q3 ●●

Elysia chlorotica is a bright green sea slug, with a soft leaf-shaped body. It has a life span of 9 to 10 months. This sea slug is unique among sea slugs as it is able to survive on solar power. *E. chlorotica* acquires chloroplasts from the algae it eats and stores them in the cells that line its digestive tract. Young *E. chlorotica* fed with algae for two weeks can survive for the rest of their lives without eating.

a Name the product of photosynthesis that provides the energy that enables *E. chlorotica* to survive for so long without eating.   1 mark

_____

The product of photosynthesis must undergo a three-stage process for the slug to access the energy in the product.

b Name and give a brief description of each of these stages.   3 marks

_____

_____

_____

_____

_____

A watery environment can have a low concentration of dissolved gases.

c Explain how having chloroplasts allows *E. chlorotica* to overcome this disadvantage.   1 mark

_____

_____

Question 7 (6 marks) ©VCAA VCAA 2008 (1) SB Q3 (adapted) ●●●

The following diagrams show:

Graph one: the rate of photosynthesis in a green plant at different wavelengths of light

Graph two: the estimated absorption of the different wavelengths of light by the different plant pigments

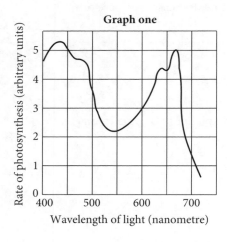

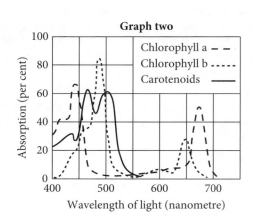

a Explain the trends in the two graphs.   2 marks

_____

_____

_____

b Explain the function of chlorophyll and carotenoid pigments in the process of photosynthesis.   2 marks

_____

_____

_____

c A person hypothesised that placing their indoor plants under a blue light at 450 nm would
result in the highest rate of photosynthesis out of any wavelengths or colours. Would this
assumption be correct?   2 marks

_____

_____

_____

Question 8 (4 marks) ⬤⬤⬛

The graph below shows the change in the partial pressure (concentration) of oxygen in the air
surrounding an experimental plant kept in bright light.

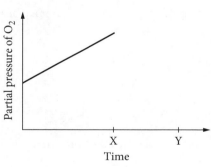

a Explain why the partial pressure of O_2 has increased.   1 mark

_____

_____

The same plant was kept in the dark for the period of time X–Y.

b **i** On the axes above, show what will happen to the partial pressure of O_2 in the time
period X–Y.   1 mark

ii Explain why you have drawn the graph the way you have.   2 marks

_____

_____

_____

Question 9 (7 marks)

A group of students ran an experiment to test the effect of light intensity on the rate of photosynthesis.

Bicarbonate solution
Spinach disc

The students cut out 50 identical discs from spinach and removed all the gases from the spinach discs. They placed 10 spinach discs into each of five beakers. The discs sunk because the gases in them had been removed. Beakers 1–4 were exposed to different light intensities and beaker 5 was left in the dark for 6 hours.

a Explain why one beaker was placed in the dark.   1 mark

b Explain how light intensity can act as a limiting factor in this experiment.   2 marks

c Provide an expected set of results for the experiment and the five beakers.   2 marks

d Identify a weakness of this experiment and state an improvement to make the results more reliable.   2 marks

9780170479448

Question 10 (5 marks)

Because of climate change and unpredictable weather conditions, plants are being exposed to extreme temperatures that they are not adapted to. Plants placed in extremely high temperatures for long periods have been found to have a decreased rate of glucose production. As the temperature lowers, it takes several hours for the rate to return to optimum.

a   Heat stress affects the set of reactions that occur in the stroma. Explain how this will affect the rate of the light-dependent reaction.   2 marks

_____

_____

_____

b   Rubisco is an enzyme that is part of the cyclic set of reactions in photosynthesis. State what occurs when an enzyme is above its optimal temperature.   1 mark

_____

_____

_____

c   Explain why the rate of glucose production increases only slowly, even when the temperature is reduced back to the plant's optimal temperature.   2 marks

_____

_____

_____

Test 5: Cellular respiration as an example of biochemical pathways

Section A: 20 marks. Section B: 82 marks. Total marks: 102 marks.
Suggested time: 60 minutes

Section A: Multiple-choice questions

Instructions to students
- For each question, circle the multiple-choice letter to indicate your answer.

Question 1

In the process of cellular respiration, the absence of which cellular molecule slows down the production of ATP?

A CO_2

B O_2

C H_2O

D NADPH and ATP

Question 2

The electron transport stage of cellular respiration involves the production of which waste product?

A oxygen

B carbon dioxide

C water

D ATP

Question 3

The photograph is of a cell organelle.

This organelle is called a

A mitochondrion and is the site of aerobic respiration.

B mitochondrion and is the site of anaerobic respiration.

C chloroplast and is the site of photosynthesis.

D chloroplast and is the site of glycolysis.

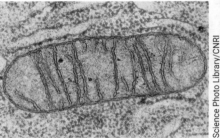

Science Photo Library/CNRI

Question 4

Where does glycolysis occur?

A In the cytosol of cells

B Within the mitochondrial matrix

C On the surface of mitochondrial membrane

D In the stroma of the chloroplast

Question 5

The aerobic pathways involved in ATP production

A produce ATP without oxygen.

B occur in the cytoplasm of the cell.

C produce ATP without glucose.

D produce more ATP than anaerobic pathways do.

Question 6 ▨▨▨

A cell was isolated from an organism and the mitochondria were removed. Only one process of cellular respiration could take place. The products of this process would be

A carbon dioxide, water and ATP.

B pyruvate and ATP.

C alcohol and ATP.

D ATP only.

Question 7 ▨▨▨

The cycling of coenzymes occurs in living cells. During cellular respiration, and in particular the Krebs cycle, the reaction involves the reduction of NAD^+ to form NADH. A possible reaction for this is

A $NAD^+ + 2e^- + H^+ \rightarrow NADH$

B $NADPH + H^+ \rightarrow NADP^+ + 2e^- + 2H^+$

C $NADP^+ + 2e^- + 2H^+ \rightarrow NADPH + H^+$

D $NADH \rightarrow NAD^+ + 2e^-$

Question 8 ▨▨▨

Aerobic respiration is a three-phase process. Which of the following molecules is a by-product of the electron transport stage of respiration?

A CO_2 **C** ATP

B H_2O **D** NADH

The following diagram relates to Questions 9 and 10.

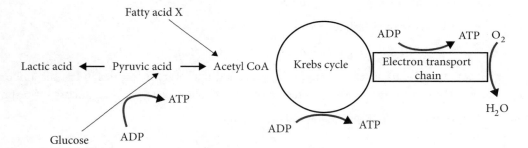

Question 9 ©VCAA VCAA 2014 SA Q12 ▨▨▨

If there is insufficient glucose for cellular respiration, fatty acids can be changed to acetyl CoA. Each fatty acid X molecules produces eight molecules of acetyl CoA. The diagram summarises the pathways for the breakdown of fatty acid X and glucose.

The number of molecules produced in each step is **not** shown.

Referring to the information above and your knowledge of cellular respiration, which one of the following conclusions can be made?

A Most of the ATP is made in the Krebs cycle.

B Pyruvic acid is converted into acetyl CoA under anaerobic conditions.

C No ATP can be formed from the breakdown of glucose under anaerobic conditions.

D One fatty acid X molecule produces more ATP in aerobic conditions than one glucose molecule does.

Question 10 ●●

Referring to the diagram above. Which molecule(s) is produced as a waste product in the Krebs cycle?

A NADH and FADH$_2$

B oxygen

C carbon dioxide

D glucose

The following information relates to Questions 11 and 12.

Rotenone is a chemical compound that is used as an insecticide and a piscicide (a substance that kills fish). The rotenone molecule disrupts the electron transport chain in animal cells by interfering with one of the essential reactions within the electron transport chain.

Question 11 ©VCAA VCAA 2015 SA Q9 ●●●

Which one of the following statements best explains the effect of rotenone in causing death in insects and fish?

A The rate of glycolysis would increase.

B ATP would accumulate in the mitochondria.

C Aerobic respiration in the mitochondria would be disrupted.

D The plasma membrane would no longer be permeable to oxygen.

Question 12 ©VCAA VCAA 2015 SA Q10 ●●●

In the past, people sometimes put extracts containing rotenone into rivers to poison the fish, allowing them to be more easily caught. When rotenone-poisoned fish are eaten by people, no effect is observed in the humans. Which of the following statements best explains this observation?

A Rotenone is not absorbed through the plasma membranes of people who have eaten poisoned fish.

B Rotenone is not absorbed by fish tissue and remains dissolved in water.

C Human cell metabolism does not involve the electron transport chain.

D Rotenone only affects organisms that respire anaerobically.

The following information relates to Questions 13 and 14.

An experiment was set up with different amounts of yeast at different temperatures to test the amount of ethanol produced. The yeast was mixed in a solution of 5 mL apple juice to provide glucose for respiration.

Below are the results of the amount of ethanol produced.

Mass yeast (g)	Volume of ethanol produced (mL)				
	Test 1 0°C	Test 2 15°C	Test 3 25°C	Test 4 35°C	Test 5 45°C
5	1	3	6	8	5
10	2	5	12	20	8
15	3	8	19	28	7
20	4	11	28	35	9

Question 13

Which of the following would increase the rate of the reaction of reaction in Test 4?

A Increase oxygen availability.

B Increase glucose availability.

C Decrease temperature.

D Increase carbon dioxide concentration.

Question 14 ⬤⬤⬤

From the data above and using your biological knowledge, which of the following graphs accurately depicts the data for change in temperature?

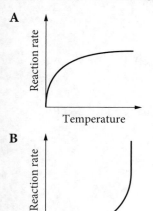

A

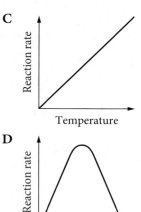

C

B

D

Question 15 ⬤⬤▢

Glycolysis is the first stage of cellular respiration. The diagram below shows a summary of the process of glycolysis. Using your knowledge of glycolysis and the diagram below you can conclude that

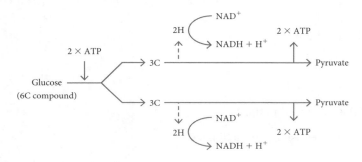

A pyruvate is a 3-carbon compound.

B pyruvate cannot be broken down any further.

C for each molecule of glucose broken down by glycolysis there is a net production of 4 ATP.

D for each molecule of glucose broken down by glycolysis 2 ADP and Pi are used.

Question 16 ⬤⬤▢

The Krebs cycle is a series of reactions that produces electron carriers for the electron transport chain. Which of the following would **not** increase the rate at which $FADH_2$ and NADH are produced?

A Increase oxygen availability.

B Increase carbon dioxide concentration.

C Increase production of pyruvate.

D Increase NAD^+ and FAD^{2+} availability.

Question 17 ⬤◯◯

Anaerobic respiration occurs in animal and yeast cells in the absence of oxygen. Glucose is partially broken down into pyruvate, releasing 2ATP. A difference between anaerobic respiration in animals and yeast is

A yeast breaks down the pyruvate into monomers and release O_2.

B animal cells produce lactic acid and CO_2, whereas yeast cells produce acetyl CoA.

C animal cells become more acidic and yeast forms ethanol and CO_2.

D animal cells produce 4ATP with each pyruvate molecule broken down and yeast produce 2ATP.

Question 18 ©VCAA VCAA 2017 SA Q10 ⬤◯◯

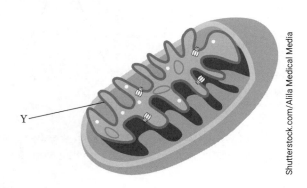

Y

Shutterstock.com/Alila Medical Media

The structure labelled Y is where

A glucose enters glycolysis.

B NAD^+ is converted into NADH.

C the majority of ATP is produced in the cell.

D pyruvate is broken down, releasing carbon dioxide.

Question 19 ⬤⬤◯

The reaction $ADP + P_i \rightarrow ATP$

A occurs in the third stage of cellular respiration as the electrons and hydrogen ions from NADH and $FADH_2$ facilitate the conversion of $ADP + P_i$ into ATP.

B occurs in the first stage of anaerobic respiration from the break down of pyruvate into lactic acid.

C occurs in the Calvin cycle during a series of chemical reactions catalysed by enzymes.

D occurs during the break down of NADPH and $FADH_2$ during the Calvin cycle.

Question 20 ©VCAA VCAA 2002 (1) SA Q13 ⬤⬤◯

Many plants naturally contain bitter-tasting chemicals called cyanogenic glycosides. In animals, these chemicals interfere with the electron transport stage of cellular respiration. The effect on animals that eat these plants is

A a decrease in ATP production.

B an increase in energy storage.

C a decrease in oxygen production.

D an increased level of aerobic respiration.

Section B: Short-answer questions

> **Instructions to students**
> • Answer all questions in the spaces provided.

Question 1 (7 marks) ●●●

The following diagram is a simplified outline of cellular metabolism, showing how NAD$^+$ and NADH link the citric acid cycle and the production of ATP.

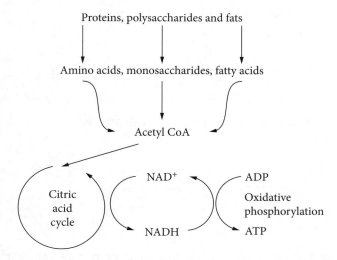

a By what other name is the citric acid cycle known?   1 mark

_____

b Describe the role of acetyl CoA in the above diagram.   1 mark

_____

_____

c Why are the arrows of the citric acid cycle linked with the arrows of NAD$^+$/NADH and ADP/ATP?   2 marks

_____

_____

d Describe the role of ATP in cells.   1 mark

_____

_____

e In the following table, state the amount of ATP that is formed in each stage.   2 marks

Stage	Number of ATP molecules
Formation of acetyl CoA from glucose	
Citric acid cycle	
Oxidative phosphorylation	

Question 2 (10 marks) ⬤⬤◼

A biologist carried out a series of experiments to investigate the rate of anaerobic respiration in yeast cells under different conditions. The reactions of fermentation are catalysed by about 12 enzymes. In addition to these enzymes, yeasts also contain the enzyme invertase, which converts sucrose into glucose and fructose. These can then be used in fermentation. Five flasks were set up with the following contents and conditions.

Flask number	Contents of flask	Temperature (°C)	pH
1	100 mL sucrose solution, 0.5 g yeast	38	7
2	100 mL sucrose solution, 0.5 g yeast	38	5
3	100 mL sucrose solution, 0.5 g yeast	15	7
4	100 mL sucrose solution, 0.5 g yeast	15	5
5	100 mL sucrose solution, 0.5 g yeast	75	7

The rate of fermentation was measured by recording the volume of gas given off from each flask.

a   What **two** variables are being investigated in this series of experiments?   1 mark

b   State what gas is being produced in this set of experiments.   1 mark

The volume of gas produced was recorded at 10-minute intervals over the course of an hour. The results are shown below.

Time (min)	Volume of gas (mL)				
	Flask 1	Flask 2	Flask 3	Flask 4	Flask 5
0	0	0	0	0	0
10	0.5	0	0.1	0	0.1
20	1.1	0	0.3	0	0.1
30	3.8	0	0.8	0	0.1
40	5.6	0	1.1	0	0.1
50	6.0	0	1.5	0	0.1
60	6.0	0	1.7	0	0.1

c   On the basis of these results, what conditions were most favourable for fermentation?   1 mark

d   How could you explain the lack of any gas production in flask 2?   1 mark

e   i   Describe the effect of temperature on the rate of fermentation.   2 marks

ii Outline an explanation to account for the differences in fermentation rates in
flasks 1, 3 and 4.

2 marks

f The experiment was repeated with different disaccharides from sucrose. Lactose (glucose + galactose)
and maltose (glucose + glucose) were used. These were placed in separate flasks with 0.5 grams of
yeast, as in the initial experiment. Both flasks were adjusted to a pH of 7 and kept at 38°C. To the
experimenter's surprise, no fermentation occurred in these flasks. Give **two** possible explanations to
account for the lack of fermentation in these flasks.

2 marks

Question 3 (9 marks)

The break down of complex molecules to produce an energy source for reactions is vital for an
organism's survival.

a Write the word equation for aerobic cellular respiration.

2 marks

b A biologist measured the CO_2 levels in a muscle cell of a mouse to test the rate of anaerobic
respiration versus aerobic and concluded that they were occurring at a similar rate because of the
release of CO_2. Explain an issue with this type of measurement and the biologist's conclusion
about the rates of both reactions.

2 marks

c Anaerobic respiration occurs in all organisms in the absence of oxygen. Explain how
anaerobic respiration alters the pH in animal cells and the effect this could have on the cells'
functioning.

2 marks

The movement of molecules through the semipermeable membrane from the extracellular fluid enables the reactions in cells to occur.

d Glucose is a large polar molecule required for cellular respiration. Explain how the glucose enters the muscle cells.   2 marks

e Explain how carbon dioxide and oxygen enter and leave the cell.   1 mark

Question 4 (6 marks) ⬤⬤⬤

A drug company is developing a type of drug that aims to decrease the build-up of lactic acid in muscle cells through acting as an alternative source of oxygen during strenuous exercise. Before the drug can be tested on participants, the company will test the drug on muscle cells in a culture in the lab.

a Propose a hypothesis for the drug and its effectiveness.   1 mark

b Design an experiment based on your hypothesis to determine how the company could determine if the drug is effective and the best dosage for the maximum effect.   4 marks

c State a set of results that would support your hypothesis.   1 mark

Question 5 (10 marks)

a Explain how the products of the Krebs cycle are involved in the electron transport chain in the mitochondria.   2 marks

A mutation occurred in the gene that codes for the enzyme mitochondrial malate dehydrogenase (MDH) in the biochemical pathway of the Krebs cycle. The mutation alters the enzyme and inhibits its function.

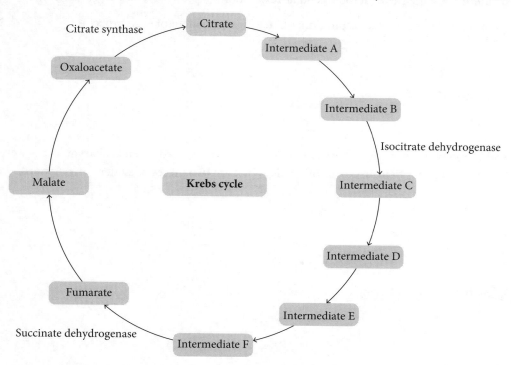

b What would happen to the biochemical pathway in the diagram if the malate dehydrogenase was no longer functional?   2 marks

_____

_____

_____

_____

c How would the mutation in the DNA affect the malate dehydrogenase on a molecular level?   3 marks

_____

_____

_____

_____

_____

d How would lowering the temperature of the cell affect the Krebs cycle and the electron transport chain?   2 marks

_____

_____

_____

e Other than ATP, what are the molecule(s) produced during this stage?   1 mark

_____

Question 6 (7 marks)

The mitochondria are the site of aerobic cellular respiration.

a   Compare aerobic to anaerobic respiration, discussing **one** benefit of each process.   2 marks

_____

_____

_____

b   The endosymbiotic theory hypothesises that mitochondria evolved when prokaryotic cells were
engulfed by other cells to form eukaryotic cells. What are **two** pieces of evidence to support
this theory?   2 marks

_____

_____

c   Complete the table by referring to the diagram of a mitochondrion.   3 marks

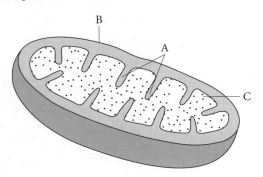

	Name of structure	Process or role in cellular respiration
A		
B		
C		

Question 7 (11 marks)

An experiment was designed to test the rate of cellular respiration at three temperatures.

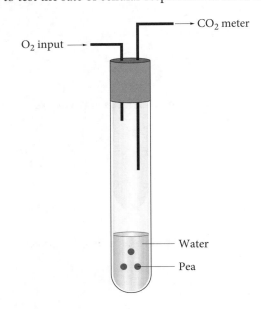

Each experimental set-up included one test tube containing 50 mL of water and 15 germinating peas. The test tube was sealed but provided with different amounts of O_2. CO_2 was measured to determine the relative rate of cellular respiration.

The results of the experiment are shown in the following graph.

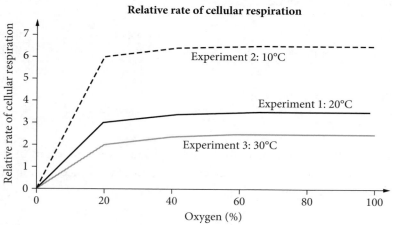

a Identify the limiting factor at 20% oxygen for all three reactions.   1 mark

_____

b For Experiment 1, identify a limiting factor other than temperature at 80% oxygen.   1 mark

_____

c Referring to the data above, explain how the temperature affects the rate of cellular respiration.   2 marks

_____

_____

_____

d Which stage of cellular respiration uses oxygen and how does it enable the stage to occur?   3 marks

_____

_____

_____

_____

e A student hypothesised that an experiment at 50°C would have a greater rate of cellular respiration. Would this hypothesis be supported?   2 marks

_____

_____

_____

f Identify **two** improvements that could be made to the experiment.   2 marks

_____

_____

Question 8 (10 marks) ⬤⬤▮

An experiment was carried out to determine the rate of yeast fermentation in a sealed container. Apple juice (to act as a glucose source), buffer solution and yeast were added to a conical flask and the concentrations (arbitrary units) of ethanol, oxygen and carbon dioxide were analysed over 180 minutes. The results are shown in the table below.

Measurement	Time (min)						
	0	**30**	**60**	**90**	**120**	**150**	**180**
Ethanol concentration	0	0	2	4	6	9	9
Oxygen concentration	5	3	1	0	0	0	0
CO_2 concentration	8						

a   The student completed the measurements for the ethanol and oxygen but didn't include the information for the concentration of CO_2. What would occur to the concentration of CO_2 over the 180 minutes? Explain why.   2 marks

_____

_____

b   Which processes would be occurring between 30 and 60 minutes?   2 marks

_____

_____

c   Compare the ATP production per glucose molecule at 30 minutes and 150 minutes.   1 mark

_____

_____

d   Complete the following graph with the concentration of ethanol.   2 marks

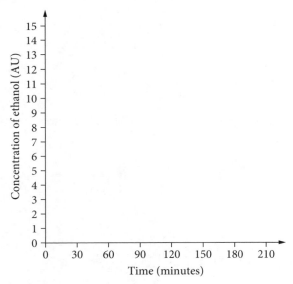

e Explain what would occur if there was an unlimited supply of glucose within the conical flask and the experiment was continued for a couple of days.   1 mark

_____

_____

f Why does the cell produce ethanol from pyruvate?   2 marks

_____

_____

_____

Question 9 (6 marks) ©VCAA VCAA 2016 SB Q2 ○ ● ●

Plant materials containing cellulose and other polysaccharides are reacted with acids to break them down to produce glucose. This glucose is then used by yeast cells for fermentation.

a Why is fermentation important for yeast cells?   1 mark

_____

b What are the products of fermentation in yeast cells?   1 mark

_____

A by-product of the acid treatment of plant materials is a group of chemical compounds called furans. It has been observed that as the concentration of furans increases, the rate of fermentation decreases. The enzyme alcohol dehydrogenase is required for the process of fermentation.

c Design an experiment to test the hypothesis that one of the furans, called furfural, is an inhibitor of the enzyme alcohol dehydrogenase. Assume that the experiment will be repeated many times and that environmental factors are kept constant.   4 marks

_____

_____

_____

_____

_____

_____

_____

_____

_____

Question 10 (6 marks) ©VCAA VCAA 2016 SB Q4 ●●●

The apparatus shown below was used in a series of experiments to study aerobic respiration.

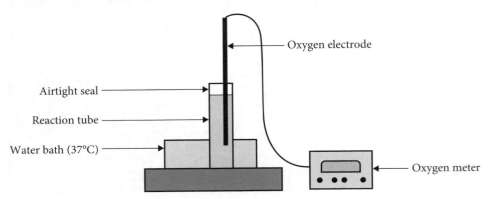

In three different experiments, the reaction tube initially contained the following:

1   suspension of mitochondria

2   cytosol of cells from which the mitochondria had been removed

3   suspension of mitochondria and cytosol of cells.

The temperatures and pH of the mixtures within the reaction tubes were carefully controlled so as not to damage the mitochondrial structure or any of the enzymes. In each experiment, a solution containing glucose was first added to the mixture in the reaction tube and the oxygen concentration was measured for 3 minutes. Then, a pyruvate solution was added and the oxygen concentration was measured again for 3 minutes.

Using your knowledge and understanding of aerobic respiration and mitochondria, complete the tables with your prediction of the change in oxygen concentration of the mixture in the reaction tube after the addition of each substance and give a reason for your prediction.

Experiment 1 – Suspension of mitochondria   2 marks

Suspension added	Change in oxygen concentration (increase/decrease no change)	Reason
Glucose		
Pyruvate		

Experiment 2 – Cytosol of cells from which the mitochondria would have been removed   2 marks

Suspension added	Change in oxygen concentration (increase/decrease/no change)	Reason
Glucose		
Pyruvate		

Experiment 3 – Suspension of mitochondria and cytosol of cells   2 marks

Suspension added	Change in oxygen concentration (increase/decrease/no change)	Reason
Glucose		
Pyruvate		

Test 6: Biological applications of biochemical pathways

Section A: 10 marks. Section B: 40 marks. Total marks: 50 marks.
Suggested time: 35 minutes

Section A: Multiple-choice questions

Instructions to students
- For each question, circle the multiple-choice letter to indicate your answer.

Question 1 ◉◯◯

Which of the following is correct about the production of biofuel?

A A non-renewable energy source with the addition an organic compound forms the biofuel.

B Lactic acid can be isolated for a high-energy compound during the fermentation of corn with yeast.

C Fermentation of bacteria in culture mediums can produce H_2 used for a biofuel.

D Biogas and natural gas are composed of natural compounds formed through the biofuel production.

Question 2 ◉◉◯

Both biofuels and fossil fuels release carbon dioxide into the atmosphere when burnt. Which one of the following statements reflects the climate-friendly reason for using biofuels rather than fossil fuels?

A Fossil fuels burn cleaner that biofuels.

B Fossil fuels are much cheaper to produce than biofuels.

C Biofuels draw down carbon dioxide from the existing atmosphere whereas fossil fuels contain carbon dioxide from ancient atmospheres.

D Biofuels use up animal and plant waste that would decompose and produce methane whereas fossil fuels is made up of animal and plant material that decomposed millions of years ago.

Question 3 ◉◉◯

Social factors can influence how people respond to bioethical issues. An example of this is

A politicians using their power to legislate against the continued mining of coal.

B people reducing their carbon footprint in response to the changes in climate.

C banks refusing to lend money to companies to search for more fossil fuels.

D mining companies being taken to court for the damage they are doing to the environment.

Question 4 ◉◉◯

A scientist used CRISPR to alter a gene in a somatic cell of a tomato plant with the aim of decreasing the rate at which the fruit matured, allowing farmers an extended picking period and time on the shelf. The gene modification would be present in

A all flowers produced by the plant.

B the tomatoes produced by the offspring of the plant.

C the gametes of the plant.

D the cells and cells produced from these in the modified plant only.

Question 5 ●●

Which of the following is **not** a disadvantage of using biofuels?

A Change in the ecosystem with the use of more fertiliser and pesticides

B Requires large amount of land and water to produce the crops

C Modification of engines for the use of pure bioethanol

D Limited amount of available resources for the production of biofuels

Question 6 ●●●

Bioethanol can be produced by fermenting sugar rich fruit and vegetables such as beet, sugarcane and maize in a fermenter. You would expect the conditions inside the fermenter to include

A temperature of 100° C and pH of 2.

B temperature of 89° C and pH of 7.

C temperature of 20° C and pH of 14.

D temperature of 36° C and pH of 5.

Question 7 ●●●

CRISPR interference is a type of application developed from the original CRISPR technology. In this technique, the Cas9 is inactivated and denatured to prevent cleavage of DNA when the guide RNA recognises the specific sequence. Transcription rates can be altered by preventing the enzymes responsible for binding to the promoter region, inhibiting any expression of the gene. A second technique is RNA interference, which breaks down the RNA produced from the genes of interest. Which of the following is an advantage of the CRISPR interference over the original CRISPR technique?

A RNA interference cuts at more sites in the DNA genome at its specific sequence.

B There are fewer offsite effects and risk of mutations in the organism's genome.

C There is more specific targeting of the gene.

D Fewer dosages of the CRISPR system is required in the target cells.

Question 8 ●●●

Scientists are currently looking at the potential use of seaweed and algae for the production of biofuels to reduce the amount of land mass being used for biofuel crops. Which of the following would be the most relevant considerations when selecting the species to be used?

A Seasonal growth period, energy production, labour

B Photosynthetic rate, chlorophyll pigment, cell density

C Chromosome complexity, reproductive rate, energy requirement

D Impact on ecosystem, infection rate, replication style

Question 9 ●●●

Which of the following are waste products of fermentation in yeast?

A ATP and heat

B Carbon dioxide and ATP

C Ethanol and carbon dioxide

D Pyruvate and ATP

Question 10 ●●●

To reduce the amount of crops being used for biofuels instead of for food, some farmers are now using the discarded plant material after crops have been harvested. These complex starchy parts of the plant contain lignocellulosic material (material composed of lignin, cellulose and hemicellulose). This adds a level of complexity to the process because yeast is only able to ferment glucose.

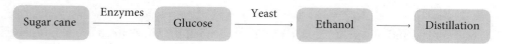

The enzymes α-amylase and glucoamylase are added with the sugar cane. The function of these enzymes is to

A increase the activation energy and facilitate the formation of glucose from cellulose and lignin.

B be involved in a series of chemical reactions to break up the glucose in the presence of cellulose.

C facilitate the formation of glucose through a biochemical pathway, digesting the cellulose and lignin.

D provide the active site for the yeast to use glucose in their fermentation.

Section B: Short-answer questions

> **Instructions to students**
> • Answer all questions in the spaces provided.

Question 1 (9 marks) ●●

Yeast are single-celled, microscopic organisms in the fungi kingdom that can undergo aerobic or anaerobic respiration. These organisms undergo fermentation in the absence of oxygen. It is this process that scientists have been investigating to isolate the end product for use in biofuels.

a In the space below, write the word equation for fermentation by yeast.   2 marks

b Name the product from fermentation that is isolated and used in biofuels.   1 mark

c Why is there a limit to how much product can be formed in a particular batch of yeast?   1 mark

d Why is it more complicated to use starchy crop waste for biofuel production than use the crop itself (e.g. corn or sugar cane)?   2 marks

e State the advantage of adding ethanol to petrol.

1 mark

_____

_____

_____

f State one economical and one political factor that could affect the widespread use of biofuels for transport.

2 marks

_____

_____

_____

_____

Question 2 (8 marks)

Some archaea and bacteria can undergo anaerobic respiration to produce a compound that can be used in biofuel. These bacteria or archaea can be grown in controlled cultures.

a Appropriate conditions are required for bacteria to undergo anaerobic fermentation. Explain why.

1 mark

_____

_____

b Explain **two** conditions that would need to be controlled in the culture medium to ensure the products were being produced and why.

2 marks

_____

_____

c In a fermentation, substrate was added continuously to the bacteria. In the space below, draw a labelled graph that shows the rate of reaction over time.

2 marks

d What would the bacteria produce under anaerobic conditions that would be useful in biofuel production?   1 mark

_____

e How can the product of bacterial fermentation be used for energy production?   2 marks

_____

_____

_____

Question 3 (7 marks) ⬤⬤⬤

In 2020, the world's population was 7.8 billion people. The United Nations expects the population to exceed 9.7 billion in 2050, with more than half of the global population growth predicted to occur in Africa. Such a substantial increase in population is expected to result in food shortages, land shortages and an increase in morbidity and mortality. For these reasons, scientists are looking to biotechnology to change how we farm.

a State **two** advantages of using CRISPR in crop production compared to selective breeding.   2 marks

_____

_____

_____

b Compare the use of transgenic organisms to the use of CRISPR in crops. Include an advantage and a disadvantage.   3 marks

_____

_____

_____

_____

_____

c What are the functions of the Cas9 and single guide RNA?   1 mark

_____

_____

_____

d How is observing a knock-out function for a gene useful for crop yield?   2 marks

_____

_____

_____

_____

_____

Question 4 (8 marks) ⬤⬤⬤

Scientists have performed many studies on different species of crops to increase yield, alter flowering progression and increase disease and pathogen tolerance. For example, inducing a mutation in the _GmFT2a_ gene of soybeans resulted in later flowering and a larger seed size. This gene is involved in the photoperiod flowering pathway of the soybean.

a State the type of cut Cas9 endonuclease would produce.   1 mark

_____

b What is a technique that the scientists would have used for cloning large amounts of the mutation before use in the CRISPR-Cas9 process?   2 marks

_____

_____

_____

_____

c Create a diagram explaining the steps involved in inserting the mutated _GmFT2a_ gene into the soybean genome.   3 marks

d Single guide RNA is generally 20 base pairs long. Why is this important?   1 mark

_____

_____

_____

e How does variation in a species enable CRISPR to be used?   1 mark

_____

_____

_____

Question 5 (7 marks)

a Define 'biofuel'.   1 mark

_____

b Compare the uses of marine and terrestrial plants in biofuel production.   2 marks

_____

_____

_____

_____

_____

_____

c Explain the benefit of using genetically modified organisms in the production of biofuel.   2 marks

_____

_____

_____

_____

_____

d Discuss **two** potential barriers to using biofuels to replace traditional fuel production.   2 marks

_____

_____

_____

_____

UNIT 4
AREA OF STUDY 1

Test 7: Responding to antigens

Section A: 20 marks. Section B: 66 marks. Total marks: 86 marks.
Suggested time: 40 minutes

Section A: Multiple-choice questions

Instructions to students
- For each question, circle the multiple-choice letter to indicate your answer.

Question 1

The role of dendritic cells in the mammalian immune system is to

A process antigens and present them on their cell surfaces to T helper cells.

B release histamine in response to a foreign antigen.

C migrate to areas of infection or injury and stimulate inflammation.

D enhance chemotaxis and cell lysis of pathogens.

Question 2

Non-specific defences in mammals include the

A mucous lining in the respiratory tract.

B production of memory cells.

C attachment of cytotoxic T cells to infected cells.

D production of antibody molecules by plasma cells.

Question 3

Which of the following acts as a chemical barrier to the entry of pathogens into body tissues?

A Interferon produced by virus-infected cells

B Histamine produced by mast cells

C Complement proteins in the blood

D Hydrochloric acid in the stomach

Question 4

Which cell of the mammalian immune system possesses an MHC II marker and interacts with lymphocytes?

A neutrophil

B macrophage

C mast cell

D natural killer cell

Question 5 ©VCAA VCAA 2015 SA Q18 ●●●

A girl is carrying a piece of wood. A small piece breaks off and becomes embedded in her finger. The next day, she notices an inflammatory response occurring in her finger. In the region around the small piece of wood embedded in her finger

A mast cells would release antibodies.

B the skin tissue would become pale and cold.

C the capillaries would become more permeable.

D red blood cells would leave the blood vessels and engulf foreign material.

Question 6 ●●●

During an inflammatory response, cytokines are released to signal to the different cells in the immune system to initiate an immune response. Which of the following does **not** occur as a direct response to the release of cytokines?

A Vascular permeability and vasodilation

B Migration of neutrophils and macrophages to the site of infection

C Increase in body temperature

D Activation of the humoral and cell-mediated responses

Question 7 ●●●

The lines of defence consist of three stages. An example of a first line of defence mechanism against a pathogen is

A a cytotoxic T cell releasing perforins to kill infected cells.

B natural flora in the digestive and respiratory tract.

C complement protein activation.

D release of histamine from damaged mast cells to attract phagocytes.

Question 8 ●●●

Sam tried a peanut butter sandwich at a friend's house for the first time. The following week he bought a jar and had peanut butter on toast for breakfast. Shortly after, his chest began to constrict, and he had trouble breathing. An inflammatory response was initiated in Sam's throat and he began to struggle to breathe. The peanut butter acted as

A an antigen.

B a pathogen.

C an allergen.

D an IgE antibody.

Question 9 ●●●

An increase in body temperature at the hypothalamus is due to which chemical signalling molecule?

A interleukin

B histamine

C interferon

D neurotransmitter

Question 10 ●●●

A virally infected cell releases a chemical to surrounding cells. Which of the following is the correct chemical and effect?

	Chemical released	Effect
A	Interferon	Builds resistance in the surrounding cells through antiviral proteins.
B	Interleukin	Stimulates the infected cell to undergo apoptosis.
C	Chemokines	Builds resistance in the infected cell.
D	Histamine	Vasoconstriction of the blood vessels to increase specific immune cell movement.

Question 11 ●●●

Bovine spongiform encephalopathy, or mad cow disease, is a neurodegenerative disease in cows that is caused by a prion. The immune system does not defend against prion infection. This is probably because

A the defective proteins are not recognised as self.

B prion infection is caused by an intracellular obligate parasite.

C there is no immediate innate inflammatory response to stimulate the humoral or cell-mediated response.

D mad cow disease occurs in the neurons where the immune system is unable to recognise foreign pathogens.

Question 12 ●●

Chemotaxis is the process whereby

A antigen-presenting cells migrate away from an infected area to the lymph nodes for humoral activation.

B leukocytes migrate from low to high concentrations of chemokine release to areas of infection.

C vascular membranes increase in permeability and vasodilate.

D chemicals are released post infection to clot the blood.

Question 13 ●●

A person became ill from an unknown pathogen. The pathogen contained no genetic material, was non-cellular and replicated when it encountered similar structures. The pathogen would be classified as

A bacteria.

B prion.

C virus.

D protozoa.

Question 14 ●●

These graphs show the number of pathogens present in the blood of organisms infected with two infectious diseases. The different growth patterns shown in these graphs could be due to the fact that

A viruses regulate their reproductive rate.

B viruses mutate faster than the immune system is able to respond.

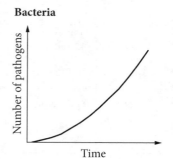

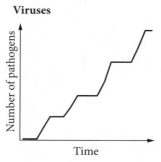

C bacteria grow in the extracellular environment, whereas viruses multiply within the cell.

D the immune system takes longer to recognise bacterial infection.

Question 15 ●●□

An innate immune cell responsible for detecting the absence of an MHC I marker on the surface of a cancer or foreign cell and releasing pore-forming proteins and granzymes (cytotoxic enzymes) is a

A natural killer cell.

B cytotoxic T cell.

C neutrophil.

D dendritic cell.

Question 16 ●□□

Chemical barriers that are involved in the resistance of plants to pathogens include

A the vertical alignment of leaves.

B hairs and thorns.

C a thick waxy cuticle.

D sap and natural secretions.

Question 17 ©VCAA VCAA 2016 SA Q20 ●□□

The inflammatory response is a defence mechanism that evolved in higher organisms to protect them from infection and injury. This response

A includes phagocyte migration to the site of the injury.

B is part of the adaptive immune system.

C is specific to the type of foreign body.

D involves the production of lymphocytes.

Question 18 ©VCAA VCAA 2010 (1) SA Q3 ●□□

One of the similarities between the defence mechanisms of a plant and an animal includes the

A production of memory cells.

B release of immune cells through a circulatory system.

C use of an epidermal layer to inhibit the invasion of pathogens.

D production of salicylic acid to warn cells of an invading pathogen.

Question 19 ©VCAA VCAA 2010 (1) SA Q4 ●●□

William has an allergy that causes problems. Eight skin-prick tests were carried out to find the source of the problem. A portion of skin was divided into eight squares and a skin-prick made with an extract from each of the possible causes. An inflamed area indicates a positive reaction. William's skin-prick test results are shown below.

Horse	House dust mite	Mould	Histamine
●	●		●
Cat	Guinea pig	Dog	Saline
	●	●	

● Indicates positive reaction

From the results it is reasonable to assume that

A William is allergic to saline.

B histamine is a negative control.

C William is allowed to keep his cat.

D the same allergen is found in a horse and a house dust mite.

Question 20 ●●●

A person who suffered from leukemia underwent a bone marrow transplant, receiving the cells from a relative with a similar tissue type. Thirty days later the person experienced graft versus host disease. The donor cells differentiated into leukocytes and began lysing and initiating apoptosis in the recipient's healthy cells. This attack on the recipient's cells would have occurred because the

A recipient's cells and the donor cells would be producing similar intracellular proteins.

B recipient's cells and the donor cells would recognise different surface receptors.

C recipient's cells and the donor cells would have different MHC II markers on the surface.

D donor cells would reject the surrounding tissue and over-replicate to outcompete with the recipient's cells.

Section B: Short-answer questions

Instructions to students
• Answer all questions in the spaces provided.

Question 1 (7 marks) ●● ▪

a Which physical barrier has been disrupted in the diagram above?   1 mark

_____

b Explain the steps that would follow the diagram above.   3 marks

_____

_____

_____

_____

_____

c Explain **two** functions of complement proteins in an infection.   1 mark

_____

_____

d As demonstrated in the diagram, pus forms at the wound. What is the composition of the pus?   2 marks

_____

_____

_____

Question 2 (6 marks)

While Julia is sitting in a park, she begins to experience an allergic reaction, starting with swelling and redness on her arms.

a Name the molecule responsible for initiating this response.   1 mark

_____

b Explain how Julia became sensitised to this molecule.   3 marks

_____

_____

_____

_____

_____

_____

c What are the steps in the subsequent exposures that led to the response Julia is experiencing?   2 marks

_____

_____

_____

_____

Question 3 (6 marks)

a Compare an antigen to a pathogen.   1 mark

_____

_____

_____

b Compare the roles of MHC I and MHC II in the innate immune system.   2 marks

_____

_____

_____

_____

c Complete the following table with the types of barriers involved in the first line of defence in plants and animals.

3 marks

	Barrier example		
	Physical	Chemical	Microbiological
Animals			
Plants			

Question 4 (9 marks) ⬤⬤⬤

Chickenpox is a viral infection caused by the varicella-zoster virus. It most cases, it causes an itchy, blistering rash across the skin along with a mild fever. However, in newborns or immunocompromised individuals, chickenpox can develop with severe symptoms and in some cases become fatal. In most children, recovery occurs in two weeks. In older people, the virus can reappear, developing into shingles after being reactivated in nerve tissue. Since the implementation of the National Immunisation Program in 2005, cases and hospitalisations for chickenpox have significantly declined. In 2017, 93% of 2-year-old children were vaccinated.

A family living in the suburbs of Melbourne presented at a local clinic with one of their four children and the father showing symptoms of the varicella virus (chickenpox). Although other family members had been in close contact, none was experiencing symptoms. It was determined that the mother, currently breastfeeding the youngest child, had suffered chickenpox when she was younger.

a How does chickenpox cause disease within the body?

3 marks

b Explain the role of **two** innate immune cells and how they would be involved in initiating a response after the virus enters the body.

2 marks

c Compare the types of immunity of the mother and the breastfeeding child.   2 marks

....

....

....

d Explain the role an infected cell would have in the body for reducing the spread of
this virus.   2 marks

....

....

....

Question 5 (5 marks)   ©VCAA   VCAA 2004 (1) SB Q8 (adapted)   ● ●

Ticks are a group of parasitic spider-like arthropods which live on the outside of an animal. They
attach to the skin of an animal such as a dog and suck blood from the host.

a   **i** Describe a factor with which an external parasite such as a tick has to cope in order
to survive, which is not a problem for an internal pathogen.   1 mark

....

....

ii Explain how an external parasite could overcome this problem.   1 mark

....

....

Investigations were carried out to monitor the patterns of growth of bacteria and viruses once
they had infected an organism. Results were presented graphically as follows.

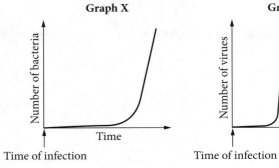

b Identify the type of infection in: 1 mark

Graph X: ....

Graph Y: ....

c Explain why there is a difference between the patterns of growth after infection of an organism.

2 marks

_____

_____

_____

_____

_____

_____

_____

_____

_____

_____

Question 6 (5 marks) ●●●

This diagram represents a virus that causes influenza. In response to the presence of a virus, interferon is synthesised and secreted into the tissue fluid by many different cell types, including macrophages and lymphocytes.

The viral core contains nucleic acid. The outer layer is composed of a lipid studded with protein molecules of two types: neuraminidase and haemagglutinin. These proteins trigger an immune response in humans and other vertebrates.

Neuraminidase

Haemagglutinin

Nucleic acid

a Why is interferon considered to be part of the innate, non-specific immune system?

1 mark

_____

b How does interferon act to limit the spread of viruses in the body?

1 mark

_____

_____

_____

c Viruses trigger a specific immune response. What is the advantage of having interferon as well?

1 mark

_____

_____

_____

d New strains of influenza appear almost every year. These new strains are distinguished by changes in the viral nucleic acid, which lead to changes in the structure of the surface protein molecules. Small changes result in proteins, which some immune cells are still able to recognise and respond to. If re-infected with the new strain, previously infected individuals may still get influenza, but it will be milder. However, occasionally, the viral proteins undergo major structural change, resulting in severe pandemics (worldwide outbreaks) of influenza. Explain how a change in viral nucleic acid leads to a change in the surface proteins of the virus.

2 marks

_____

_____

_____

_____

Question 7 (9 marks)

a State **one** circumstance when the skin is not able to provide an effective barrier.

1 mark

_____

b In parts of the body where skin is not present, the body relies on chemical barriers to prevent pathogen entry. Name and briefly explain **two** chemical barriers of the body.

2 marks

_____

_____

_____

c **i** After a short time an infected area can become swollen. Why does swelling occur?

2 marks

_____

_____

_____

ii Name **two** other symptoms you would expect to see in the affected area. For each, explain why that symptom would be present.

4 marks

Symptom 1: _____

Reason: _____

Symptom 2: _____

Reason: _____

Question 8 (7 marks)

The defence barriers in plants are a coordinated system of molecular, cellular and tissue-based responses to pathogen attack. Protection from a pathogen's initial invasion is achieved through defences such as physical and chemical barriers.

a Why do plants need a defence system?   1 mark

_____

_____

b Name **two** physical barriers protecting the plant from invasion.   2 marks

_____

_____

c Name and outline the actions of **two** chemical barriers that protect the plant from invasion.   2 marks

_____

_____

d When would the cellular and tissue-based responses to pathogens attack be activated?   1 mark

_____

e Explain **one** other response that plants can use to limit damage after pathogenic invasion.   1 mark

_____

_____

Question 9 (7 marks)

The bacterium *Staphylococcus aureus* (or golden staph) is responsible for many pus-forming infections, particularly boils and infections in open wounds.

a How may the bacteria gain access to the body?   1 mark

_____

Many strains of this bacterium have evolved a resistance to antibiotics. Outbreaks of severe infections regularly occur in many large hospitals. In Japan in 1997, a strain of *S. aureus* was found that was resistant to all antibiotics, including methicillin and vancomycin, the antibiotics of last resort. This resistance is a genetically determined trait in the bacteria.

b i Outline how the build-up of the resistance could have occurred in the *S. aureus* bacteria.   2 marks

_____

_____

_____

_____

ii Why would hospitals provide a suitable environment for the rapid evolution of bacteria?   2 marks

c Outline **two** features of bacteria that contribute to the rapid evolution of new strains.   2 marks

Question 10 (5 marks) ©VCAA VCAA 2012 (1) SB Q2 (adapted) ●●

Mast cells are found in a number of different tissues and play an important role in allergic reactions. Consider the following diagram.

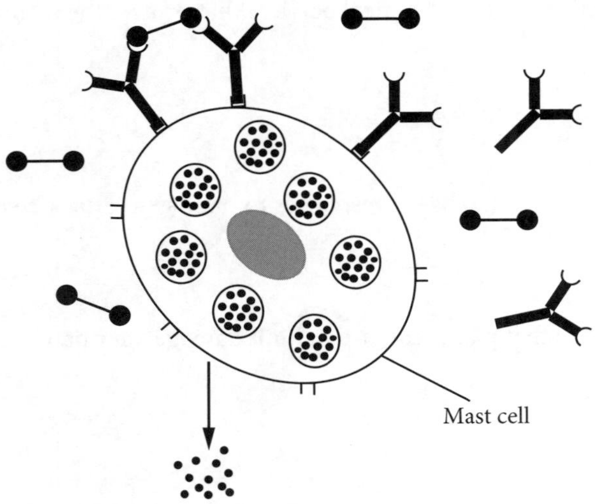

Mast cell

a State the name of a similar cell which circulates in the body.   1 mark

b Label at least **three** appropriate parts of the diagram related to the activation and action of a mast cell. Referring to the diagram and labels you have entered, outline the order of events that occur during an allergic reaction.   4 marks

Test 8: Acquiring immunity

Section A: 20 marks. Section B: 75 marks. Total marks: 95 marks.
Suggested time: 55 minutes

Section A: Multiple-choice questions

Instructions to students
- For each question, circle the multiple-choice letter to indicate your answer.

Question 1

Long-term immunity may be provided by

A an injection of antivenom after a spider bite.

B ingestion of antibodies in breast milk.

C vaccination with a solution containing killed viruses.

D antibodies that cross the placenta during foetal development.

Question 2

The lymphatic system consists of the lymph nodes together with a network of lymphatic vessels that carry
a clear fluid called lymph. The lymphatic nodes

A are where all leukocytes reside when no infection is present.

B are reduced in size during an infection.

C contain both red and white blood cells.

D are the major sites of B and T cells.

Question 3

The following diagram represents a bacterium capable of causing disease.

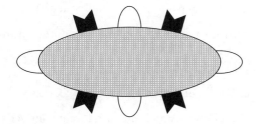

Which of the following diagrams represents an antibody that would be effective against this bacterium?

A   B 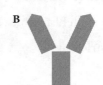 C  D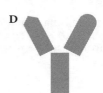

Question 4 ⬤⬤⬤

Measles is an infectious disease caused by a virus. Symptoms usually appear 6–10 days after exposure to the virus and include a rash, headache and fever. In some cases, the patient is affected by secondary bacterial infections, especially of the lungs. Antibiotics are often prescribed.

When suffering from measles and a secondary bacterial lung infection, the patient's immune cells produce antibodies. Which of the following statements is true of these antibodies?

A The same antibody is capable of attacking the measles virus and the bacteria infecting the lung.

B The antibodies that attack the virus are produced by T helper cells, whereas the antibodies that attack the bacteria are produced by B cells.

C Viral antibodies and bacterial antibodies are produced by the same plasma cells.

D Large quantities of antibodies specific to the measles virus remain in the circulation for a short time.

Question 5 ⬤⬤

The following graph shows the amount of circulating antibody in two patients exposed to the measles virus at the same time.

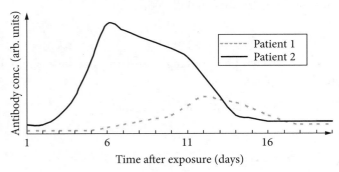

Blood antibody concentration

On the basis of this information, it is reasonable to conclude that

A patient 2 had been immunised against measles, whereas patient 1 had contracted measles naturally.

B patient 2 will suffer from measles for a shorter time than patient 1.

C on day 12, patient 1 had more circulating antibodies than patient 2.

D patient 1 is likely to suffer the same symptoms from measles in the future.

Question 6 ⬤⬤

Which of the following best describes the function of cytotoxic T cells?

A They produce antibodies, which circulate and bind with foreign antigens.

B They recognise foreign particles in extracellular fluid through MHC restriction.

C They are important in the inflammatory response and migrate through histamine release.

D They are activated through antigen-presenting cells and MHC restriction.

Question 7 ⬤

As part of the body's specific immune response, human cells infected by viruses may be killed by

A antibodies.

B helper T cells.

C complement proteins.

D cytotoxic T cells.

Question 8 ⊙⊙

Which of the following is correct when comparing long-term and short-term immunity in response to a toxin or pathogen?

	Long-term immunity	Short-term immunity
A	Cytotoxic T cells and B plasma cells produced	Larger and more rapid response on subsequent exposure.
B	Increase in inflammation and swelling on subsequent exposures to pathogen	An injection of antivenom after exposure
C	Apoptosis of some B and T cells post-infection. Memory cells reside in lymph nodes.	Second line involved in the response to assist the antibodies after neutralisation of the toxin/pathogen
D	A course of broad-spectrum antibodies	Vaccination with a solution containing virus protein coat injected

Question 9 ⊙⊙⊙

Lymphocytes

A are responsible for cell-mediated immunity but not humoral immunity.

B mature and undergo self-tolerance in the thymus.

C circulate in the blood but not in the lymph.

D develop from bone marrow cells.

Question 10 ⊙

In humans, specific (adaptive) immune responses include

A production of antibodies by plasma cells.

B production of histamine by mast cells.

C phagocytosis by monocytes.

D neutralisation of bacteria by complement proteins.

Question 11 ©VCAA VCAA 2016 SA Q24 ⊙

In the search for a malaria vaccine, scientists have focused on a protein called circumsporoite protein (CSP). CSP is secreted by the malaria parasite and is present on its surface. For the vaccination to work, the scientists want CSP to act as

A an antigen.

B an allergen.

C an antibody.

D a complement protein.

Question 12 ⊙⊙

After an antigen has reached the lymph node and a T helper cell has been activated, clonal selection and expansion of the humoral response produces B plasma cells and B memory cells. The B plasma cells would contain a large amount of which organelle?

A nucleoli

B mitochondria

C centrioles

D free-floating ribosomes

Question 13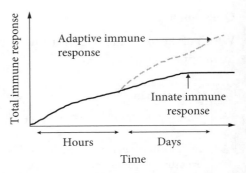

A person was exposed to the varicella virus (chickenpox) for the first time. The graph at right shows the total rate and size of the innate and adaptive immune responses.

A couple of years later, the person was exposed to the varicella virus again at the local childcare centre where they worked. Which of the statements about the immune response is correct?

A The innate and adaptive immune responses would occur at the same rate as the first exposure.

B Both the innate and adaptive immune responses would be faster and larger than the responses after the first exposure.

C The innate immune response would remain the same, whereas the adaptive immune response would be faster and larger than after the first exposure.

D The innate immune response would be faster as the individual aged and the adaptive immune response would increase due to the memory cell formation on first exposure.

Question 14

There are more than 40 species of funnel web spiders in Australia, with *Atrax robustus* (Sydney funnel web) being linked with the most human fatalities if antivenom is not received shortly after the bite. A gardener presented to a hospital's emergency department with a funnel-web spider bite and received antivenom, preventing development of any side effects. A year later, the gardener was bitten again. The gardener

A would be protected with memory cells from the antivenom.

B would still contain the antibodies from the previous antivenom protection.

C received artificial passive immunity and would need to be reinjected with antivenom.

D received natural active immunity and had formed their own antibodies over the previous year.

Question 15

The lymphatic system is composed of lymph vessels, primary and secondary organs, and filters fluid throughout the body to help rid it of foreign antigens. Which of the following is **not** a correct description of this system?

A Unidirectional fluid flow with the assistance of valves

B Fluid enters through the afferent lymph vessel and leaves the node through the efferent lymph node

C Circulates the macrophages, neutrophils, mast cells and platelets throughout the body

D Facilitated in movement through the contraction of muscles

Question 16

An antigen-presenting cell engulfs a foreign pathogen and migrates to a lymph node. Which of the following is the correct series of steps for the activation of a T helper cell?

A Presents on the MHC I marker, activating the T helper cell to release interferons to activate naive B and T cells

B Presents on the MHC II marker, activating the T helper cell to release interleukins to activate naive B and T cells

C Presents on the MHC II marker, activating the T helper cell and binds onto naïve B and T cell to activate via a contact-dependent signal

D Releases the antigen into the lymph fluid, which binds onto the MHC II marker on the T helper cell, which activates naive B and T cells through release of interleukins

Question 17 ●●

Bacillus anthracis is a bacterium that can be contracted from infected animals or from bacterial spores in soil. These bacteria produce an anthrax toxin that consists of a cell-binding protein and two enzymes that can enter cells through endocytosis. These proteins can inhibit a range of biochemical pathways and can be fatal without treatment.

Which of the following would be most effective against this specific anthrax toxin on entry into the cells?

A T helper cells

B cytotoxic T cells

C antibodies

D natural killer cells

Question 18 ●●

A child receives a vaccine for whooping cough (pertussis) in a five-dose schedule at 2, 4, 5 and 18 months and at 4–5 years. The type of immunity acquired from this vaccine would be best described as

A naturally acquired, passive immunity.

B naturally acquired, active immunity.

C artificially acquired, passive immunity.

D artificially acquired, active immunity.

Question 19 ©VCAA VCAA 2013 SA Q18 ●●

DiGeorge syndrome is a rare, congenital disease that can disrupt the normal development of the thymus gland. This disease would result in the

A swelling of lymph nodes.

B overproduction of B cells.

C reduced production of T cells.

D release of histamines from mast cells.

Question 20 ©VCAA VCAA 2012 (1) SA Q24 ●●●

A diagnostic test for HIV infection includes the following steps.

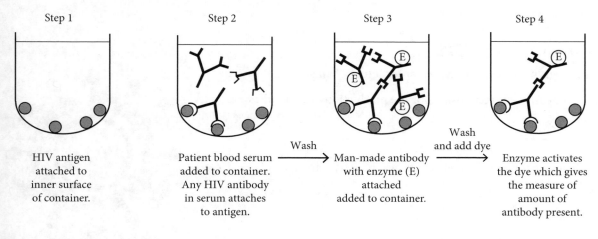

Step 1	Step 2	Step 3	Step 4
HIV antigen attached to inner surface of container.	Patient blood serum added to container. Any HIV antibody in serum attaches to antigen.	Wash → Man-made antibody with enzyme (E) attached added to container.	Wash and add dye → Enzyme activates the dye which gives the measure of amount of antibody present.

The test for HIV is reliable because the

A dye reacts with the patient's blood serum.

B enzyme has an active site for the HIV antigen.

C man-made antibody has the same shape as the HIV antigen.

D HIV antigen has a complementary shape specific to the HIV antibody.

Section B: Short-answer questions

Instructions to students
- Answer all questions in the spaces provided.

Question 1 (7 marks)

Vaccinations have been providing people with protection against pathogens since the 1940s.

a Each year, a new vaccination is required for the flu. Why is this necessary? 1 mark

_____

_____

b How is the adaptive immune response activated through vaccination? 3 marks

_____

_____

_____

_____

_____

c A vaccine is composed of which elements? 1 mark

_____

_____

d State **two** advantages of vaccinating individuals against cellular and non-cellular pathogens. 2 marks

_____

_____

_____

_____

Question 2 (6 marks)

The diagram at the right shows the structure of an antigen.

a Name the cells responsible for the production of antibodies. 1 mark

_____

b State what the monomer that an antibody is composed of. 1 mark

_____

c Draw a labelled antibody specific to the antigen shown in part b.   2 marks

d Explain **two** functions of antibodies.

2 marks

Question 3 (5 marks) ©VCAA VCAA 2015 SB Q5   ●●●

Consider the following diagram of a lymph node.

Anatomy of a lymph node

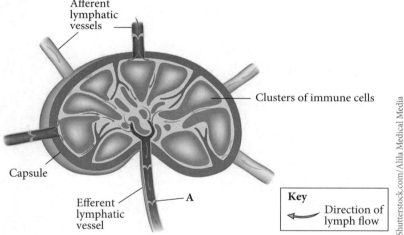

a Describe the role of the structure labelled A, found within the efferent lymphatic vessel.   1 mark

Immune cells are clustered within the lymph node (see diagram). There is more than one type of immune cell within each of these clusters.

b Name and describe the role of **one** type of immune cell found within these clusters that is involved in the innate immune response.   2 marks c   Another of the immune cell types found within these clusters has a large nucleus and extensive rough endoplasmic reticulum and plays an important role in an adaptive immune response.

Name this cell type and explain how the extensive rough endoplasmic reticulum assists this cell to perform its function.   2 marks

_____

_____

_____

_____

Question 4 (7 marks)

The following diagram represents an antibody.

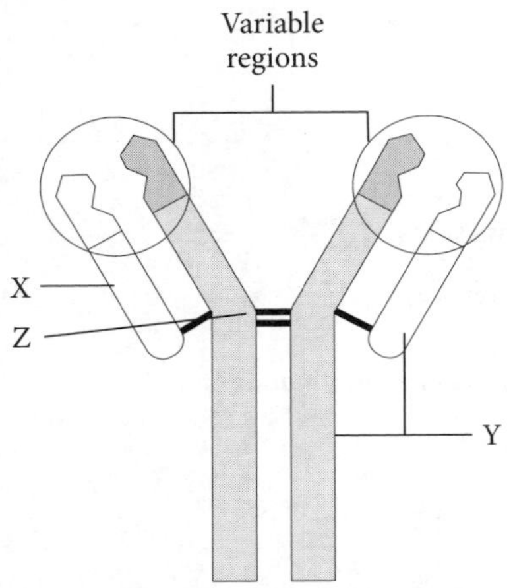

a   Name the labelled parts of the antibody.   3 marks

   i X

   _____

   ii Y

   _____

   iii Z

   _____

b   **i** What is the function of the variable regions?   1 mark

   _____

   _____

   ii State the significance of the fact that these regions are variable.   1 mark

   _____

   _____

   _____

c Explain **two** ways in which antibodies contribute to the removal of pathogens.

2 marks

_____

_____

_____

Question 5 (9 marks) ⬤⬤⬤

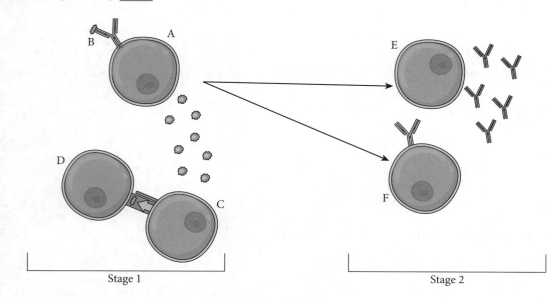

Stage 1   Stage 2

a Name the type of response being activated in the diagram.   1 mark

_____

b Name the two stages of activation.

2 marks

Stage 1 _____

Stage 2 _____

c Referring to A, B, C and D, explain the process occurring in stage 1.   4 marks

_____

_____

_____

_____

_____

_____

d Explain the process of molecule being released from cell E.   2 marks

_____

_____

_____

Question 6 (8 marks) ⬤⬤⬤

The human immunodeficiency virus (HIV) is a type of retrovirus that infects human cells, inserting its DNA into the human genome for continued infection. HIV is a progressive disease that decreases the function of the immune system over time, eventually progressing to AIDS, when the T helper (CD4) cells decrease to below 200 cells/mm^3. There is currently no cure and individuals often die due to opportunistic infections which would otherwise be non-life threatening.

a What is the role of the T helper cells in the adaptive immune response?   2 marks

_____

_____

b Why would people suffering from HIV/AIDS die from an illness such as the common cold?   2 marks

_____

_____

_____

_____

c State the type of drug used to treat a viral infection.   1 mark

_____

In the acute stage of HIV (first few weeks of infection) flu-like symptoms are present, including fever, rash, aches and swollen glands.

d State what is occurring in the body during this stage.   3 marks

_____

_____

_____

Question 7 (7 marks) ⬤⬤

The immune system is composed of three different stages to increase the protection against foreign materials. A bushwalker falls over a rock, grazing their leg and hand.

a What is the first line of defence that has been breached?   1 mark

_____

b Explain the immediate response by the body.   2 marks

_____

_____

_____

c Compare and contrast the second line of defence and the first line of defence.   2 marks

_____

_____

_____

_____

d Explain **two** reasons why it is important to have multiple lines of defence.   2 marks

_____

_____

_____

_____

_____

Question 8 (11 marks) ⬤⬤⬤

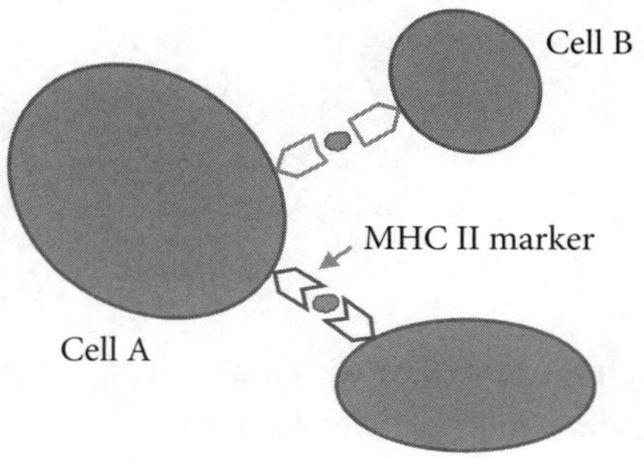

Cell B

MHC II marker

Cell A

a Where in the body does T cell maturation occur?   1 mark

_____

b Name cell A.   1 mark

_____

c Compare MHC I and MHC II markers in the activation of the cell-mediated response.   2 marks

_____

_____

_____

d Name and explain the function of the two cells produced after clonal expansion.   2 marks

_____

_____

_____

A person suffers from chronic glomerulonephritis, which causes continual inflammation of the kidney, eventually leading to significant scarring and inhibits the filtering system of the kidneys. This condition results in kidney disease and eventually kidney failure, requiring a kidney transplant.

e In relation to the adaptive immune system, why would immune-suppressant drugs be required post-surgery?   2 marks

_____

_____

_____

_____

f Compare the immune response initiated for a kidney transplant to a mature naïve B or T cell that has failed to undergo self-tolerance.   3 marks

_____

_____

_____

_____

_____

_____

Question 9 (8 marks)

Polio is a viral infection for which children receive vaccinations at 2, 4 and 6 months and 4 years. Influenza is a virus for which yearly vaccinations are required.

a Why do these viruses require different vaccination strategies?   2 marks

_____

_____

_____

b When the polio antigen enters the body in the vaccine, an adaptive immune response is initiated. Draw and label a diagram of the process of antigen presentation.   2 marks

c Explain the role of memory cells in the adaptive immune response.   2 marks

_____

_____

d Create a graph to show the first and second immune responses to the influenza vaccination.   2 marks

Question 10 (7 marks) ©VCAA VCAA 2015 SB Q4 (adapted) ●●●

In 2014, an outbreak of Ebola virus disease (EVD) occurred in West Africa. Humans may contract the virus from infected animals or from an infected person. In 2014, there was no safe and effective vaccine available to prevent EVD.

a Explain how an effective vaccine could provide long-term immunity to EVD.   4 marks

_____

_____

_____

_____

_____

b Scientists developing new vaccines for EVD are conducting trials in animal subjects. To evaluate the effectiveness of a new vaccine, both humoral and cell-mediated responses are measured in the animal subjects. Explain how these two immune responses are different. Give **two** differences in your answer.   2 marks

_____

_____

_____

_____

_____

c Explain how the implementation of a vaccination program would protect the community.   1 mark

_____

_____

Test 9: Disease challenges and strategies

Section A: 20 marks. Section B: 50 marks. Total marks: 70 marks.
Suggested time: 35 minutes

Section A: Multiple-choice questions

Instructions to students
· For each question, circle the multiple-choice letter to indicate your answer.

Question 1 ●●

With the European colonisation of Australia in the 1700s, Aboriginal and Torres Strait Islander populations were exposed to pathogens they had never been exposed to before. This resulted in epidemics that caused many deaths and a drastic decrease in the population numbers of many communities. Which of the following was **not** one of the initial diseases the Europeans bought with them, resulting in these epidemics?

A influenza

B smallpox

C lung disease

D measles

Question 2 ○●●

Bacterial gastroenteritis is an infection that causes inflammation of the lining of the stomach and small intestines. It can cause vomiting, diarrhoea, pain and nausea. Which one of the following is an intervention that would prevent the spread of bacterial gastroenteritis in the community?

A Hand washing

B Face masks

C Physical distancing

D Covering your cough

Question 3 ●●●

Measles is an infectious disease caused by a virus. Symptoms usually appear 6–10 days after exposure to the virus and include a rash, headache and fever. In some cases, the patient is affected by secondary bacterial infections, especially of the lungs. Antibiotics are often prescribed.

Antibiotics are likely to

A decrease the rate of DNA synthesis by the measles virus.

B reduce fever by resetting the body's thermostat.

C inhibit the replication of the viral RNA.

D disrupt the synthesis of new bacterial cell components.

Question 4 ●●●

The most precise method of identifying a microbial pathogen involves

A growing the pathogen on a nutrient agar plate.

B sorting the microbes by gel electrophoresis.

C isolating antibodies from the infected individual and determining the amino acid sequence.

D sequencing the microbial genome and identifying proteins.

Question 5 ◖◗◗

Herd immunity is when a large proportion of the population is immune to a disease then there are too few susceptible individuals to sustain the spread of the disease. One of the drawbacks of herd immunity is

A vaccination can be expensive.

B people are too busy to get vaccinated.

C that people who have negative beliefs about vaccinations often live in the same area.

D the percentage of vaccinated people required for herd immunity is different for different diseases.

Question 6 ◖◗◗

Modern antiviral drug design is based on identifying viral proteins or parts of proteins that can be disabled. The target proteins need to be unlike any proteins or parts of proteins in humans to avoid undesirable side effects. Proteins that are possible targets do not include

A receptor proteins on the outside of cells, which viruses use to enter the cell.

B antibodies that are specific to the virus.

C enzymes that facilitate the release of new viruses from the host cell.

D the enzyme that integrates the synthesised DNA into the host cell genome.

Question 7 ◖◗◗

The terms 'epidemic' and 'pandemic' are used to describe widespread outbreaks of a disease. Which of the following best explains the differences?

A An epidemic occurs across continents, whereas a pandemic occurs within a country.

B A pandemic occurs in a state, whereas an epidemic occurs within a country.

C A pandemic spreads across continents, whereas an epidemic is more localised.

D A pandemic occurs in all countries, whereas an epidemic occurs in one country.

Question 8 ◖◗◗

The number of people who die from a pandemic does **not** depend on the

A severity of the disease.

C effectiveness of preventative steps.

B vulnerability of affected populations.

D type of infective agent (virus or bacteria).

Question 9 ◖◗◗

Which of the following is likely when monoclonal antibodies attach to a cancer cell? The monoclonal antibodies

A increase the production of T cells.

B make the cancer more visible to the immune system.

C stimulate the production of memory B cells.

D stimulate the cancer cell to divide.

Question 10 ●●●

The detection of pathogens has become an important part of research in many fields, such as food safety, clinical research, forensics and drug discovery. An effective technique for unambiguous identification of a pathogen is

A growth on nutrient agar.

B identification of a unique DNA or RNA sequence coding for a specific protein.

C amplification of the pathogen by PCR.

D classification into a taxon by identifying under a microscope.

Question 11 ●○○

An advertisement on television aims to educate people about the correct use and administration of antibiotics. The main reason for the advertisement is to

A limit the use of antibiotics to bacterial infections only.

B save money by encouraging fewer prescriptions of antibiotics.

C prevent the build-up of resistance in the population.

D encourage people to take antibiotics each time they become unwell.

Question 12 ●○○

A global pandemic occurred in 2020 with the virus SARS-CoV-2. This virus spread rapidly through air particles and when individuals were in close contact. Which of the following is effective in preventing the onset of this infection in humans?

A Taking vitamins and probiotics   **C** Applying antiseptics

B Taking antibiotics   **D** Getting vaccinated

Question 13 ©VCAA VCAA 2016 SA Q22 ●●

S. anatum is not a common cause of food poisoning. Data has been collected and analysed for the occurrence of illness caused by this organism in Queensland over a five-year period. The graph below displays the average monthly notification rate per 100 000 of the population for the illness caused by *S. anatum*.

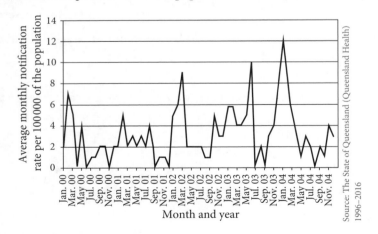

It can be concluded from the data that

A there are four periods in which the notification rate is greater than six per 100 000.

B the notification rate is always lowest during September of each year.

C the notification rate is fairly steady over the five-year period.

D the notification rate in 2001 was highest in May.

Question 14 ⬤⬤⬜

Viruses mutate in a short period of time and can cross between species. One such virus is the avian influenza H5N1 virus (which causes bird flu), which is transmitted to humans through direct contact with infected live or dead birds or their faeces. This strain of bird flu has a high mortality rate with 50% of affected individuals dying. Which of the following is not a biosecurity method that could slow or prevent the progression of these type of diseases into Australia?

A Quarantine of individuals during the pathogen's incubation period

B Giving probiotics before entering the country

C Prevention of live import of birds or any poultry products (feathers)

D Restrictions of animals and workers between poultry farms within Australia

Question 15 ⬤⬤⬤

Monoclonal antibodies are produced by scientists in the laboratory to assist in the treatment of cancer. To be most effective, the antigen-binding site on the antibody would need to be complementary to

A a protein on the outside of the cancer cells.

B the virus that induced the cancer.

C white blood cells, which scavenge up the marked cells.

D enzymes upregulated in the cytosol of the cell.

Question 16 ⬤⬤⬤

Whooping cough is a highly contagious respiratory disease that is spread through air particles and is caused by the gram-negative bacterium *Bordetella pertussis*. Scientists isolated the bacterium and identified a range of molecules to develop a vaccine to reduce morbidity in the population. Which of the following would be the most effective antigen in the vaccine for an adaptive immune response to be initiated and memory cells to form?

A The operon on the bacterial chromosome with the pertussis toxin

B Outer-membrane proteins pertactin and filamentous heamagglutinin

C Cell wall peptidoglycan components from a gram-negative bacterium

D Lactose-metabolising enzyme β-galactosidase

Question 17 ©VCAA VCAA 2013 SA Q16 ⬤⬤⬜

Ross River fever is caused by a virus that lives in kangaroos and wallabies. When a female mosquito bites an infected animal, it picks up viral particles. When the mosquito bites a human, the virus enters the bloodstream. The virus then reproduces in blood cells, resulting in fever, rashes and joint pain. Using the information given, it can be concluded that

A the viral vector is a mosquito.

B the virus is a cellular pathogen.

C Ross River water transmits the virus.

D many kangaroos and wallabies would be killed.

Question 18 ● ●

To create monoclonal antibodies, a B plasma cell extracted from a mouse that has been exposed to the specific antigen for the cancer or autoimmune disorder and fused with a myeloma cell is used to produce a hybridoma cell. These hybridoma cells are created because

A B plasma cells are only short lived and are cleared through apoptosis after an infection.

B the myeloma cell exposes a B cell to its specific antigens to enable antibodies to be complementary and specific to its antigens.

C they enable a greater range of antibodies with different antigen-binding sites to be produced from the original isolated B plasma cell.

D they can be inserted into the patient for continual treatment of their disease through constant antibody release.

Question 19 ©VCAA VCAA 2017 SA Q25 (adapted) ● ● ●

Multiple sclerosis (MS) is an autoimmune disease. In sufferers of MS, the myelin coating of nerve cell axons is damaged. This damage results in poor transmission of nerve messages between the brain, the spinal cord and the rest of the body. One aspect of MS diagnosis is imaging the brain to detect visible areas of demyelination, called plaques. Scientists are investigating factors that increase the likelihood of developing MS. Recently, the 'hygiene theory' has been considered a possible factor. This theory proposes that, if a child's environment is overly hygienic and does not allow sufficient exposure to a wide range of non-self antigens, an overactive immune system will result later in life.

A recent study tested for the presence of antibodies to the bacteria that cause stomach ulcers, *Helicobacter pylori*, in the blood of 550 MS patients and 299 healthy people. Both groups of people had the same proportion of each gender and were of similar age. Exposure to *H. pylori* usually occurs by the age of two years. The results of the antibody testing showed that the rate of *H. pylori* infection was 30% lower in the women with MS than in the healthy women or healthy men. The findings of this study are consistent with the suggestion that

A monoclonal antibodies could be used to treat MS.

B males are affected by MS 30% more often than females.

C suffering from a stomach ulcer is a common symptom of MS.

D in females, childhood exposure to *H. pylori* helps to protect against MS.

Question 20 ● ● ●

Monoclonal antibodies are antibodies that can be produced in a lab and are specific to an antigen present in higher concentrations on the surface of cancer cells. While these antibodies can be used in isolation for the identification of the cells by the body's own immune cells, they can also be combined with other treatments to provide targeted radiotherapy or chemotherapy to the cancer cells (conjugated monoclonal antibodies). Ibritumomab tiuxetan is a monoclonal antibody with a radiotherapy drug which targets the CD20 antigen on the B cells in non-Hodgkin's lymphoma.

The radiotherapy drug would be bound to which part of the antibody?

A The CD20 variable region of the monoclonal antibody

B The antigen-binding site on both sides of the monoclonal antibody

C The CD20 antigen-binding site of the cancerous B cell

D The constant region of the heavy chain of the monoclonal antibody

Section B: Short-answer questions

> **Instructions to students**
> - Answer all questions in the spaces provided.

Question 1 (5 marks) ●●●

Breast cancer affects around 18 000 Australians every year. Individuals with mutations in their *BRCA1* and *BRCA2* genes have an increased risk of developing breast cancer because the mutated gene cannot produce a protein that can repair DNA damage during cell division. Traditionally, cancer has been treated with chemotherapy, which kills off cells that are rapidly dividing. However, this treatment has side effects such as hair loss and digestive issues because the treatment is not specific to the cancer cells. In recent years, monoclonal antibodies that can target specific proteins on the outside of cancer cells have been produced. One frequently used in breast cancer treatment is trastuzumab, which binds to the HER2 protein when the individual has HER2-positive breast cancer.

a Explain the process of making the monoclonal antibodies. 3 marks

b Identify **one** way in which these monoclonal antibodies could kill off these cancer cells. 1 mark

c Identify a limitation with the use of monoclonal antibodies. 1 mark

Question 2 (4 marks) ●○○

Complete the following table for the identification of a pathogen under investigation. 4 marks

	Bacteria	Virus	Prion	Fungus
Contains genetic material				
Prokaryote, eukaryotic or neither				
Replication via				
Treatment				

Question 3 (6 marks) `●●`

Polio is an infectious disease caused by a virus that attacks the nervous system and leads to symptoms within a few hours. The virus can be transmitted by person-to-person contact, by the faecal–oral route and through contaminated water or food. The disease can result in a range of symptoms but more than half of people that survive this disease are left with permanent paralysis. By mid-2020, polio had been eradicated from Africa after 4 years of no reported cases and was only found in Pakistan and Afghanistan. Although there is no cure for polio, a vaccine has been used since the 1960s with individuals having four injections in early childhood.

a What would the vaccine consist of?   1 mark

_____

b State the type of immunity a vaccine initiates.   1 mark

_____

c Draw a graph of the antibody response on the first exposure to the vaccine and the booster vaccine.   2 marks

d State **two** reasons polio is still found in Pakistan and Afghanistan.   2 marks

_____

_____

Question 4 (6 marks) `©VCAA` `VCAA 2012 (1) SB Q7` `●●`

Yellow fever is caused by a virus transmitted through the bite of a particular species of mosquito.

a Would you describe a virus as a cellular or non-cellular pathogen? Justify your answer.   2 marks

_____

_____

b An unvaccinated person travelled to an area where yellow fever virus existed and became exposed to the virus. Describe **two** ways in which the first line of defence of their body would protect against an infection by this virus.   2 marks

_____

_____

_____

_____

c As a requirement for re-entry, travellers returning to Australia from Africa and South America must have proof of vaccination against yellow fever. Explain why this precaution is taken and what course of action Australian authorities may take for an unvaccinated person wanting to re-enter Australia.   2 marks

_____

_____

_____

Question 5 (5 marks) ●●

A person presents to a doctor with a range of symptoms. The doctor collected a sample and sent it off to the lab for testing to determine which type of pathogen the person is infected with.

a Identify a technique that could be used to determine if the pathogen was bacterial and not viral.   1 mark

_____

_____

b The lab ran a series of tests and eventually determined that the pathogen was a type of virus. However, to treat the virus, the lab needed to know the strain. State and explain a process that could be completed for this.   2 marks

_____

_____

_____

_____

c What type of treatment could the person be provided with and how could it inhibit the viral infection?   2 marks

_____

_____

_____

Question 6 (4 marks)

The Ebola virus outbreak of 2014 created a major health scare for many countries. The spread of the virus is shown on the map below.

Tracking Ebola worldwide

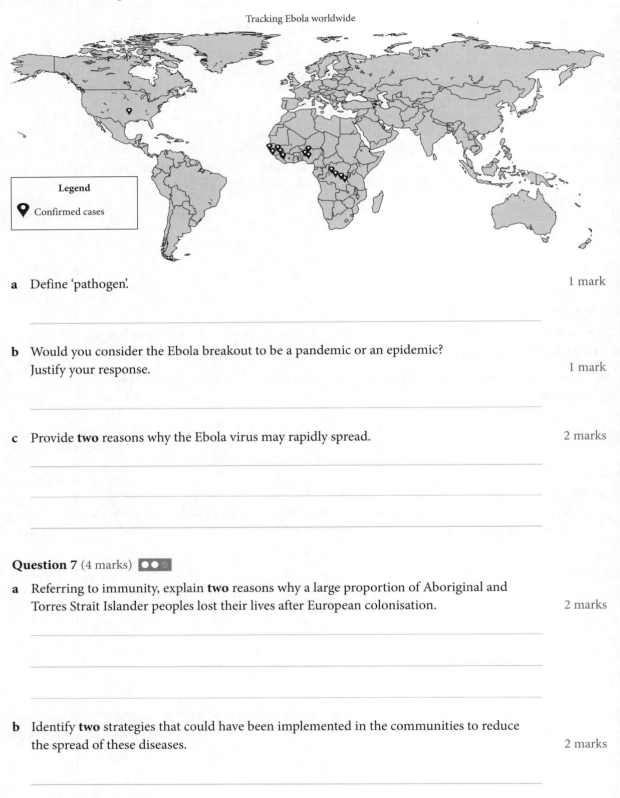

Legend

📍 Confirmed cases

a Define 'pathogen'.   1 mark

_____

b Would you consider the Ebola breakout to be a pandemic or an epidemic?
Justify your response.   1 mark

_____

c Provide **two** reasons why the Ebola virus may rapidly spread.   2 marks

_____

_____

_____

Question 7 (4 marks)

a Referring to immunity, explain **two** reasons why a large proportion of Aboriginal and
Torres Strait Islander peoples lost their lives after European colonisation.   2 marks

_____

_____

_____

b Identify **two** strategies that could have been implemented in the communities to reduce
the spread of these diseases.   2 marks

_____

_____

Question 8 (5 marks) ○▢▢

Tuberculosis is a communicable disease caused by the bacterium *Mycobacterium tuberculosis*. This bacterium infects the lungs and is spread by person-to-person contact through droplets in the air (coughing, sneezing etc.). Although many countries had controlled or eradicated tuberculosis in the past, the disease is re-emerging in some locations.

a   State **two** reasons that could account for the increase in cases in developed countries.   2 marks

_____

_____

b   State the type of drug used to treat *Mycobacterium tuberculosis*.   1 mark

_____

c   Explain how the most effective drug can be determined.   2 marks

_____

_____

_____

_____

_____

_____

Question 9 (4 marks) ●●●

The Wurundjeri have been the Traditional Owners of the Birrarung (Yarra River) area since before European colonisation and still occupy the land today. Traditionally, the tribes and communities of the Wurundjeri were small and moved to new areas regularly where adequate food was available and to allow the current area to regenerate. The movement of the camps also prevented human waste building up. The transition to more permanent settlements increased the number of people in a particular area and altered the natural food availability. The adoption of Western practices, such as European dress, decreased the amount of skin exposed to the sun – this had previously aided in the destruction of pathogens. The transition to a more Westernised lifestyle negatively affected the health of the Wurundjeri through the spread of infection.

Analyse and explain **two** reasons why the changes described could have influenced the emergence of infectious diseases and death.   4 marks

_____

_____

_____

_____

_____

Question 10 (7 marks) ●●

The bacterium *Staphylococcus aureus* (or 'golden staph') is responsible for many pus-forming infections, particularly boils and infections in open wounds.

a How could the bacterium gain access to the body?   1 mark

_____

Many strains of this bacterium have evolved a resistance to antibiotics. Outbreaks of severe infections regularly occur in many large hospitals. In Japan in 1997, a strain of *S. aureus* was found that was resistant to all antibiotics, including methicillin and vancomycin, the 'antibiotics of last resort'. This resistance is a genetically determined trait in the bacteria.

b **i** Outline the important steps in natural selection that have led to the evolution of
resistant strains of *S. aureus*.   2 marks

_____

_____

_____

_____

_____

_____

ii Why would hospitals provide a suitable environment for the rapid evolution of bacteria?   2 marks

_____

_____

_____

c Outline **two** features of bacteria that contribute to the rapid evolution of new strains.   2 marks

_____

_____

Test 10: Genetic changes in a population over time

Section A: 20 marks. Section B: 65 marks. Total marks: 85 marks.
Suggested time: 45 minutes

Section A: Multiple-choice questions

Instructions to students
- For each question, circle the multiple-choice letter to indicate your answer.

Question 1

Which of the following does **not** affect the gene pool of a population?

A migration

B death

C mutation

D cell division

Question 2

Which of the following is **not** an example of a selection pressure operating in a population?

A Pollution, which gives dark-coloured moths an advantage

B A pesticide that selects for resistance in an insect population

C A mutation that causes a colour change in insects which blends with their surroundings

D An antibiotic that selects for resistance in a bacterial population

Question 3

Consider the diagram below showing the recovery of a population after experiencing a bottleneck effect. The allele represented by the letter D represents a broad phenotype.

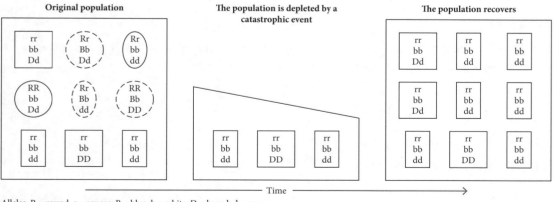

Alleles: R = round, r = square; B = blue, b = white; D = broad, d = narrow

How did the frequency of allele D change from the original population to when the population recovers?

A Frequency of allele D was 7% in the original population compared to 4% in the recovered population.

B Frequency of allele D was 39% in the original population compared to 22% in the recovered population.

C Frequency of allele D was 59% in the original population compared to 32% in the recovered population.

D Frequency of allele D was 89% in the original population compared to 41% in the recovered population.

Question 4 ☐●●●

If the gene pool of a population is large, this indicates that

A there is a large number of organisms in the population.

B there is extensive genetic diversity in the population.

C the organisms in the population are fit.

D there is a high chance of extinction.

Question 5 ☐☐●●

Two species of wallaby inhabit parts of the northern region of South Australia. Similarities in their anatomy have led researchers to conclude that they share a common ancestor. The most likely order of events that have resulted in the evolution of these two species of wallaby from a common ancestor is

A reproductive isolation, natural selection, genetic drift.

B natural selection, geographical isolation, reproductive isolation.

C geographical isolation, natural selection, reproductive isolation.

D geographical isolation, genetic drift, natural selection.

Question 6 ☐☐●●

The northern elephant seal is a large marine predator that lives in the waters of the north eastern Pacific Ocean. It was extensively hunted during the late 19th century to such an extent that by the end of the century, only about 20 remained. Hunting restrictions have led to a resurgence in seal numbers, and present populations number over 30 000. This species has much less genetic variation than the southern elephant seal, which was not hunted to the same extent.

The loss of genetic variation in the northern elephant seal is an example of

A the founder effect.

B a bottleneck.

C a mass extinction.

D natural selection.

Question 7 ☐☐●●

Two species are reproductively isolated from each other due to prezygotic or post-zygotic barriers. Which of the following is an example of a post-zygotic isolating mechanism?

A Differences in reproductive seasons

B The production of sterile hybrids

C Differences in seasonal mating calls

D Differences in habitats

Question 8 ☐☐●●

In order for natural selection to operate on a population of organisms, the population must have individuals with different

A phenotypes.

B reproductive seasons.

C mating calls.

D mutation rates.

Question 9

Both genetic drift and natural selection can alter the frequency of alleles in a population. One difference between these agents of change is that

A genetic drift acts on populations while natural selection acts on individuals.

B natural selection may be random, whereas genetic drift is directional.

C natural selection has more effect in smaller populations and genetic drift has more effect in larger ones.

D the changes caused by genetic drift may not be adaptive.

Question 10

A farmer isolates strawberry plants that produce the largest strawberries and cross-pollinates these plants for the next season's seeds. This type of breeding

A increases genetic variation in the plants.

B increases the rate of mutations in the growth genes.

C decreases genetic variation.

D simulates natural selection.

Question 11

In humans, the Y chromosome carries the *SRY* (sex-determining region Y) gene that codes for the production of the male hormone testosterone and determines maleness. Occasionally, a mutation occurs resulting in the *SRY* gene being deleted from the Y chromosome and transferred to the X chromosome, as seen in individuals A and B in the diagram below.

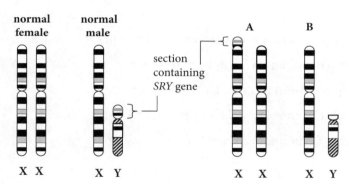

This type of mutation is known as a

A trinucleotide repeat.

B translocation.

C point mutation.

D frameshift insertion.

Question 12

The diagram below depicts the change that occurred in the frequency of phenotypes in a beetle population over a 10-year period.

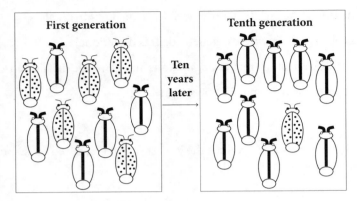

The selection pressure that was operating over that time was likely to have been

A a disease.

B the availability of food.

C a predator.

D the climate.

Question 13

In animal populations, gene flow is the same as

A migration.

B evolution.

C selection.

D mutation.

Question 14

In which of the following populations would you expect to find the most genetic variation?

A Asexually reproducing organisms in a stable environment

B Sexually reproducing organisms in a variable environment

C Sexually reproducing organisms in a stable environment

D Asexually reproducing organisms in a variable environment

Question 15

Many biologists think that the human population survived a severe evolutionary bottleneck in the relatively recent past (about 70 000–50 000 years ago). It has been suggested that this was due to a huge volcanic eruption and subsequent 'volcanic winter', which cooled and darkened Earth for many years.

Evidence that would support the hypothesis that the human species had suffered a severe bottleneck would include

A human remains found in volcanic ash deposits.

B low genetic diversity among present-day human populations.

C high levels of diversity in mitochondrial DNA among people from different continents.

D the presence of human fossil remains in deposits older than 50 000 years.

Question 16 ●●

The Amish population of eastern Pennsylvania, USA, carry an unusually high frequency of a gene mutation that causes a number of inherited disorders, such as polydactyly (additional fingers or toes) and a hole in the heart. The conditions are very rare in the general population. The Amish population stemmed from about 200 Swiss immigrants. The most likely explanation for the high frequency of these disorders in the Amish population is

A the founder effect.

B artificial selection.

C natural selection.

D random mating.

Question 17 ●●●

Two related species of plant that inhabit the same area were examined. The two species have very similar phenotypes and have the same flowering season. They are also pollinated by the same species of insect. The following diagrams show the chromosomes present in the somatic cells of these two species.

Species A

Species B

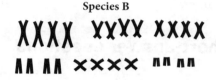

On the basis of the information above, it is reasonable to conclude that

A species A and B are reproductively isolated due to prezygotic mechanisms.

B species A and B evolved into two species because of the presence of a geographical barrier.

C both species A and species B are able to produce viable gametes.

D the ancestor of these species would have contained three homologous pairs.

Question 18 ●●

Selective breeding is the traditional method for improving crops and livestock. The list below contains the steps in a selective breeding program.

I Select the best offspring from parents to breed for the next generation.

II Choose the desirable characteristics.

III Repeat the process over a number of generations.

IV Find parent organisms that show the characteristics.

The correct order of steps is as follows:

A I → II → III → IV

B II → IV → I → III

C III → II → IV → I

D IV → I → III → II

Use the following information to answer Questions 19 and 20.

In populations of fruit flies, there are individuals that are resistant to the effects of insecticides. Insecticide-resistant fruit flies arose as a result of a mutation. In normal insecticide-susceptible fruit flies, a specific section of mRNA has the sequence GCU, whereas in the insecticide-resistant fruit flies the sequence is UCU.

Question 19 ©VCAA VCAA 2014 SA Q22 ●●

Considering the mRNA sequence of the insecticide-resistant fruit flies, the corresponding sequence of nucleotides on the individuals, DNA is

A AGA.

B AUA.

C CGA.

D TGT.

Question 20 ●●

The above mutation is an example of a

A chromosomal deletion.

B nucleotide substitution.

C nucleotide insertion.

D frameshift mutation.

Section B: Short-answer questions

Instructions to students
- Answer all questions in the spaces provided.

Question 1 (7 marks) ●●

The Galapagos Islands off the coast of South America were extensively explored by Charles Darwin and many biologists after him. The islands are relatively young, having arisen from volcanic activity under the ocean.

Giant tortoises inhabit seven of the islands; on four more islands, the populations have been exterminated by humans in the last 150 years. There is only one species of tortoise, *Chelonoidis nigra*, but 15 distinct subspecies that differ in shell shape, colour, thickness and neck length. These differences between subspecies indicate that divergent evolution has occurred in tortoises. Although not usually swimmers, the tortoises do float and have been known to survive at sea for several days. It is believed that movement of tortoises between islands has maintained gene flow between the populations.

a What is meant by 'gene flow'?   1 mark

b Why are the 15 subspecies of giant tortoises not considered to be separate species?   1 mark

Long-necked tortoises inhabit the more arid islands where they reach up to feed on shrubs and prickly pear cacti. On these islands, the prickly pear plants have thick trunks and sharp spines (see below). On islands with no tortoises, these plants grow low to the ground and have soft spines. Short-necked tortoises inhabit the wetter islands where groundcover plants are abundant.

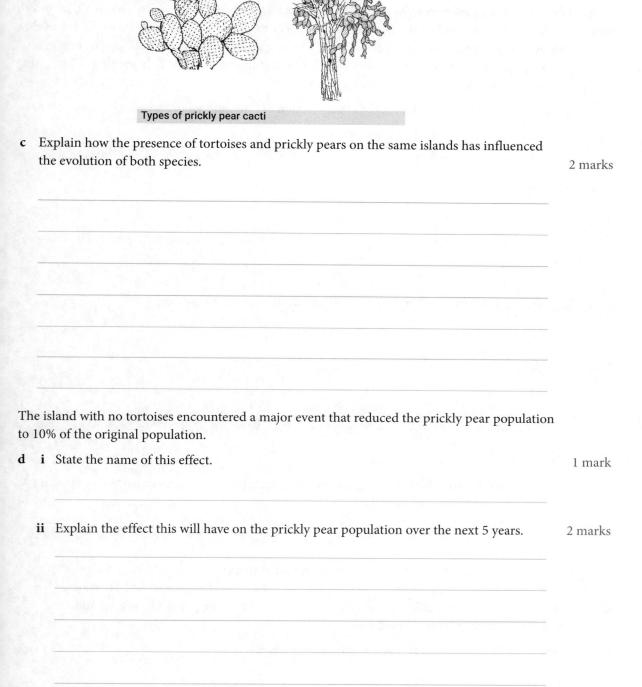

Islands with no tortoises   Islands with long-necked tortoises

Types of prickly pear cacti

c Explain how the presence of tortoises and prickly pears on the same islands has influenced the evolution of both species.   2 marks

_____

_____

_____

_____

_____

_____

The island with no tortoises encountered a major event that reduced the prickly pear population to 10% of the original population.

d i State the name of this effect.   1 mark

_____

ii Explain the effect this will have on the prickly pear population over the next 5 years.   2 marks

_____

_____

_____

_____

Question 2 (7 marks) ●●

Many animals, from grasshoppers to lions, have evolved so that their colours make them inconspicuous in their environments. These camouflaged organisms may look like twigs, leaves or a rocky background. The opposite phenomenon also occurs in nature. Many species of birds, especially the males, have very bright plumage, making their presence obvious to any potential predator or prey. The huge, brightly coloured tail of a peacock was surely no advantage in the struggle for existence! Charles Darwin proposed a theory of sexual selection. Females of certain species selectively mate with the more glamorous males.

Even more perplexing are the bright colours of many caterpillars, which have no sex life at all. Many of these brightly coloured caterpillars are also poisonous or taste particularly bad for predators such as birds. It was Darwin's contemporary, Alfred Russel Wallace, who suggested that this type of colouring acts as a warning to other species such as birds that they should attack at their own peril. This warning colouration can be seen in many species of insects, frogs and fish.

a   Explain how natural selection could lead to the evolution of camouflaged organisms.   2 marks

b   Explain, in genetic terms, how bright plumage in male birds may have evolved.   2 marks

c   For poisonous caterpillars, what advantage is there in being brightly coloured, rather than relying on their unpleasant taste to deter predators?   1 mark

While working in the field, a biologist came across a caterpillar that closely resembled a poisonous species. Both species have bright red bands along the length of their bodies with spikes. The biologist hypothesised that the non-poisonous species had evolved to resemble the poisonous species and thus gained a selective advantage to avoid predators.

d   What would birds be acting as for the caterpillars in the area?   1 mark

e What further information would support the biologist's hypothesis that the colouration had given the non-poisonous species a selective advantage? 1 mark

_____

_____

_____

Question 3 (7 marks) ⬤⬤⬤

The Galapagos Islands are an archipelago of 19 volcanic islands that are home to an array of unique species found only in this region. The marine iguana (*Amblyrhynchus cristatus*) is thought to have reached the islands by chance after being washed up from the mainland about 10 million years ago. These iguanas are the only known marine lizard in existence, and differ from land lizards by having blunt snouts and sharp claws, which allow them to access the algae off underwater rocks, and a flattened tail for swimming. Their diet of seaweed causes a large amount of salt to build up in their tissues. To manage this, the iguanas have a gland in their nostrils to sneeze out the excess salt.

a The gland for expelling salt is not present in the mainland population. Explain how this feature could have appeared and remained in the population. 1 mark

_____

_____

_____

b Scientists identified a new marine iguana on an island in the Galapagos that looked similar to the other marine iguanas. How could scientists determine if they were the same species? 2 marks

_____

_____

_____

_____

Every 7 years, a weather change (El Niño) occurs in the Galapagos Islands that alters the sea currents and, as a consequence, increases the sea temperature. This reduces the amount of algae and green seaweed growth and increases brown seaweed, which the tortoises cannot digest.

c What is the selection pressure described? 1 mark

_____

_____

d Explain the impact of this selection pressure on the marine iguana population on the islands. 3 marks

_____

_____

_____

_____

_____

Question 4 (6 marks) ©VCAA VCAA 2016 SB Q8 ●●●

Two species of *Cryptasterina* sea stars are found in coastal Queensland. *Cryptasterina pentagona* is found in warmer water further north, while *Cryptasterina hystera* is found further south in cooler water.

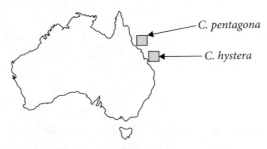

Researchers have concluded that these two species arose from a recent common ancestor via natural selection. They believe that, over thousands of years, the sea environment has changed, with the boundary line between cold water and warm water moving further north. They have found that water temperature and predation of sea star larvae by cold-water predators are important selection pressures for these sea stars.

a Using the information above, explain how natural selection can lead to differences in phenotypes between these two sea star species.   4 marks

b One of the phenotypic differences between these two species of sea stars is their method of reproduction. *C. pentagona* reproduces sexually and its sperm and eggs are free-floating in the ocean. *C. hystera* self-fertilises and its fertilised eggs are kept within the sea star until maturity. The researchers found that one species of *Cryptasterina* has a significantly higher diversity of alleles in its gene pool than the other species.

Using this information about reproduction strategies, which species of *Cryptasterina* would you expect to have the highest diversity of alleles? Explain your answer.   2 marks

Question 5 (7 marks) ⬤⬤⬤

Over the last decade, the bacterium *Neusseria gonorrhoeae* (gonorrhoea) has become resistant to penicillin, tetracycline and cuprofloxiacin, raising concerns that the disease will be completely untreatable in the near future. Doctors are currently prescribing a combination of antibiotics to treat gonorrhoea.

a What are **two** ways that antibiotics work?   1 mark

b Explain how bacteria can become resistant to an antibiotic in a population.   2 marks

c What are **two** mechanisms that can be used within communities to reduce continued antibiotic resistance?   2 marks

d On the agar plates below, draw the bacterial growth showing the antibiotic disc with tetracycline infused when it was first introduced into the population and in today's bacteria population.   2 marks

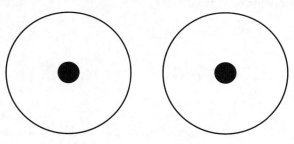

Question 6 (6 marks) ⬤⬤◻

The bread wheat, *Triticum aestivum* is hexaploid meaning it has six sets of chromosomes (6n). This provides the wheat with greater yield, drought tolerance, disease resistance and nutrient-use efficiency, all desirable qualities in a commercially grown crop. There are other examples of polyploidy crops such as peanuts, oats, bananas and coffee.

a Polyploids can arise naturally, what event would give rise to polyploids?   1 mark

b Commercial crops exhibiting polyploidy have been selected for by farmers for their desirable qualities. Explain the process that farmers would have undertaken to select these crops.   3 marks

_____

_____

_____

_____

_____

_____

c Explain the effect on genetic variation and the wheat's survival under changing conditions.   2 marks

_____

_____

_____

_____

Question 7 (7 marks) ●●●

Influenza is the type of virus responsible for seasonal flu epidemics each year around the world. People suffering from the flu experience fever, headaches, muscle pain, cough, sore throat and a runny nose. Influenza is moderate to highly infectious and consequently able to spread rapidly through populations through air droplets when a host is coughing, sneezing or talking. Each year, the flu vaccine is produced to reduce morbidity and mortality within the population.

a What are the main components of a virus?   1 mark

_____

b Define 'antigenic drift'.   1 mark

_____

c A person received their 2023 flu vaccination containing a new H1N1 strain and H3N2 strain. Explain whether this person would require the flu vaccination in 2024. Justify your response.   2 marks

_____

_____

_____

_____

d How can a virus cause disease within an individual?   2 marks

_____

_____

_____

_____

e Other than vaccination, what are **two** mechanisms that can be implemented in the
community to reduce the number of individuals suffering from the flu?   1 mark

_____

Question 8 (6 marks) ⬤⬤⬤

a Why is it advantageous to have a large gene pool?   2 marks

_____

_____

b Compare the rates at which migration and mutations alter the gene pool.   2 marks

_____

_____

_____

c Name the two types of genetic drift and identify a difference.   2 marks

_____

_____

_____

_____

_____

Question 9 (6 marks) ©VCAA VCAA 2017 SB Q5 ⬤⬤⬤

The rufous bristlebird (*Dasyornis broadbenti*) is a ground-dwelling songbird. The rufous bristlebird is
found in gardens near thick, natural vegetation and builds nests in shrubs close to the ground. The rufous
bristlebird feeds on ground-dwelling invertebrates. It is a weak flyer and is slow to go back to areas from
which it has been previously eliminated. Two distinct populations of rufous bristlebird exist in Victoria.
The distribution of each population is shown on the map of Victoria below. The distance between
Population A and Population B is over 200 km.

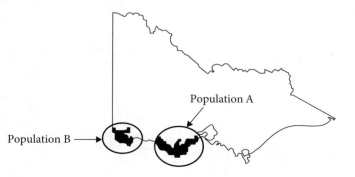

Source: Flora & Fauna Guarantee Action
Statement, 1993, no. 49; © The State of
Victoria, Department of Sustainability and
Environment, 2003

a Define the term 'gene flow' and explain whether gene flow is likely to occur between these two populations.

3 marks

_____

_____

_____

_____

b Both of the rufous bristlebird populations in Victoria are small. Referring to the theory of natural selection, explain why the rufous bristlebird is at risk of extinction.

3 marks

_____

_____

_____

_____

_____

_____

_____

Question 10 (6 marks) ⬤⬤⬤

Ellis–van Creveld syndrome is a rare genetic condition first discovered in 1940. Equal numbers of males and females were affected and sufferers were of normal intelligence. However, all were moderately to severely dwarfed and had malformations of the hands, including polydactyly (extra fingers) and bone deformities that made it impossible to make a fist. There were also characteristic facial malformations, including a short upper lip and dental problems. In 50% of cases, there were also cardiovascular problems. It was these cardiovascular problems that led to a high mortality in childhood. In the 1960s, more cases of Ellis–van Creveld syndrome were found in Lancaster County in the USA than had been reported in the world to that date. All of these cases were found among a group of Amish. The ancestors of today's Amish population fled religious persecution in Switzerland between 1720 and 1770. In all, some 200 people settled in Lancaster County and remained reproductively isolated from the rest of the American population. There are more than 7000 Amish people in this area today.

Of the initial settlers, either one or two individuals carried the allele for Ellis–van Creveld syndrome. This is a clear example of the founder effect in a human population.

a   i What is meant by the founder effect?

1 mark

_____

ii How has the founder effect influenced gene frequency in the Amish population?   1 mark

_____

_____

_____

_____

b Remaining reproductively isolated has meant that there has been a lot of inbreeding in the Amish population of Lancaster County. How has inbreeding influenced the prevalence of this disease?   1 mark

_____

_____

c As stated, there were one or two individuals with the allele in the initial population of 200. How has it become more common in this population, despite the obvious negative effects of this condition?   1 mark

_____

_____

_____

d Until recently, Ellis–van Creveld syndrome was often fatal in childhood due to the number of cases in which there were cardiovascular problems. This condition is now treatable and many more affected children are surviving well into adulthood. What effect is this likely to have on the future population of Lancaster County Amish?   2 marks

_____

_____

Test 11: Changes in species over time

Section A: 20 marks. Section B: 66 marks. Total marks: 86 marks.
Suggested time: 45 minutes

Section A: Multiple-choice questions

Instructions to students
· For each question, circle the multiple-choice letter to indicate your answer.

Question 1

There are 14 different species of finches that originated from the Galapagos Island finches, and 1478 species of freshwater cichlid fish. What kind of evolution are these an example of?

A adaptive radiation

B convergent evolution

C bottleneck effect

D genetic drift

Question 2

The absolute dating of organisms assigns a supposed date to an artefact based on many factors found at the archaeological site. Which of these is used as an absolute dating technique?

A index fossils

B rock levels in Earth

C tree-ring dating

D radiometric dating

Question 3

The evolution of life on Earth occurred in several discrete steps, which are detailed below.

1 Development of multicellular organisms

2 Formation of eukaryotic cells

3 Ability to live on land

4 Formation of prokaryotic cells

Arrange these steps to indicate the order in which evolution occurred.

A 1, 3, 2, 4

B 2, 4, 3, 1

C 3, 1, 4, 2

D 4, 2, 1, 3

The following information relates to Questions 4 and 5.

The length of time it takes for half of the atoms in a sample of radioactive material to decay is called the half-life of the radioactive substance. The half-lives of four radioactive materials are shown in the table.

Radioactive material	Half-life (years)
Uranium-235	704 million
Potassium-40	1.25 million
Carbon-14	5573
Argon-39	269

Fossil evidence shows that in the Pleistocene period (from 2.6 million years ago to 12 000 years ago), Australia was home to several species of megafauna: huge marsupials and flightless birds. They are all thought to have been extinct by about 46 000 years ago. Smaller marsupials, such as kangaroos and wombats, now dominate the Australian fauna.

Question 4 ⬤⬤⬤

What is the best explanation for this change in Australian fauna?

A Increased food supply caused the megafauna to decrease in size.

B The mass extinction of the megafauna was caused by a comet striking Earth.

C Smaller marsupials were better able to survive climatic change and warmer temperatures.

D Predators introduced through land bridges during the ice age wiped out the megafauna.

Question 5 ⬤⬤⬤

Which of the following radioisotopes could be used to measure the age of a fossil of the last of the giant marsupials?

A uranium-235

B potassium-40

C carbon-14

D argon-39

The following information relates to Questions 6 and 7.

The following diagram shows samples of fossils found in sedimentary deposits from three different locations in Australia.

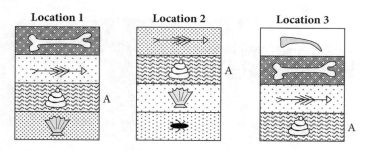

Question 6 ⬤⬤⬜

The fossil found at each of the locations in the sediment layer marked 'A' is the shell of a marine mollusc. Which of the following conclusions can you draw about the molluscs?

A The molluscs in each of these locations were all deposited at different time periods.

B The molluscs were deposited before the evolution of vertebrates.

C The molluscs are found in deposits of approximately the same age in each of the locations.

D The molluscs are now extinct, because they are not seen in later deposits.

Question 7 ⬤⬤⬜

Observing the rock strata of the three locations, we can assume that

A each rock stratum would be the same thickness and would have been formed in a similar period.

B location 1 contains the youngest rock stratum.

C each rock stratum would contain the same fossils in each location.

D location 2 contains the oldest fossils in the rock strata.

Question 8 ⬤⬜⬜

The graph below shows the decay of carbon-14. Carbon-14 has a half-life of 5730 years.

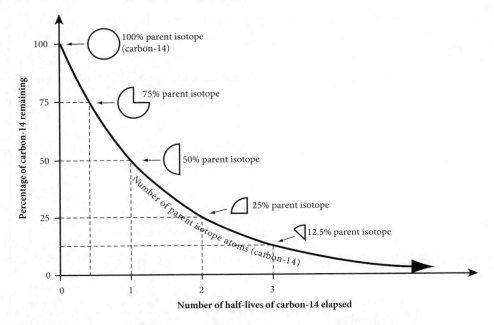

Using the information in this graph, if a sample contains 40% of the original amount of carbon-14 then you can conclude that the age of the sample is

A 4300 years.

B 7100 years.

C 11 400 years.

D 17 200 years.

The following information relates to Questions 9 and 10.

The palms on Lord Howe Island off the coast of New South Wales have been highly studied in terms of speciation. The island formed 6.9 million years ago and the palms diverged after formation. *Howea forsteriana* and *H. belmorena* are two species on the islands that developed with the lack of a geographic barrier.

Question 9

Speciation of these palms could have occurred through

A a mountain range separating the two populations on either side.

B the crossing over of flowering seasons.

C the altitude that the palms grew from the ocean.

D the rate of photosynthesis.

Question 10

A new palm was discovered in a small section of the Lord Howe Island that had not previously been a habitat for palms, owing to the pH of the soil. The plant contained longer, thinner palm leaves and more closely resembled a cactus. Scientists attempted to cross-pollinate this species with *H. belmorena* but were unsuccessful. The plant produced from the cross-pollination with *H. forsterianna* grew but was infertile. From this information you could assume

A *H. forsterianna* and the unknown palm are the same species.

B post-zygotic factors such as the pH of the soil prevented cross-pollination from naturally occurring.

C speciation has occurred from *H. belmorena*.

D the new palm is more closed related to *H. forsterianna* than it is to *H. belmorena*.

The following information relates to Questions 11 and 12.

In 2019, a major natural disaster occurred in Victoria and South Australia. The Black Summer bushfires dramatically decreased Australia's koala population as many areas of the bushland became uninhabitable and koala populations became isolated in small pockets of unburnt eucalyptus.

Question 11

Koalas cannot travel large distances because of the energy expenditure required, and they often stay within a 6–10 km radius of their habitat. What effect would this have on the populations over many generations?

A The koalas would be exposed to similar selection pressures due to the fires and adapt similarly.

B The koalas would adapt to their environment and may evolve into separate species over many generations.

C Gene flow will occur between the populations, ensuring the populations remain the same species.

D Sympatric speciation will occur where the koalas evolve into separate species.

Question 12

Wildlife rescue groups are implementing programs to rotate koalas from different populations into other areas. What is the purpose of this?

A To enable selective breeding to increase desired characteristics within the koala population

B To decrease the gene pool of the new populations to increase survival

C To reduce the impact on the eucalyptus tree ecosystem

D To increase variation within the population to increase survival

The following information relates to Questions 13 and 14.

A fossil of the ornithopod dinosaur *Diluvicursor pickeringi* was found in a 113-million-year-old rock formation in western Victoria. The fossil included a tail, a partial hind limb and some vertebrae. *D. pickeringi* grew to 2.3 m long. Evidence suggests that the dinosaur was fossilised in a log-filled hollow at the bottom of an ancient riverbed. Two stratigraphically younger fossils that had been discovered previously at a nearby site were found to be closely related to *D. pickeringi*.

Question 13 ©VCAA VCAA 2018 SA Q25 ●●

It is most probable that the two stratigraphically younger fossils would have been found in a layer of rock that

A was closer to the present-day ground surface than the rock surrounding the *D. pickeringi* fossil.

B contained a smaller quantity of carbon-14 than the rock surrounding the *D. pickeringi* fossil.

C was located at a depth of 2.3 m below the ancient riverbed.

D was formed from extremely hot, volcanic lava flow.

Question 14 ©VCAA VCAA 2018 SA Q26 ●●●

Palaeontologists believe that the Victorian ornithopods shared a close common ancestor with several ornithopod fossils found in Antarctica, South America and Africa. Which one of the following is the most likely explanation for the distribution of these fossils?

A Antarctica, South America and Africa were joined to Australia in the distant past.

B The strong tails of the ornithopods enabled them to swim for sustained periods of time.

C The small forelimbs of the ornithopods suggest that they were evolving wings for flight.

D Seagoing, scavenger birds carried the fossil bones of the ornithopods to other continents.

Question 15 ●●

The first fossil of a megafauna *Euryzygoma dunense* was discovered in Queensland in 1912. This quadrupedal marsupial herbivore lived in the temperate eucalypt forests where rainfall was high and it ate a range of shrubs and leaves. Scientists think *E. dunense* was present about 25 million years ago. A relative dating technique that could be used to support this claim is

A potassium-argon dating.

B stratigraphy.

C radiometric dating.

D crystal radiation of surrounding minerals.

Question 16 ©VCAA VCAA 2016 SA Q38 ●●●

In India, a group of scientists was studying fossils from a coal deposit formed during the Permian period (290–245 million years ago). They found three fossil species from the same genus in different levels (strata) of the coal. When radiocarbon dating on these fossils was performed, it showed exactly the same levels of carbon-14 in all three fossil species. The data is summarised in the table below.

Fossil species	Depth at which fossil was found in the coal deposit (m)	Proportion of carbon-14
Gangamopteris major	6.2	0.0001
Gangamopteris obliqua	8.1	0.0001
Gangamopteris clarkeana	4.7	0.0001

Which one of the following is the correct conclusion to draw from these findings?

A There is no evolutionary relationship between these three fossil species.

B *G. clarkeana* is the common evolutionary ancestor of *G. major* and *G. obliqua*.

C As carbon dating is a more reliable dating technique than analysis of strata in coal deposits, the fossils of *G. major*, *G. obliqua* and *G. clarkeana* are all the same age.

D An analysis of strata in coal deposits is a more reliable dating technique than carbon dating for Permian fossils; the fossil of *G. major* is younger than the fossil of *G. obliqua*.

Question 17

Ammonites were a species of mollusc found within a relatively short period during the Mesozoic era from 245 to 65 million years ago. This species can help scientists identify the age of other specimens. The ammonites are classified as

A transitional fossils.

B indirect fossils.

C index fossils.

D carbon fossils.

Question 18

A person said that 'The fossil record can't be used for ecological analysis because it provides an incomplete image of the environment and species present in each rock stratum'. Which of the following is **not** a supporting argument of this statement?

A Not all fossils have been located and analysed.

B Fossilisation conditions are not always optimal.

C Decomposition occurs quickly after death in exposed areas.

D Exoskeletons would not have fossilised.

Question 19

Which of the following describes the carbon-14 dating technique?

A Dates material found in rocks and minerals

B Measures the ratio of carbon-12 to carbon-14 in organic remains

C Is a technique used for fossils up to 50 000–60 000 years old with a half-life of 30 000 years

D Is a type of relative dating

Question 20 ©VCAA VCAA 2015 SA Q35

Potassium-40 has a half-life of 1.25 billion years. In igneous rocks closely associated with a fossil layer, the ratio of potassium-40 to its radioactive breakdown product, argon-40, is approximately 1:1. The age of the fossils in the fossil layer would be close to

A 125 million years.

B 310 million years.

C 1.25 billion years.

D 2.5 billion years.

Section B: Short-answer questions

Instructions to students
- Answer all questions in the spaces provided.

Question 1 (5 marks)

The age of Earth and its inhabitants has been determined through two complementary lines of evidence: relative dating and numerical dating. Only some materials are suitable for numerical dating. Two types of index fossils are located in the rock strata shown in diagram at the right: mollusc shells, which were exoskeletons for soft-bodied organisms, and trilobites, which were hard-shelled insects.

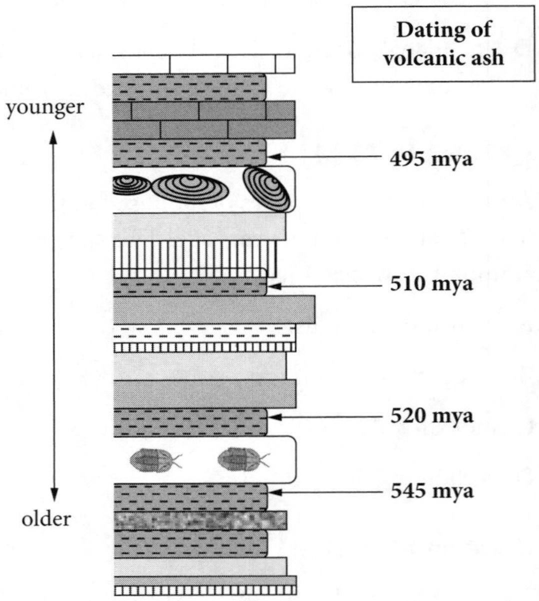
Dating of volcanic ash
younger
older
495 mya
510 mya
520 mya
545 mya

a What is the age of the trilobite fossil in the diagram? 1 mark

____

b What technique do scientists use to numerically date a sample? 1 mark

____

c The mollusc fossil was probably deposited more recently than the trilobite fossil. What information from the relative dating results can be used to substantiate this claim? 1 mark

____

____

d What is **one** problem that can interfere with the reliability of relative dating? 1 mark

____

____

e What could be done to verify the results from relative dating? 1 mark

____

____

Question 2 (7 marks)

The following table shows the radioactive decay of carbon-14 over time.

Time (years)	Amount of carbon-14 remaining (%)
0	100
5000	55
10 000	30
15 000	17
20 000	9
25 000	5
30 000	3
35 000	1.6

a Graph the data on the grid provided. Label the axes clearly.   2 marks

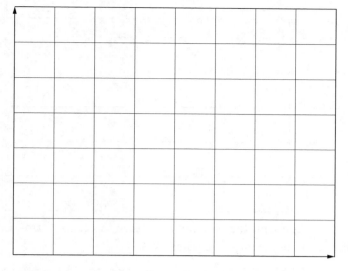

b A fossil was found and dated by the carbon-14 method. What is the approximate age of the
fossil if it contained 20% of its original carbon-14?   1 mark

_____

c **i** State **one** reason why carbon-14 dating is frequently used to date animal remains.   1 mark

_____

ii What is meant by an isotope's half-life?   1 mark

_____

d A mineralised partial skull of a primate, thought to be several million years old, is submitted
for dating. Explain **two** reasons why carbon-14 dating might not be a useful technique in
this instance.   2 marks

_____

_____

_____

Question 3 (7 marks)

Several events were involved in the evolution of life on Earth. These events can be summarised by the
following.

V = endosymbiosis of mitochondria

W = origin of multicellular eukaryotes

X = endosymbiosis of chloroplast

Y = origin of bacteria

Z = origin of mammals

a Complete the table with the correct order of events and give a different reason for each choice.

6 marks

Order	Choice	Reason
1st		
2nd		
3rd		
4th		
5th		

b The following cladogram represents the evolutionary links between life forms. Complete the table with the correct names of the major domains (A–E).

2 marks

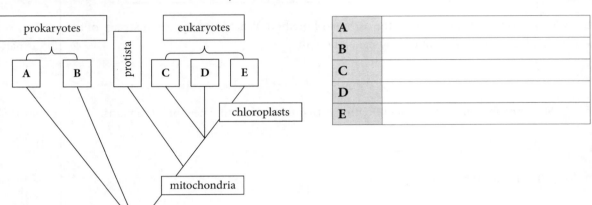

A	
B	
C	
D	
E	

Question 4 (7 marks)

Fourteen closely related finch species are currently living on the Galapagos Islands. The diagram depicts the relative beak sizes and shapes along with the type of food eaten by each species.

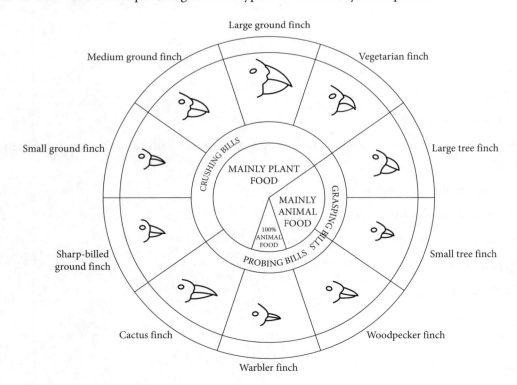

a State the type of speciation that has occurred within the population of Galapagos finches.   1 mark

_____

b Why do the different species have such different beak shapes?   1 mark

_____

_____

c The greater the gene activity of the beak protein in the developing embryo, the wider and deeper the beak. Which of the above finches would have the largest amount of this gene activation?   1 mark

_____

d A major weather change is experienced in the Galapagos and kills off the small bushes that produce soft fruit, but favours the large spiky bushes that have large hard seeds. Explain what would occur within the population of small tree finches.   3 marks

_____

_____

_____

_____

e Which of the birds is the ancestral species?   1 mark

_____

_____

Question 5 (7 marks) ⬤⬤⬤

Lord Howe Island off the coast of New South Wales is home to a family of palms, with two of the palms, _Howea forsteriana_ and _H. belmorena_, becoming established after the formation of the land 6.9 million years ago. The palms are located throughout the island but favour particular altitudes and soil pH levels.

a Explain the difference between prezygotic and post-zygotic isolation mechanisms. Provide an example of each in your answer.   2 marks

_____

_____

_____

_____

b   What kind of speciation has occurred in the palms on Lord Howe Island?   1 mark

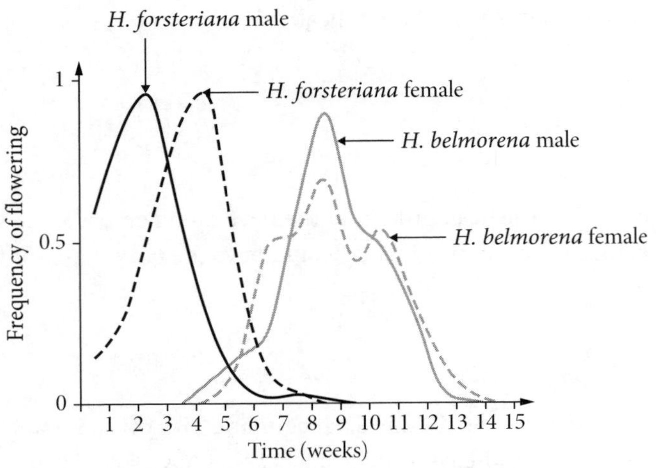

_____

c   Describe the data in the graph and explain how this could have affected the formation
of the species.   2 marks

_____

_____

_____

_____

H. forsteriana tends to grow in a more basic soil (pH 8–9) and a habitat of 0–60 metres in altitude, whereas
H. belmoreana grows in a more acidic soil (pH 5–7).

d   Some botanists isolated a group of *H. forsteriana* palms with a range of alleles and placed
them in similar conditions to *H. belmoreana*. Explain the potential outcome for these plants
and their offspring.   2 marks

_____

_____

_____

Question 6 (6 marks)

The family Paradisaeidae (birds of paradise) is a group of more than 36 bird species located in New
Guinea and surrounding areas. They live in the rainforests with plentiful food, habitats and mates.

a   What is meant by 'divergent evolution'?   1 mark

_____

b What evidence would you need to support the hypothesis that two species of Paradisaeidae were the result of divergent evolution?   1 mark

c With plentiful food, no predators and a stable environment, these birds may not be exposed to any selection pressures. What effect would this have on the gene pool of the population?   2 marks

Male birds of paradise have large brightly coloured feathers and perform extravagant courtship dances for the females, who are often a dull brow colour.

c   i What is the selection pressure described?   1 mark

ii A small population of the bird of paradise species *Lophorina superba* preferred to mate in the evening, whereas most of the species mated in the early morning, meaning that the smaller and larger populations were not interacting. What would happen to the two populations if this behaviour continued?   1 mark

Question 7 (7 marks) ●●●

Stygofauna are an ancient group of animals that reside in ground water, with some originating back to when land masses were connected. Because they have evolved in the dark, they lack pigmentation and eyes, and instead have complex feelers and highly developed sensory skills. One species located in Western Australia is the blind cave eel (*Ophisternon candidum*). Changes resulting from the colonisation of continents, such as changes in water quality, removal of water through bores and wells, and compaction, means underground water systems that were once connected are now isolated from each other and are subject to different types of water and environments.

a What effect could this have on the blind cave eel species?   4 marks

b If Western Australia experiences continuous high rainfall for 5 years and the water systems return to their original state, what are **two** things that might occur?   2 marks

_____

_____

_____

c One of the populations that became isolated consisted of five blind cave eels. These eels were separated from the original main waterways that contained 900–1000 organisms at any given time. What would be the effect of this phenomenon?   1 mark

_____

_____

Question 8 (5 marks)

A rare fossil believed to be a common ancestor of bears and pandas has recently been discovered in lowland China. It was dated at 40 million years before present.

a Describe **two** ways in which the age of the rare ancestral bear fossil could have been established.   2 marks

_____

_____

_____

_____

b i What conditions are necessary for the formation of fossils?   2 marks

_____

_____

_____

ii Why would fossils of giant pandas be relatively rare?   1 mark

_____

_____

_____

Question 9 (6 marks) ©VCAA VCAA 2012 (2) SB Q6 (adapted) ●●●

One form of dating the age of a fossil is by radioactive carbon dating. The ratio of carbon-14 to nitrogen-14 (^{14}C:^{14}N) in the fossil is analysed and compared with the ratio of these elements in an organism living today. The graph below shows the rate of decay for carbon-14.

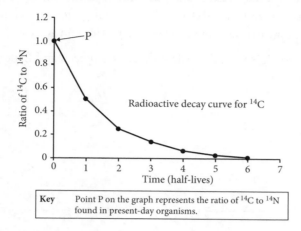

Key   Point P on the graph represents the ratio of ^{14}C to ^{14}N found in present-day organisms.

A fossil kangaroo skull was found in a limestone cave. The skull's C^{14}:N^{14} ratio was analysed and found to contain one-quarter (¼) of the carbon-14 of a kangaroo that died in 2012.

a **i** Place an X on the curve to show the fossil's C^{14}:N^{14} ratio.   1 mark

_____

ii Given the half-life of carbon is approximately 5700 years, what is the approximate age, in years, of the kangaroo skull?   1 mark

_____

Carbon-dating analysis is not always possible, and the age of the fossil can be estimated by dating the rock in which it is found.

b **i** Why is carbon-dating analysis not always possible?   1 mark

_____

ii Name another absolute dating technique that can determine the age of the rock surrounding a fossil.   1 mark

_____

c What is a relative dating technique that could be used to determine the age of the unknown fossil?   2 marks

_____

_____

_____

Question 10 (8 marks) 〇〇

Fossilisation is a rare process in which animals become preserved after their death. Two species present in the Mesozoic era were *Glossopteris* and *Cynognathus*.

Fossils of a *Glossopteris*, a fern that existed 200–300 million years ago, have been discovered in South America, Africa, India, Antarctica and Australia. The *Cynognathus* was a four-legged land animal existing 251–245 million years ago, in both South America and Africa.

a Explain how it would be possible to find fossils such as *Glossopteris* and *Cynognathus* in many places around the world. 2 marks

_____

_____

_____

b Outline the steps for fossilisation that would have occurred to preserve *Cynognathus* if it died in a riverbed. 2 marks

_____

_____

c A scientist found *Cynognathus* footprints in an old riverbed. What type of fossil would this be classified as? 1 mark

_____

d Ammonites were found in the same rock stratum as the *Cynognathus*. Ammonites are classified as index fossils. What are **two** characteristics of an index fossil? 2 marks

_____

_____

_____

e A fossil located more recently than the fern and *Cynognathus* is the *Archaeopteryx*. Explain how fossils such as *Archaeopteryx*, a transitional fossil, provide evidence in the evolution of species. 1 mark

_____

_____

_____

Test 12: Determining the relatedness of species

> Section A: 20 marks. Section B: 60 marks. Total marks: 80 marks.
> Suggested time: 40 minutes

Section A: Multiple-choice questions

> **Instructions to students**
> - For each question, circle the multiple-choice letter to indicate your answer.

Question 1

A scientist identified a new species and compared a conserved gene with a similar species. The gene had five differences in the amino acid sequence, but the nucleotide sequence difference was higher at 12 differences. This can be accounted for by

A DNA being universal.

B the degenerate code.

C different conserved genes being analysed.

D differences in pre-transcriptional modification.

Question 2

The wing of a bird and the wing of an insect are

A homologous structures.

B analogous structures.

C derived from a common ancestor.

D the result of divergence.

Question 3

The following diagram shows the evolutionary relationship between six species, A–F.

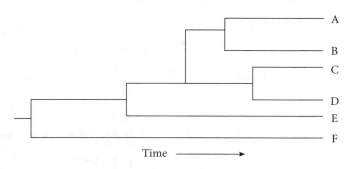

Using the information contained in this diagram, which two species are the most closely related?

A D and F

B D and E

C B and D

D A and E

Question 4 ●○○

An ancestor of whales, *Pakicetus* was a land-dwelling animal that inhabited Earth 50 million years ago. It hunted small land animals and fish. It lived on riverbanks and in lakes. Over time, these animals evolved to the whales we see today. The pelvis of *Pakicetus* is still present in whales today but is no longer functional. This is classified as

A a homologous structure.

B an analogous structure.

C a vestigial structure.

D a transitional structure.

Question 5 ●●○

Which of the following pieces of information would be least useful when determining the relationship between two species of organisms?

A DNA base sequences of both species

B Details of analogous structures found in the two species

C Amino acid sequences of related proteins

D Analysis of fossil remains

Question 6 ●●○

The following diagrams show the forelimb bones of a dog and a mole.

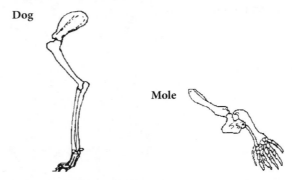

Which of the following statements best explains the similarities and differences in the anatomy of these two limbs?

A They are homologous and modified by divergent evolution.

B They are analogous and modified by divergent evolution.

C They are homologous and modified by convergent evolution.

D They are analogous and modified by convergent evolution.

Question 7 ●●○

The beak of the finch is the terrestrial analogue of the jaw of the cichlid fish. Both evolved in order to

A exploit available niches.

B facilitate eating of bigger prey.

C escape predation.

D avoid extremes of climate.

Question 8 ●○○

Even though snakes cannot walk, they still have tiny hind leg bones buried in their muscles towards their tail end. These leg bones are thought to be leftovers from their lizard ancestors.

The snake leg bones are an example of

A convergent evolution.

B divergent evolution.

C vestigial organs.

D adaptive radiation.

Question 9 ○●●

Which of the following multicellular organisms was the first to evolve?

A mammals

B reptiles

C birds

D flowering plants

Question 10 ○●●

Embryos of the Phylum Chordata such as fish, amphibian, reptiles, birds and mammals all show evidence of pharyngeal slits, a dorsal notochord and a tail that extends past the anus at some stage of their development. The similarities between these embryos suggest

A vestigial structures.

B analogous structures.

C a shared common ancestor.

D they have evolved the same features in response to a common selection pressure.

Question 11 ●●●

Two fossils were located in the same rock strata in the same region that was once a riverbed. Both fossils were in the class Osteichthyes. These bony fish had similar bone structures and were similar sizes. Which of the following could be determined by this information?

A The fossils were from the same species of fish.

B The structures were analogous because of similar selection pressures.

C The fish both contained vestigial organs which were used earlier in evolution.

D The fish were present in the same time period.

Question 12 ●●●

The mantella frogs from Madagascar and the poison dart frog from the tropical central and south America independently evolved to have bright yellow and black colouring to protect against predators and produce similar toxic chemicals on the surface of their skin. The similarity in their defensive function is evidence of

A homologous structures.

B analogous structures.

C allopatric speciation.

D gene flow between populations.

The following information relates to Questions 13 and 14.

Question 13 ©VCAA VCAA 2016 SA Q39 ○●●

Cytochrome c is a protein that consists of 104 amino acids. Many of these 104 sites on cytochrome c contain exactly the same amino acids across a large range of organisms. There are, however, some differences at certain sites. It is hypothesised that different organisms, all containing cytochrome c proteins, descended from a primitive microbe that lived over 2 billion years ago. The table below uses the three-letter codes for various amino acids found at specific sites for each organism.

Molecular homology of cytochrome c

Organism	Site 1	Site 4	Site 11	Site 15	Site 22
Human	Gly	Glu	Ile	Ser	Lys
Pig	Gly	Glu	Val	Ala	Lys
Dogfish	Gly	Glu	Val	Ala	Asn
Chicken	Gly	Glu	Val	Ser	Lys
Drosophila	Gly	Glu	Val	Ala	Ala
Yeast	Gly	Lys	Val	Glu	Lys
Wheat	Gly	Asp	Lys	Ala	Ala

Using only the data for the molecular homology of cytochrome c, which one of the following organisms is most closely related to the dogfish?

A *Drosophila*

B chicken

C human

D yeast

Question 14 ⬤⬤⬤

Using only the data for the molecular homology of cytochrome c, which pair of organisms is most distantly related to wheat?

A dogfish and *Drosophila*

B *Drosophila* and yeast

C *Drosophila* and pig

D human and yeast

The following table relates to Questions 15 and 16.

Differences in the amino acid sequences and DNA mutations of the haemoglobin of a number of species are shown in the table.

Species compared	Number of amino acid differences	DNA mutational differences
Human vs chimpanzee	0	1
Human vs gorilla	2	2
Human vs monkey	12	16
Monkey vs chimpanzee	12	12
Monkey vs gorilla	14	17

Question 15 ⬤⬤⬤

From the information in the table, it can be concluded that

A the minimum number of DNA mutations to alter an amino acid is two.

B the monkey is more closely related to the human than the chimpanzee.

C a change in the DNA will always result in a change in the amino acid coded for.

D the numbers in the table account for the minimum number of mutations that have occurred between the species.

Question 16 ⬤⬤

There are 12 differences between the amino acid sequences of the monkey and the human for the haemoglobin protein, and 16 mutational differences. What is a possible explanation for this difference?

A There are fewer bases in the DNA of the human than in the monkey.

B The human gene would have more introns than the monkey gene.

C More than one DNA triplet can code for each amino acid.

D One DNA triplet can code for multiple amino acids.

The following information relates to Questions 17 and 18.

Consider the following phylogenetic tree for different species of lice. The tree has been constructed from molecular and morphological data.

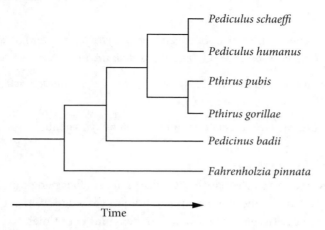

Time

Question 17 ©VCAA VCAA 2016 SA Q29 ●●

This information suggests that

A *Pedicinus badii* shares a more recent common ancestor with *Pthirus gorillae* than with *Fahrenholzia pinnata*.

B *Pediculus humanus* is more closely related to *Pedicinus badii* than it is to *Pthirus pubis*.

C the six species of lice would have evolved by convergent evolution.

D *Pediculus schaeffi* is the ancestor of *Pediculus humanus*.

Question 18 ●●●

A conserved gene with 3000 amino acids was observed between the different species of lice. *Pediculus humanus* and *Pthirus gorillae* contained 5 different amino acids between them and *Pthirus gorillae* and *Fahrenhoizia pinnata* contained 11 amino acid differences. From this information it can be assumed that

A only 11 nucleotides would differ between *Pthirus gorillae* and *Fahrenhoizia pinnata* in the conserved gene.

B the minimum number of differences in the nucleotide sequence of *Pediculus humanus* and *Pthirus gorillae* is 5.

C *Pediculus humanus* and *Pthirus gorillae* diverged further back than *Fahrenhoizia*, resulting in the 5 amino acid differences.

D *Pthirus gorillae* would have more amino acids in common with *Pediculus humanus* than with *Pthirus pubis*.

The following information relates to Questions 19 and 20.

Penguins are all part of the same family of *Spheniscidae*. There are currently 18 species of penguins, which are mainly located in the southern hemisphere. All penguins had a common ancestor 40 million years ago. Emperor penguins (*Aptenodytes forsteri*) diverged from king penguins (*Aptenodytes patagonicus*) 10 million years ago; African penguins (*Spheniscus demersus*) and Galapagos penguins (*Spheniscus mendiculus*) diverged 5.3 million years ago.

Question 19

Which of the following statements is correct?

A *Aptenodytes forsteri* is more closely related to *Aptenodytes patagonicus* than *Spheniscus mendiculus* is to *Spheniscus demersus*.

B There would be fewer genetic differences between *Aptenodytes forsteri* and *Spheniscus demersus* than between *Aptenodytes patagonicus* and *Spheniscus mendiculus*.

C There would be more amino acid differences between *Aptenodytes patagonicus* and *Spheniscus demersus* than the two in the *Spheniscus* genus.

D The *Spheniscus* genus are less closely related than the *Aptenodytes* genus.

Question 20

The *Aptenodytes* are large penguins, whereas the *Spheniscus* are smaller penguins. The two penguins originated from the same common ancestor around 20 million years ago. Which of the following could account for the change in size between the two genera of penguins over time?

A Spending a large percentage of their lifespan in the ocean

B Different mating calls and breeding partners

C Differences in environmental temperatures and habitat

D Laying eggs that hatch to produce individual offspring

Section B: Short-answer questions

> **Instructions to students**
> • Answer all questions in the spaces provided.

Question 1 (7 marks)

Unlike birds, whose anatomy is adapted for flight in nearly all species, mammals have evolved into a wide variety of forms. Some groups took to the air, while some, against the long-term trend of evolution, returned to the sea. The diagrams below show part of the anatomy of three of these diverse mammals.

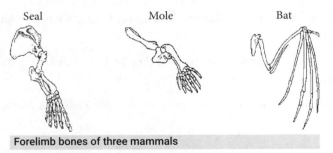

Seal   Mole   Bat

Forelimb bones of three mammals

a What evolutionary process has led to the adaptations in the limbs of the above mammals?   1 mark

b What do biologists call these similar structures?   1 mark

c What evidence is there from the images above that the three mammals have a common ancestor?   1 mark

_____

_____

_____

d How would Charles Darwin have explained the change in shape of the above structures?   2 marks

_____

_____

_____

_____

e The bone structure of bats' wings are considerably different from that of non-flying mammals. Surprisingly, the coding regions of many of the genes involved in bone growth in bats are very similar to the coding regions of other mammals. How could the same structural genes produce a short leg in an animal like a mouse and the elongated bones of a bat's wing?   2 marks

_____

_____

_____

_____

_____

Question 2 (6 marks) ⬤⬤⬤

Palaeomastodon is thought to be the ancestral species of all elephants. This elephant evolved into _Primelephas,_ the common ancestor of elephants. The African elephant diverged 5 million years ago from _Primelephas_ and subsequently diverged into the African savanna elephant and the African forest elephant between 5 million and 2 million years ago. The mammoth diverged 4 million years ago. The mammoth and Asian elephants are more closely related than the African and Asian elephants, which developed 2.5 million years ago. The only elephants still existing today are the African savanna elephant, the African forest elephant and the Asian elephant.

a Form a phylogenetic tree from the information provided above including the mammoth, African elephants and the Asian elephant.   3 marks

b Explain how two different cladograms could be formed from information about a group
of organisms.   1 mark

_____

_____

_____

c Another group of scientists hypothesised that the mammoth was the same species as the Asian
elephant and appeared structurally different because of the small number of elephants that became
isolated due to a change in sea levels from the small population. What is the event of a small
population moving to a new area classified as?   1 mark

_____

d How could the scientists find evidence to support this hypothesis?   1 mark

_____

Question 3 (6 marks) ⬤⬤⬤

The giant panda's natural distribution is limited to the bamboo forests of China. Although commonly
called a panda bear, for many decades the giant panda was believed to be a member of the raccoon family,
having many features in common with the lesser panda and the common raccoon. These features include
similar dentition, digestive tracts, habitat and feeding behaviour. These common features led to the
development of the phylogenetic tree seen below. Differences in body shape between the two pandas were
believed to be the result of divergent evolution.

Phylogenetic tree based on morphology

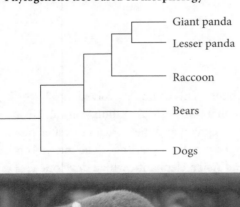

The lesser panda and the giant panda

a **i** What is meant by 'divergent evolution'?   1 mark

_____

_____

ii Explain how divergent evolution can lead to the development of a new species.   3 marks

_____

_____

_____

_____

_____

_____

_____

_____

_____

b Name a different technique that could be used to analyse the species above, which could give a different relationship. Which would be a more accurate phylogenetic tree and why?   2 marks

_____

_____

_____

Question 4 (7 marks) ●●●

The North Island of New Zealand is home to four distinct beetle species of the genus _Lissotes_. These ground-dwelling beetles have a common ancestor, which inhabited the entire island in the early Pliocene.

During the Pliocene, rising sea levels broke up the island into a series of smaller islands separated by straits of sea water. A long period of geographic isolation and divergent evolution followed, in which the beetles evolved into four distinct species. These species vary in size, and particularly in the shape and position of their mouth parts. Even though the sea level has long since dropped again, reuniting the islands, the beetles remain classified as four distinct species.

a What criteria would a biologist use to determine that the beetles are four distinct species?   1 mark

_____

_____

_____

b Explain what is meant by 'geographic isolation'.   1 mark

_____

c Explain how a long period of isolation may lead to the evolution of new species.   3 marks

_____

_____

_____

_____

d Consider the four beetle species as species A, B, C and D. Scientists think that species A most closely resembles the common ancestor of the four species and that species C and D have diverged most recently. Draw a branching diagram that shows the relationship between these four species.   2 marks

Question 5 (7 marks) ●●●

The table at the right contains the number of amino acid differences between the beta chain of human haemoglobin and the haemoglobins of other species. The haemoglobin chain consists of 146 amino acids. The average rate of evolutionary change between haemoglobins has been shown to be one amino acid every 3.5 million years.

Species	Differences
Human	0
Gorilla	1
Rhesus monkey	8
Dog	15
Frog	67

a i Which of the above species had a common ancestor with humans most recently?   1 mark

_____

ii Explain your answer to part **i**.   1 mark

_____

_____

_____

b What conclusions can be made about the genetic code of the human and the gorilla haemoglobin gene compared to the amino acid chain?   2 marks

_____

_____

_____

c According to the data in the table, how many years ago did the rhesus monkey and humans share a common ancestor?

1 mark

_____

d State **two** different types of evidence that can be used to determine evolutionary relationships.

2 marks

_____

_____

Question 6 (8 marks)

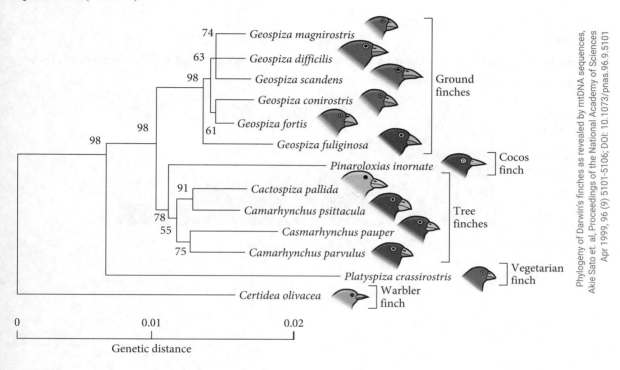

Phylogeny of Darwin's finches as revealed by mtDNA sequences, Akie Sato et. al, Proceedings of the National Academy of Sciences Apr 1999, 96 (9) 5101-5106; DOI: 10.1073/pnas.96.9.5101

a Explain the steps involved in the formation of the 14 different species of Galapagos finches from the mainland ancestral species.

3 marks

_____

_____

_____

_____

b Explain how developments in DNA analysis have influenced the way in which we draw phylogenetic trees and the relatedness between species.

2 marks

_____

_____

_____

c   The ground finches on different islands have no gene flow between them but have similar beak sizes. Explain how the similarities seen in the diagram have occurred.   2 marks

_____

_____

_____

d   Which of the finches is most closely related to *G. conirostris* and when did they diverge?   1 mark

_____

Question 7 (4 marks)

The gene for cytochrome c, which is conserved in many species, was analysed and the numbers of amino acid differences were recorded in the table below.

	Human	Whale	Snake	Rabbit	Elephant
Human		2	26	13	5
Whale			22	11	4
Snake				2	11
Rabbit					7
Elephant					

a   Draw a phylogenetic tree from the information above.   2 marks

b   Hypothesise where a chimpanzee and yeast would be on the phylogenetic tree above. Explain your reasoning.   2 marks

_____

_____

_____

_____

Question 8 (6 marks) ©VCAA VCAA 2015 SB Q9 ●●●

A fossil of an extinct species called *Indohyus major*, found in northern India, is thought to share a recent common ancestor with the group of living organisms called cetaceans. Cetaceans include dolphins and whales.

Indohyus major

Source: Nobu Tamura
(http://spinops.blogspot.com)

a Name the type of evolution that describes the relationship between *I. major* and cetaceans.   1 mark

_____

For a long time, scientists have believed that cetaceans are related to the group of terrestrial mammals classified as artiodactyls, which includes pigs and hippopotami. The name artiodactyl refers to the shape of the feet or hooves of these animals. To work out the evolutionary relationships between *I. major* and living animals, scientists closely studied their bones and skeletal structures.

b What name is given to the study of the similarities and differences between the bones and skeletal structures of animals, including fossils of extinct species?   1 mark

_____

The table below shows a summary of the scientists' findings.

Animal	Feet	Limb bones	Inner-ear bones
Cetaceans (e.g. whales)	artiodactyl	thick	thick
Suids (e.g. pigs)	artiodactyl	thin	thin
Hippopotamids (e.g. hippopotami)	artiodactyl	thick	thin
I. major	artiodactyl	thick	thick

c **i** Using all the information provided, complete the diagram below to show the evolutionary relationships between the following animals: **A** cetaceans (e.g. whales and dolphins), **B** suids (e.g. pigs), **C** hippopotamids, (e.g. hippopotami), **D** *I. major*.

Write the corresponding letter of the animal (**A–D**) in the boxes provided.   2 marks

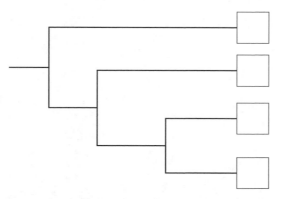

ii Explain the reasoning behind your response to part **c i**.   2 marks

_____

_____

Question 9 (4 marks) ⬤⬤⬤

DNA analysis that involves hybridisation techniques of measuring the temperature required to separate the DNA strands has been used to established the genetic make-up of some members of the Carnivora order. The results of this analysis are included in the following phylogenetic tree.

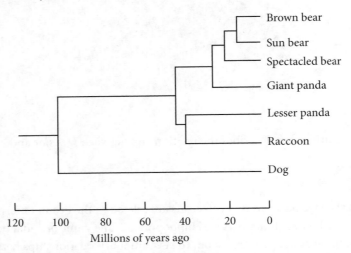

Phylogenetic tree based on DNA hybridisation

a According to the phylogenetic tree, at what point in the past did the

   i giant panda and lesser panda share a common ancestor?   1 mark

   ii giant panda and the other bears share a common ancestor?   1 mark

b Identify which species is most closely related to the raccoon. Justify your answer.   1 mark

c Given that the giant panda and the lesser panda are not closely related, how would you account for their similar features?   1 mark

Question 10 (5 marks) ©VCAA VCAA 2008 (2) SB Q5 (adapted) ●●●

Three different kinds of plants, a cactus, euphorbia and milkweed, have similar adaptations for growing in desert environments. They all have long, fleshy stems for water storage, protective spines and reduced leaves. Both the euphorbia and milkweed plants are believed to have evolved from leafy plants adapted to more temperate climates. They share a more recent ancestor than either one does to a cactus.

a   Draw an evolutionary tree (also called phylogenetic tree or cladogram) to demonstrate this
relationship.   1 mark

b   Name the process by which the ancestral leafy euphorbia plant could have given rise to the
desert-adapted species described above.   1 mark

c   What is a possible explanation for there being so few cactus fossils?   1 mark

d   A botanist located a cactus that looked similar to the cactus species but was located in a different
region. Explain a way in which the botanist could determine if the two cacti were the same
or separate species.   2 marks

Test 13: Human change over time

Section A: 20 marks. Section B: 64 marks. Total marks: 84 marks.
Suggested time: 40 minutes

Section A: Multiple-choice questions

Instructions to students
- For each question, circle the multiple-choice letter to indicate your answer.

Question 1 ●●●

Which of the following statements about the migration of *Homo sapiens* to Australia and the founding of Australia's Aboriginal population is correct?

A The Australian Aboriginal population has a closer genetic link to the *Homo sapiens* in Papua New Guinea than the founding African population.

B *Homo sapiens* migrated through Asia to Papua New Guinea 90 000–100 000 years ago.

C The Aboriginal population evolved from the *Homo erectus* population through interbreeding with the Neanderthals.

D *Homo sapiens* migrated before the continents separated from one land mass.

Question 2 ●●○

Scientists observing the fossils of the *Homo* ancestors noticed gradual changes in their anatomy as they began to undertake bipedal locomotion. Which of the following was not a selection pressure for becoming bipedal?

A To observe predators in the long grassy savannas

B An increase in brain size and cranium

C Less surface area exposed to the sun leading to increasing body temperature regulation

D Energy efficient for locomotion compared to walking on all four limbs

Question 3 ●●○

Which of the following aligns with the classification of hominins versus primates?

	Hominin	Primate
A	Modern and extinct great apes	Apes, monkeys, humans and all their ancestors
B	Specialised tool use	Centralised foramen magnum
C	Bipedal locomotion	Stereoscopic colour vision
D	Large relative brain size	S-shaped spine

Question 4

The following diagram represents the relationship between three hominin species as determined by DNA sequencing.

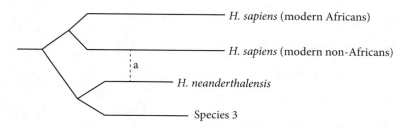

The species name missing from this diagram (species 3) is:

A *H. erectus*   **B** *H. denisovans*   **C** *A. africanus*   **D** *A. afarensis*

Question 5

Mitochondrial DNA (mtDNA) is passed from a mother to her young during sexual reproduction. There is little change in the mtDNA from one generation to the next. mtDNA

A has a high rate of crossing over at meiosis.

B can be used to trace the paternal line.

C is useful for researchers to trace the maternal line back in time.

D changes by 50% each generation.

Question 6

The mtDNA sequences of three modern populations of *Homo sapiens* were compared to Neanderthal and chimpanzee sequences. The results are shown in the diagram at the right.

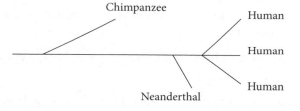

What can be concluded?

A The Neanderthal sequence branches before the divergence of the various human mitochondrial DNA lineages, but after the split from chimpanzees.

B The mtDNA sequences were conserved across the three groups.

C The Neanderthals were an ancestor of modern humans.

D The mtDNA sequences of the Neanderthal would be more similar to that of chimpanzees than to living humans' sequences.

Question 7

A newly discovered fossil primate was determined to be a member of the hominoid group. This species is now extinct. Which feature could have been used to make this determination?

A The hand showing evidence of nails, rather than claws

B Opposable thumbs and five digits

C A short bowl-shaped pelvis relative to body length

D Rotating shoulder joints

Question 8 ⬤⬤⬤

The brain size has been analysed in a range of species, with the following information collated:

Chimpanzees: 395 cm³ and quadruped

Gorillas: 490 cm³ and quadruped

Australopithecus: 440 cm³ and facultative biped

Homo habilis: 610 cm³ and obligate biped

Homo erectus: 860 cm³ and obligate biped

Homo floresiensis: 440 cm³ and obligate biped

Homo sapiens: 1350 cm³ and obligate biped.

Which of the following can be concluded from this information?

A Brain size and cranial size increased with the increase in complexity of the *Homo* genus.

B Increase in brain size enabled the species to become bipedal.

C Increased protein in the diet from hunting and running bipedally is linked with increased brain capacity.

D The older the skeleton fossil the more simplistic their behaviour, demonstrated through the brain size.

Question 9 ⬤⬤

Research published shows that modern humans and Neanderthals did interbreed, but on a very small scale. Researchers compared the genomes of modern humans from a variety of continents with Neanderthal DNA. They discovered that Europeans and Asians share 1–4% of their DNA with Neanderthals, and Africans share none. This data is consistent with which of the following hypotheses about human evolution?

A Modern humans and Neanderthals bred after modern humans left Africa but before they spread to Asia and Europe.

B Asian populations evolved from local populations of *Homo erectus*.

C Neanderthals evolved in Africa and then migrated to Europe and Asia.

D Interbreeding occurred in isolated pockets in Northern Europe.

Question 10 ©VCAA VCAA 2018 SA Q39   ⬤

Which general trend is shown by hominin fossils?

A The older the fossil, the more central the position of the foramen magnum in the skull.

B The older the fossil, the smaller the braincase that surrounds the cerebral cortex.

C The more recent the fossil, the less bowl-shaped the pelvis.

D The more recent the fossil, the larger the jaw bones.

Question 11 ⬤

Leber's hereditary optic neuropathy is a vision disorder that causes shrinking of the optic nerve. It occurs because of a defective gene in mitochondrial DNA. Which of the following statements is true of this condition?

A It is more common in women than in men.

B A man who has the condition will pass it on to his daughters but not his sons.

C A man who has the condition must have inherited it from his mother.

D A woman who has the condition will be the only one in the family affected.

Question 12 ⬤⬤

Homo denisovans is the first hominin species to be described using its DNA sequence rather than fossil evidence. In fact, only a single finger bone and tooth were found in a cave in Denisova, Siberia, but this was sufficient to obtain a DNA sequence. Examination of Denisovan DNA showed that

A Denisovans interbred with modern European populations.

B Denisovans predated *H. erectus* in Asia.

C Denisovans were more closely related to *H. neanderthalensis* than to *H. sapiens*.

D Denisovans were more closely related to modern Papuan people than to Neanderthals.

Question 13 ©VCAA VCAA 2015 SA Q38 ⬤⬤⬤

Fossil remains of a number of individuals from the genus *Australopithecus* were found at various sites in the eastern half of Africa and have been dated to between 3 and 4 million years old. These fossil remains

A are descendants of *Homo erectus*.

B represent the oldest evidence found of primates.

C show early evidence that hominins were bipedal.

D represent the earliest examples of the hominoid super-family.

Question 14 ⬤⬤⬤

Two common models of the evolution of *Homo sapiens* include the 'out of Africa' hypothesis and multiregional hypothesis. With the discovery of more fossils and the use of DNA analysis, scientists can get a clearer picture of how the migration of *H. sapiens* occurred. Which of the following supports the out of Africa hypothesis?

A High variation and a large gene pool through continuous interbreeding with populations in different regions

B Fossils of *H. sapiens* located throughout the continents aged at similar time scales

C Presence of *H. erectus* in multiple regions

D Greater variation in the mitochondrial DNA in Africa than all other continents

Question 15 ⬤⬤

Which of the following statements is true of *Homo neanderthalensis*?

A Their brains were smaller on average than the brains of modern humans.

B They had a relatively shallow and long pelvis similar to that of *Australopithecus*.

C They had relatively large brow ridges and noses.

D Their limbs had similar proportions to those of early *Australopithecus* species.

Question 16 ⬤⬤

A new hominid fossil was located in an old riverbed in Siberia. Which of the following would have contributed to the fossilisation of this specimen?

A Rapid burial by volcanic ash

B Presence of scavengers

C Anaerobic conditions

D Being washed down a turbulent riverbed

Question 17 ●●

In comparison to modern humans, members of the species *Homo erectus* had

A   proportionally much longer arms to legs.

B   the foramen magnum at the back of the skull rather than the base of the skull.

C   fully opposable toes.

D   more prominent brow ridges and muscle attachments.

Question 18 ●●●

For an organism to be classified in the primate order, they must have specific characteristics. Which of the following are **not** specific to the primate classification?

A   Hands and feet with flat nails and an opposable thumb or toe

B   Milk-producing mammary glands and thick fur over body surface

C   Large brain relative to their body size and extensive foetal brain development

D   3D colour vision with forward-facing eyes

The following information relates to Questions 19 and 20.

A hominin species, *Homo floresiensis*, was identified from fossils found on an isolated Indonesian island. These fossils were dated to be 18 000 years old. The adult skull of this upright bipedal hominin had a cranial volume less than one-third of the average cranial volume of a modern adult human. It had harder, thicker eyebrow ridges than *H. sapiens*, a sharply sloping forehead and no chin. *H. floresiensis* was just over 1 metre tall and their arm-to-leg ratio was slightly larger than that of modern humans. They weighed approximately 16 kg. The fossils were found in sediment that also contained stone tools and fireplaces for cooking. The fireplaces contained the burnt bones of animals, each animal weighing more than 350 kg. The stone tools included blades, spearheads and cutting and chopping tools.

Question 19 ©VCAA VCAA 2014 SA Q39 ●●●

The above evidence and current theories of hominin evolution indicate that

A   the very small stature of *H. floresiensis* is unexpected in a species that is geographically isolated on a small island.

B   *H. floresiensis* was a species that showed considerable social cooperation and methods for passing on knowledge.

C   the sloping forehead and absence of chin of members of the *H. floresiensis* species suggest they were members of the first migration of hominins into Australia over 50 000 years ago.

D   the skull shape and size of *H. floresiensis* suggest a close relationship to the *H. neanderthalensis* species.

Question 20 ©VCAA VCAA 2014 SA Q40 ●●

The fossils of *H. floresiensis* showed that they had opposable thumbs. The development of an opposable thumb in primate evolution

A   is used to distinguish members of the genus *Homo* from the other great apes.

B   was a necessary step in the development of bipedalism in hominins.

C   was an important anatomical development that assisted toolmaking in hominins.

D   is a significant factor in determining the arm-to-leg ratio of modern humans.

Section B: Short-answer questions

> **Instructions to students**
> • Answer all questions in the spaces provided.

Question 1 (7 marks)

The following diagram shows the evolutionary relationships between the hominoids on the basis of DNA sequencing studies. The scale represents millions of years before present.

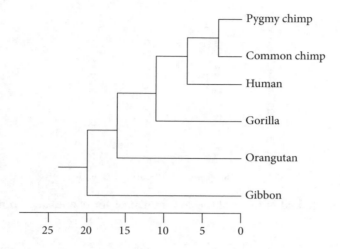

a According to the diagram, how long ago did the common ancestor of chimps and humans exist?

1 mark

b Besides humans, none of the other hominoid families are found in Australia. Orangutans and gibbons are found nearby in South-East Asia. Give **one** reason that would account for the absence of apes in Australia.

1 mark

The following table shows genetic differences among some primate species determined by DNA hybridisation.

Groups compared	Difference in DNA sequences (%)
Human and chimpanzee	1.6
Human and gibbon	3.5
Human and rhesus monkey	5.5

c i Using this table and the diagram, estimate the number of years since the existence of the common ancestor of the hominoids and the rhesus monkey.

1 mark

ii Explain the difference in DNA sequences and how this can determine an evolutionary tree. 2 marks

The following diagram shows chromosome 2 from humans and the apes. This is the only major chromosomal difference between apes and humans. Humans have a pair of chromosome 2. The ape species all have two smaller chromosomes, which have many sequences in common with human chromosome 2. They have been named 2A and 2B.

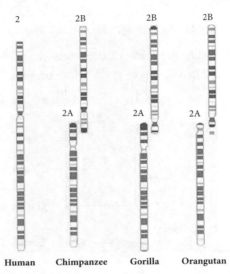

d **i** What was the likely diploid number of the common ancestor of apes and humans?   1 mark

_____

ii Using the information above, state what molecular event in the chromosomes appears to have occurred on the human evolution line?   1 mark

_____

Question 2 (6 marks)

The following diagram shows three hominin skulls.

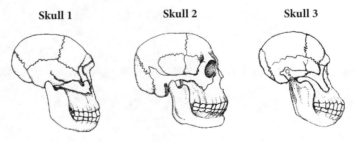

a Which of these skulls appears to be the earliest hominin?   1 mark

_____

b Give **two** reasons for your answer to part **a**.   2 marks

_____

_____

_____

c On what continent would you expect skull 3 to have been found? Explain your choice.   1 mark

_____

d The course of future human evolution is subject to speculation. One suggestion is that we should look at evolutionary changes of the last 4 million years and extend these trends. If we do this, we arrive at a huge-brained, small-faced future human.

Other suggestions about our future fate include:

The future?

- loss of useless organs such as the appendix and wisdom teeth due to lack of use

- enlargement of the eyes due to the increased importance of vision, especially reading and computer work.

- decreasing mouth and teeth size due to reliance on processed foods.

Evaluate the above theories about our future evolution. In particular, how likely are the changes suggested?   2 marks

_____
_____
_____
_____
_____

Question 3 (7 marks)

Consider the structures of the *Australopithecus* genus and *Homo* genus.

a State **two** ways in which the *Homo* genus structures altered to facilitate bipedalism.   2 marks

_____
_____
_____

b Explain an advantage of bipedalism for the *Homo* genus.   1 mark

_____
_____
_____

c Explain how brain size and bipedalism are interrelated in evolution.   2 marks

_____
_____
_____
_____
_____

d What are **two** changes that occurred to the skull structure (other than cranial size) as
species evolved bipedally and altered their diet?

2 marks

_____

_____

Question 4 (6 marks) ⬤⬤⬤

A land mass known as Sahul consisted of Papua New Guinea and Australia until about 8000 years ago
when it became separated because of a rise in sea levels. When sea levels were lower, the now-separated
continents were connected by land bridges. For example, when the sea levels were at their lowest, Timor
and Sahul were only separated by 90 km of sea. There is no evidence of boats in archaeology, probably
because of the materials used and the lack of preservation, but the earliest human evidence is located in
the Northern Territory in a rock shelter dating back 65 000 years. With a rise in sea levels, there is the
potential that many early occupation sites have been lost.

Within the sites around Australia, human skeletons have been dated to 40 000 years ago, 20 000 years
before the first evidence of large stone tools, rock art and charcoal deposits give an insight to the
Indigenous populations' behaviours. The skeletal remains show a larger and more robust _Homo_ skeleton
with a wide variation in physical characteristics.

a Suggest the genetic similarity between the two Indigenous populations of Papua New Guinea
and Australia compared to European and African populations.

1 mark

_____

_____

_____

b Why would the skeletal remains of archaic _Homo sapiens_ have been larger than the skeleton
of modern _Homo sapiens_?

2 marks

_____

_____

_____

_____

c What is the significance of discovering archaeological sites containing material other
than skeletons?

1 mark

_____

_____

_____

d What is an absolute dating technique that could have been performed to determine the age of the human fossils to 40 000 years.   2 marks

_____

_____

_____

Question 5 (6 marks) ⬤⬤⬤

a Complete the following table for the different classifications.   3 marks

Classification	Species included	Key characteristic
Primate		
Hominin		
Hominoid		

b A new fossil of a hominin is located that is determined to be _Homo florensis_. What is **one** feature that would facilitate the classification of the fossil to the _Homo_ genus and not _Australopithecus_?   1 mark

_____

_____

_____

c Explain how the human evolutionary tree has altered over time with the introduction of DNA analysis.   2 marks

_____

_____

_____

_____

_____

_____

Question 6 (7 marks) ©VCAA VCAA 2016 SB Q10 ⬤⬤

Over the past 20 years, a number of new hominin fossils have been discovered. _Homo erectus georgicus_ was found near the banks of the Black Sea in Georgia and _Homo naledi_ was found in a cave in South Africa.

a Consider the conditions that may have led to the fossilisation of members of these species. Complete the table below by identifying **one** condition in the environment of each species that will have made fossilisation possible. The same answer cannot be used for both species.   2 marks

Species	Environment	Condition
H. erectus georgicus	Near the banks of the Black Sea	
H. naledi	Cave in South Africa	

Shown right is a photograph of a skull of H. erectus georgicus. Scientists compared this skull to that of modern humans (*Homo sapiens sapiens*).

By Rama - Own work, Public Domain,https://commons.wikimedia.org/w/index.php?curid=7529356

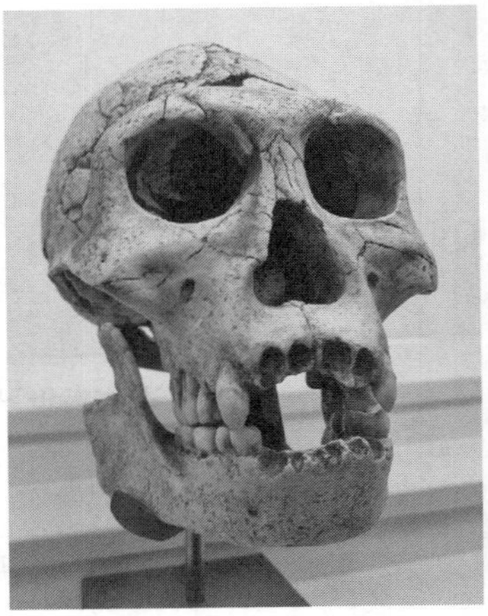

b Describe any **two** features of a skull that allowed scientists to determine that this was a much earlier species of the genus *Homo* than modern humans (*H. sapiens sapiens*). 2 marks

c Describe **one** structural feature (other than skull structure) of *H. naledi* that would indicate it is a more modern species than members of the genus *Australopithecus*. 1 mark

d Fifteen different skeletons of *H. naledi* were found in the cave. It was noted that they were all of different ages. Describe **two** pieces of evidence that scientists could have looked for in the cave to indicate cultural evolution within this species. 2 marks

Question 7 (4 marks) ●●●

Two groups of palaeontologists presented different theories on how *Homo sapiens* evolved. One group assessed the fossil evidence and location of the human remains, while the other analysed the mitochondrial and nuclear DNA to determine the path of human movement. From these analyses, two different evolutionary trees were produced.

a Outline **one** similarity and **one** difference of the 'out of Africa' theory to the multiregional theory that the palaeontologists were referring to. 2 marks

b Individuals within both of these groups have hypothesised that *Homo neanderthalensis* and the Denisovans were the same species. How could this be determined? 2 marks

Question 8 (6 marks) ©VCAA VCAA 2015 SB Q11

Fossil evidence indicates that 30 000–80 000 years ago, populations of the two hominin species – modern humans (*Homo sapiens*) and the extinct Neanderthals (*Homo neanderthalensis*) – lived close to one another in parts of the Middle East, Europe and Asia. Researchers have constructed a theory about the relationships between ancient populations. This is represented in the following diagram.

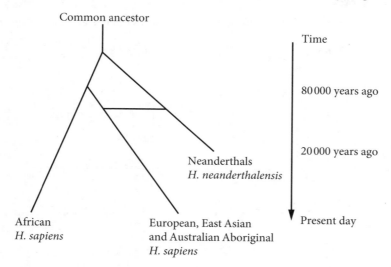

Recent DNA evidence has shown that:

- the genome of living humans of African descent does not contain Neanderthal DNA

- the genomes of living humans of European, East Asian and Australian Aboriginal descent all contain small amounts of Neanderthal DNA (1–4%).

a i Suggest how DNA from *H. neanderthalensis* entered the genome of present-day European, East Asian and Australian Aboriginal *H. sapiens*, and continues to be found in modern populations.   1 mark

_____

_____

ii What implication does this DNA evidence have for the classification of the two hominin species, *H. sapiens* and *H. neanderthalensis*, according to the common definition of a species?   1 mark

_____

b There are several theories about the geographical origins of *H. sapiens*. Scientists consider that the absence of Neanderthal DNA in present-day African *H. sapiens* lends support to one theory about the geographical origins of *H. sapiens*. Name this theory and explain how the recent DNA evidence provided above supports it.   3 marks

_____

_____

_____

_____

c What does the DNA evidence provided above suggest about the timing of the migration of the first Australian Aboriginals to arrive in Australia?   1 mark

Question 9 (6 marks)

a Complete the following table with the different characteristics of apes and *Homo sapiens*.   4 marks

Characteristic	Chimp	Homo sapiens
Pelvis structure		
Foramen magnum		
Arm to leg ratio		
Hands/feet		

b Compare *Australopithecus afarensis* to chimpanzees and *Homo sapiens*.   2 marks

Question 10 (9 marks)

Skull A represents a member of the species *Australopithecus afarensis*, which lived in Africa approximately 3.5 million years ago. It had a cranial capacity similar to that of modern apes but is classified as a hominin.

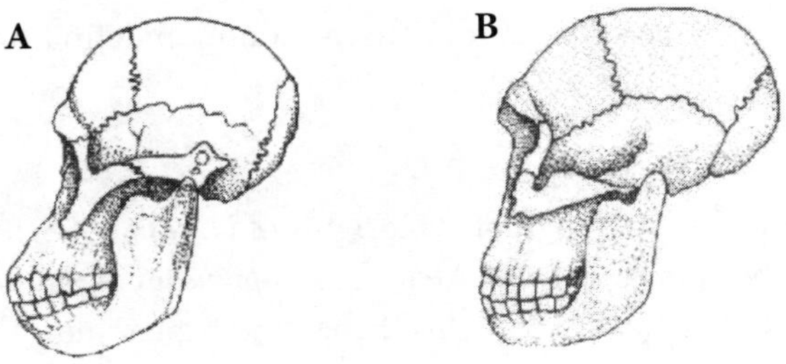

a Name **one** feature of skull A that would support its classification as a hominin rather than an ape.   1 mark

Skull B represents another African hominin species. It was discovered at about the same time as the skull above.

b   i   Is this skull representative of an earlier or a later species than the skull above?   1 mark

_____

ii   Name **two** features of the skull that support your answer to part **bi**.   2 marks

_____

In October 2004, fossils of a new hominin species named _Homo floresiensis_ were discovered on the Indonesian island of Flores. The fossils were only 18 000 years old and in many features resembled _Homo erectus_, but were smaller and had a small cranial capacity. Despite its small brain, there was evidence that it hunted, killed and butchered animals. It is widely believed that _Homo floresiensis_ evolved from an isolated _Homo erectus_ population and existed until relatively recently.

c   i   Name **two** features of a fossil skull that would support the hypothesis that _Homo floresiensis_ was related to _Homo erectus_, rather than being a member of _Homo sapiens_.   2 marks

_____

ii   What evidence would support the hypothesis that _Homo floresiensis_ was a hunter and butchered animals?   2 marks

_____

_____

_____

d   The following diagram shows part of one interpretation of the evolutionary tree of the hominins.

Show the position of _Homo floresiensis_ on the diagram.   1 marks

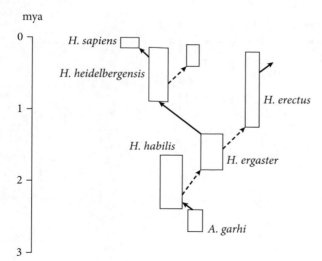

Test 14: Experimental design

Section A: 20 marks. Section B: 72 marks. Total marks: 92 marks.
Suggested time: 45 minutes

Section A: Multiple-choice questions

Instructions to students
*   For each question, circle the multiple-choice letter to indicate your answer.

Question 1

The following data was collected for an experiment trialling the effectiveness of a new antibiotic on the growth of a particular bacteria. The bacteria were cultured with different concentrations of the antibiotic-impregnated discs of filter paper on a nutrient agar plate. The plate was incubated for 24 hours at 35°C and the zone of inhibition was measured.

Concentration of antibiotic (µg/mL)	0	1	2	4	8	16
Zone of inhibition (mm)	0	0	2	5	5	5

Which of the following graphs is the best representation of the data?

A

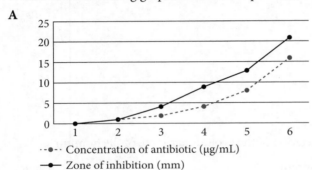

B

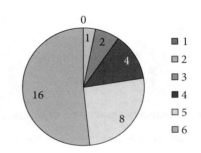

C

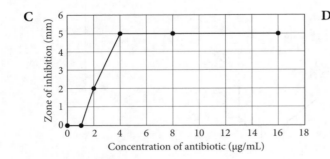

D

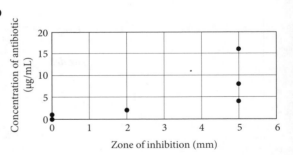

Question 2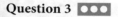

An activity was performed to measure the reaction time of students. To do this, one student held the zero end of a 1-metre ruler between the thumb and forefinger. The ruler was dropped and the distance before another student caught the ruler was measured. The results are displayed in the graph at the right.

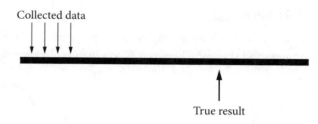

Which of the following is a conclusion that could be reasonably drawn from the graphed data?

A The reaction time of the subject improved as the trials progressed.

B The reaction time of the subject slowed as the trials progressed.

C The speed of reaction is directly proportional to the number of trials.

D The lower the ruler, the faster the reaction time.

Question 3 ⬤⬤⬤

Data was collected for a biological experiment. The data was graphed on a line to show the spread of the data. The data was compared for precision and accuracy.

Collected data

True result

The data collected can be described as having

A high precision and low accuracy.

B high precision and high accuracy.

C low precision and low accuracy.

D low precision and high accuracy.

The following information relates to Questions 4–6.

Wine vinegar is a popular ingredient in many salad dressings. The production of wine vinegar involves the use of two different microorganisms: *Saccharomyces* and *Acetobacter*. *Saccharomyces*, a yeast, is used to ferment grape juice. In the absence of oxygen, the yeast cells metabolise the grape sugar (glucose), producing alcohol and carbon dioxide:

$$C_6H_{12}O_6 \rightarrow 2C_2H_5OH + CO_2$$

Acetobacter, a bacterium, converts alcohol to acetic acid, the sour-tasting chemical that gives vinegar its distinctive taste. This enzyme-catalysed reaction requires oxygen:

$$C_2H_5OH + O_2 \rightarrow CH_3COOH + H_2O$$

At the beginning of the process, a suspension of *Saccharomyces* and *Acetobacter* is added to grape juice in a fermentation tank. The fermentation tank is deprived of oxygen for a period of time. Later in the process, the fermented juice is exposed to air to complete the production of vinegar. The yeast and bacteria are later removed from the liquid. The graph at the right shows the change in alcohol concentration during the production of wine vinegar.

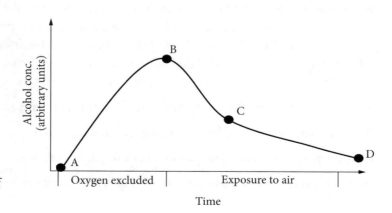

Question 4

The rate of metabolism of alcohol was greatest

A between points A and B.

B between points B and C.

C between points C and D.

D after point D.

Question 5

It is reasonable to conclude that the decline in alcohol concentration after point B is due to the

A death of the yeast cells.

B absence of any glucose in the solution.

C denaturing of yeast enzymes in the presence of oxygen.

D metabolism of alcohol by *Acetobacter*.

Question 6

Which of the following graphs best represents the concentration of glucose in the liquid over the same time period shown? The point 'X' on the graphs represents the point at which the liquid is exposed to air.

A

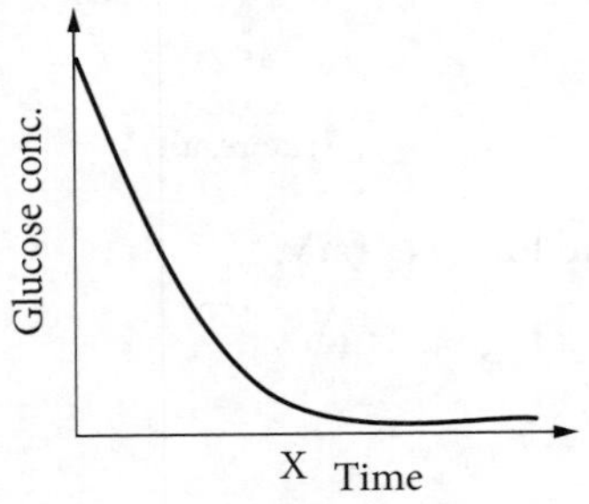

B

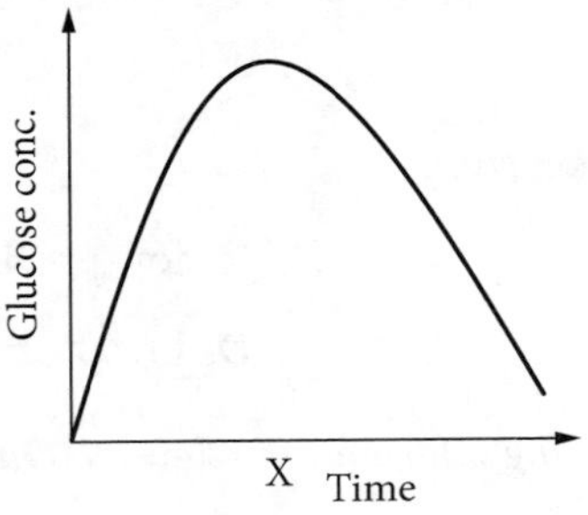

C

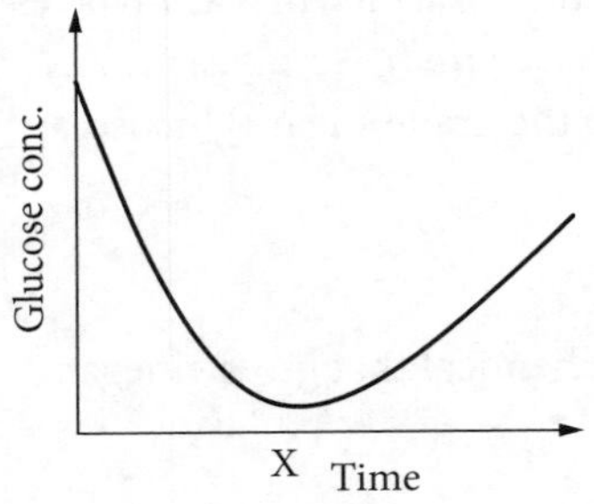

D

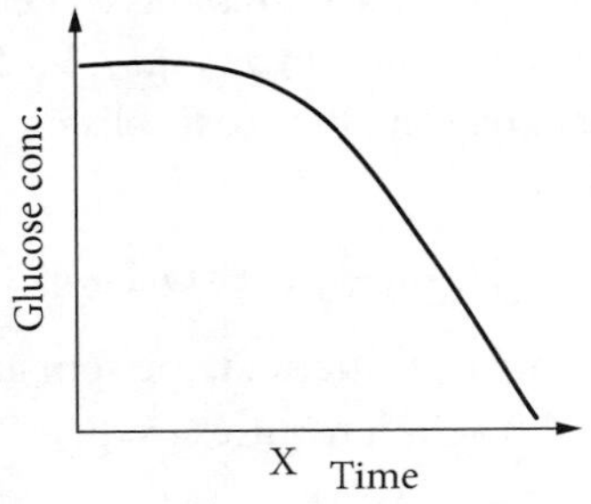

Question 7

An activity was performed to measure the rate of reaction with the enzyme amylase. A student prepared three test tubes, each containing 5 mL of amylase and 5 mL of starch. They placed each test tube at a different temperature (20, 40, 60°C) and observed the reaction from the addition of iodine into the solution after 10 minutes. All of the trials caused the iodine to change from yellow to black in colour, indicating a large amount of starch was present. A potential reason for this result is

A a systematic error with the thermometer, resulting in temperature baths being lower in temperature.

B the enzyme sample had been contaminated and denatured.

C the experimenter added less starch to one of the test tubes.

D a random error resulted in the timing of the experiment being varied.

Question 8 ▮▮▮

Experimental results may be defined in terms of accuracy and precision. Results with a high precision

A can be reproduced.

B produce data that shows a lot of variation.

C produce data with a high standard deviation.

D are the result of poor scientific method.

Question 9 ▮▮▮

Which of the following observations contains qualitative data?

A The amount of oxygen produced was 30 mL min^{-1}.

B The average height of the seedlings was 12 mm.

C A gas was produced by the reaction.

D The enzyme worked best at 37°C.

Question 10 ▮▮▮

Errors in an experiment can lead to unreliable data. Which of the following sources of error can be controlled by the experimenter?

A Vague scale on equipment

B Miscalibrated instruments

C Natural variations in measurements

D Addition of the wrong amounts of substrates

The following information relates to Questions 11 and 12.

Some scientists wanted to investigate the effectiveness of a new antibiotic on the growth of a particular bacteria. They prepared different concentrations of the antibiotic and impregnated discs of filter paper with the different concentrations. They applied the bacteria to a nutrient agar plate and carefully arranged the discs on the culture. The plate was incubated for 24 hours at 35°C.

Question 11 ▮▮▮

What is the best way of collecting data for this experiment?

A Measuring the zone of inhibition around each of the discs

B Identifying the density of the bacterial growth on the plate

C Finding the mass of the plate after incubation

D Classifying the type of bacteria that has grown

Question 12 ▮▮▮

To ensure the results obtained are reliable, which of the following would also be needed?

A Measuring the mass of the antibiotic

B Including a control disc with no antibiotic

C Growing another type of bacteria with the antibiotic

D Heating the plate at 70°C to remove contaminants

Question 13 ©VCAA VCAA 2017 SA Q8 ●●▪

During an experiment, a student measured the varying pH levels using a digital pH meter. The student calibrated the meter using a pH 7 buffer solution. The reason the student calibrated the pH meter was to

A ensure a random error would not influence the results.

B eliminate the effect of all uncontrolled variables.

C enable the use of the instrument with precision.

D allow the pH to be measured accurately.

Question 14 ©VCAA VCAA 2019 SA Q9 ●●▪

An experiment was carried out by students to test the effect of temperature on the growth of bacteria. Bacterial cells were spread onto plates of nutrient agar that were then kept at three different temperatures: –10°C, 15°C and 25°C. All other variables were kept constant. The experiment was carried out over four days. The nutrient agar was observed every day at the same time and the percentage of nutrient agar covered by bacteria was recorded.

The students wanted to check the reliability of their data. The students should

A repeat the experiment several times to find out if they would obtain the same data.

B organise their data into a different format to help identify a trend.

C change the independent variable in the experiment.

D rewrite the method for completing the experiment.

Question 15 ●▪▪

The function of a control group is to

A reduce the number of variables within an experiment.

B ensure random selection of specimens, preventing bias.

C act as a baseline to compare results to.

D observe the action of the dependent variable.

Question 16 ●●●

A student performed an experiment on the rate of oxygen production from hydrogen peroxide breakdown by the enzyme catalase at different temperatures, using the amount of bubbles produced. They repeated the experiment multiple times and found the variance in the data collected was 100 ppm, but the data was 0.1 ppm away from the true result. This experiment could be considered

A precise but not accurate.

B both precise and accurate.

C accurate but not precise.

D neither precise nor accurate.

The following information relates to Questions 17 and 18.

Four groups of students carried out an experiment in which the effect of glucose concentration on the fermentation rate of yeast was measured. The fermentation rate was determined by the rate of temperature change of the fermenting mixture. Before the experiment, each group practised measuring the temperature of water and checked their thermometer against an electronic thermometer that gave a true measure of temperature. The following results were obtained during the practice.

Group	Thermometer readings (°C) 1st measurement	2nd measurement	3rd measurement	Electronic thermometer reading (°C)
1	18.0	17.0	17.5	20.1
2	18.0	18.0	18.5	20.5
3	21.0	21.0	20.5	19.9
4	18.0	19.0	21.0	20.2

Question 17 ©VCAA VCAA 2018 SA Q11 ●●●

Which one of the following statements is correct?

A Group 1's measurements are the most accurate but the least precise.

B Group 2's measurements are accurate but not precise.

C Group 3's measurements are precise but not accurate.

D Group 4's measurements are both accurate and precise.

Question 18 ●○○

The independent variable from the experiment is

A temperature.

B glucose concentration.

C yeast concentration.

D ethanol production.

Question 19 ●●○

A carbon dioxide probe was used in a yeast fermentation experiment to determine the rate at which glucose is broken down in the absence of oxygen at different concentrations of glucose. The glucose amounts were increased in 10 mL increments and one spatula of yeast was added to each airtight container. The experiment was repeated multiple times and an average of the results was collected. A potential error that could affect the precision of the experiment is

A calibration of the carbon dioxide probe.

B the species of the yeast.

C the amount of yeast added into each chamber.

D the use of a control group.

Question 20 ©VCAA VCAA Sample Exam 2020 SA Q9 ●●○

The enzyme lactate dehydrogenase is found in a wide variety of organisms. It catalyses the conversion of both pyruvate to lactate and lactate to pyruvate. The bacterium *Thermoanaerobacter ethanolicus* lives in geothermal (hot) springs. The river buffalo (*Bubalus bubalis*) is a domestic animal common in Pakistan. Scientists studying the enzyme lactate dehydrogenase from these two organisms produced the following graphs.

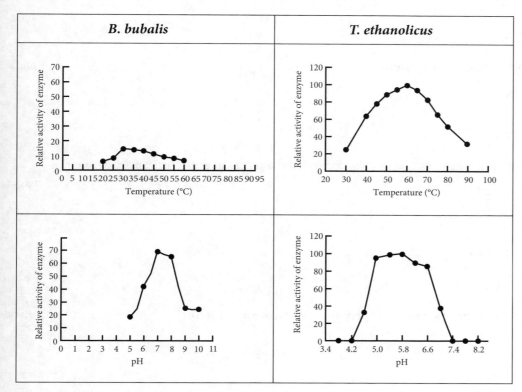

Adapted from MS Nadeem et al., 'Purification and characterisation of lactate dehydrogenase from the heart ventricles of river buffalo (Bubalus bubalis)', in Pakistan Journal of Zoology, vol. 43(2), 2011, p. 318;Adapted from Q Zhou and WL Shao, 'Molecular genetic characterisation of the thermostable L-lactate dehydrogenase gene (ldhL) of Thermoanaerobacter ethanolicus JW200 and biochemical characterisation of the enzyme', in Biochemistry (Moscow), vol. 75, no. 4, 2010, p. 529

From the graphs, it is reasonable to conclude that the

A optimum temperature at which the enzyme operates is higher in the bacteria than in the river buffalo.

B bacterial form of the enzyme would denature at temperatures below 60°C.

C bacterial form of the enzyme has an optimum pH of 7.4.

D body temperature of a river buffalo is 40°C.

Section B: Short-answer questions

> **Instructions to students**
> - Answer all questions in the spaces provided.

Question 1 (6 marks) ▮▮▮

A group of students performed an experiment with yeast in an anaerobic environment. The students aimed to determine the best temperature for fermentation to occur by testing the solution and determining the pH. Carbon dioxide forms carbonic acid when dissolved in a solution. Flasks B, C and D contained yeast and glucose solution; flask A contained glucose solution only. After 24 hours, the students tested the solutions in the sealed conical flasks with blue litmus paper, which remains blue in neutral and basic solutions but turns red in acidic solutions.

The results are shown in the table below.

Time (min)	Blue litmus paper results			
	Flask A	Flask B	Flask C	Flask D
0	Blue	Blue	Blue	Blue
20	Blue	Purple	Blue	Blue
40	Blue	Pink	Blue	Pink
60	Blue	Red	Blue	Dark pink

a Name and explain the function of flask A.   2 marks

b Form a hypothesis for the experiment.   1 mark

c Explain the type of data the students collected. What is a potential error that can arise from this type of data?   2 marks

d What change could be made to this experiment to improve its reliability?   1 mark

Question 2 (6 marks) ⬤⬤

A group of students was tasked with designing an experiment to determine factors that should be controlled for the greatest rate of photosynthesis.

a Other than light intensity, what are **two** limiting factors that could be associated with the rate of photosynthesis? 1 mark

_____

b Students formed a hypothesis that the lower the light intensity, the greater the rate of photosynthesis in tomato plants.

   i Form an experiment to test this hypothesis. 4 marks

_____

_____

_____

_____

_____

   ii Form a conclusion to refute the hypothesis. 1 mark

_____

_____

Question 3 (10 marks) ⬤⬤⬤

A scientist is testing a range of different treatments on acute myeloid leukemia cells, to find one with the highest rate of cancer removal and the lowest rate of side effects, such as killing normal, functioning cells.

A cell culture is isolated with an equal number of normal cells and acute myeloid leukemia cells. All of the myeloid leukaemia cells contained an upregulation (increase) of CD52 antigen. The cells have been radioactively tagged to identify the difference in the two types. These acute myeloid leukaemia cells can produce a protein that inhibits the function of cytotoxic T cells to override their action and the immune system.

The cell culture is separated into six sterile Petri dishes and a range of treatments are provided:

1 Cell culture untreated

2 Cell culture and cytotoxic T cells only

3 Cell culture and CD52 antigen monoclonal antibody

4 Cell culture and CD52 antigen monoclonal antibody bound with a chemotherapy drug

5 Cell culture and cytotoxic T cell + tumour protein inhibitor

6 Cell culture and chemotherapy drug

The cultures are analysed over a week to observe the rates of normal and cancerous cell growth.

a   Explain the function of Petri dish 1.   1 mark

_____

_____

b   List **two** controlled variables for this experiment.   2 marks

_____

_____

c   Identify the independent variable for this experiment.   1 mark

_____

	Petri dish					
	1	**2**	**3**	**4**	**5**	**6**
Cell growth normal cells	Normal growth no death	Normal growth	Normal growth	Normal cell growth	20% reduction in cells	60% reduction
Cell growth acute myeloid leukemia cells	High growth no death	30% reduction	No growth no reduction	80% reduction	75% reduction	88% reduction

d   Using your biological knowledge, explain the difference in the cell growth in
Petri dishes 3 and 4.   2 marks

_____

_____

_____

_____

_____

e   Which treatment is the best according to the aim of the scientist? Justify your response.   2 marks

_____

_____

_____

f   Explain why cancer cells can sometimes be overlooked by the immune system.   **2 marks**

_____

_____

_____

_____

Question 4 (4 marks) ●●●

Enzyme function is dependent on the temperature and pH of the surroundings. Catalase is an enzyme produced by the liver and which breaks down hydrogen peroxide (H_2O_2) to oxygen (O_2) and water (H_2O).

Design an experiment to determine the optimal pH or temperature of the environment for the enzyme catalase

_____

_____

_____

_____

_____

_____

_____

_____

Question 5 (5 marks) ●●●

Since antibiotics were first introduced, resistance in bacteria has been increasing, and once-effective antibiotics can now not be used in treatment. A group of scientists tested a range of antibiotics on bacteria to observe which had the greatest effect. The experiment was carried out by placing different bacterial solutions on agar plates with discs that had been infused with each of the four antibiotics. These plates were placed on the bench and observed after 48 hours. The results are in the table below.

Bacterium	Zone of inhibition (mm)			
	Penicillin	**Erythromycin**	**Cefazolin**	**Tetracycline**
Streptococcus pneumoniae	1.1	6	3.5	7.9
Mycobacterium tuberculosis	12	5	7	5
Anthrax bacterium	5	5.5	4.3	6.5
Bacillus thuringiensis	0	6	7.5	2

a   Identify **two** precautions that should be followed when using live specimens.   **1 mark**

_____

_____

_____

b Identify the dependent variable.   1 mark

_____

c Identify a weakness in the experimental design.   1 mark

_____

_____

_____

d Which antibiotic(s) would be least effective for _Mycobacterium tuberculosis_? Justify your
response.   1 mark

_____

e Which of the bacteria has low resistance to penicillin?   1 mark

_____

Question 6 (7 marks) ●●●

A group of students was investigating the action of the enzyme catalase at different temperatures. This
enzyme breaks down hydrogen peroxide into water and oxygen in the liver. The concentration of oxygen
was measured with an oxygen probe in the sealed container.

The students combined 20 mL of hydrogen peroxide measured in a beaker with 5 grams of blended
liver (for the catalase enzyme) into a conical flask and the oxygen probe was inserted into the flask.

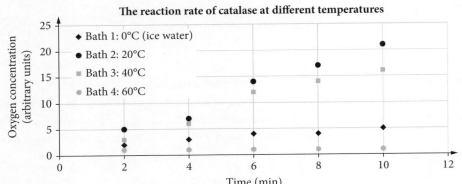

a Identify the trends in the data above.   3 marks

_____

_____

_____

_____

b How could you ensure the results are reliable?   1 mark

_____

c A student hypothesised that the data for bath 4 was low because the experiment had not been started. Is this hypothesis supported by the results?

2 marks

_____

_____

_____

d Identify a source of systemic error that could occur in the experiment. How can this be corrected?

1 mark

_____

_____

_____

Question 7 (4 marks)

A group of Biology students wished to investigate the effects of deficiencies of various elements on the growth of the aquatic plant *Spirogyra*.

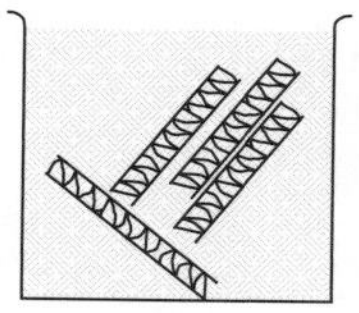

They set up four flasks similar to the one shown on the right. Each contained a nutrient solution and a quantity of the *Spirogyra* plant.

Each of the four flasks was deprived of one nutrient – iron, potassium, magnesium or nitrate. The nutrient solution contained all other plant nutrients. The plants were placed in a well-lit part of the laboratory and their growth was observed after 2 weeks.

a What hypothesis might the students have been testing?

1 mark

_____

b What results would support this hypothesis?

1 mark

_____

_____

While being impressed with the students' initiative, their Biology teacher criticised the experiment's design because it did not include a control.

c **i** What is meant by a 'control'?

1 mark

_____

_____

ii How would you improve the design of the experiment to include a control group?

1 mark

_____

Question 8 (14 marks) ©VCAA VCAA 2017 SB Q11 (adapted) ●●●

Matthew investigated how changes in
environmental temperature affected
oxygen (O_2) and carbon dioxide (CO_2)
levels in the air around a cockroach.
He used three digital probes linked to a
computer, a closed animal chamber and
a heat lamp in the experimental set-up
shown.

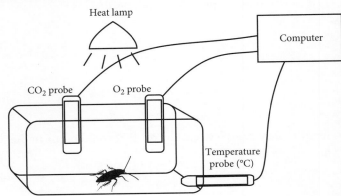

a Name the cellular process being investigated in Matthew's experiment.   1 mark

b Identify the dependent variables and the independent variable.   2 marks

Before placing the cockroach in the chamber, Matthew decided to measure the temperature, and carbon
dioxide and oxygen levels for 4 minutes. The following results were recorded.

Time (minutes)	CO_2 (%)	O_2 (%)	Temperature (°C)
0	0.04	22.3	29.5
1	0.04	22.1	29.8
2	0.04	22.0	30.0
3	0.04	22.0	30.0
4	0.04	22.0	30.0

c Explain why Matthew recorded the data for 4 minutes and not just 1 minute.   1 mark

After the initial 4-minute period, Matthew quickly placed the cockroach in the chamber and began
recording the data from the digital probes. After 10 minutes, he placed ice packs around the sides of the
animal chamber to slowly bring the temperature of the chamber down to 10°C. He recorded the data
using the digital probes for a further 20 minutes. He repeated the experiment once every day for the next
six days with the same cockroach. At all times, he took care to ensure that the cockroach showed no signs
of stress.

d Other than repeating the entire experiment, identify **two** control measures Matthew should have
included in his experimental design. Explain how each of these control measures could affect the
results if not kept constant.   4 marks

Matthew constructed the following graphs from the averaged results of the seven experiments.

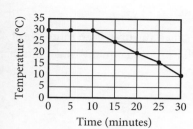 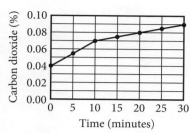 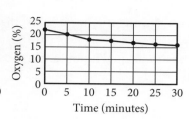

e **i** Using the graphical data, describe the changes in the levels of carbon dioxide and oxygen
when the temperature in the chamber was kept constant compared to when the temperature
was decreasing.   2 marks

_____

_____

_____

_____

ii What conclusion do you think Matthew can draw from his investigation?   4 marks

_____

_____

_____

_____

_____

_____

_____

_____

Question 9 (8 marks) ©VCAA VCAA 2016 SB Q2 ●●●

Plant materials containing cellulose and other polysaccharides are reacted with acids to break them
down to produce glucose. This glucose is then used by yeast cells for fermentation.

a Why is fermentation important for yeast cells?   1 mark

_____

b What are the products of fermentation in yeast cells?   1 mark

_____

A by-product of the acid treatment of plant materials is a group of chemical compounds called furans. It has been observed that as the concentration of furans increases, the rate of fermentation decreases. The enzyme alcohol dehydrogenase is required for the process of fermentation.

c Design an experiment to test the hypothesis that one of the furans, called furfural, is an inhibitor of the enzyme alcohol dehydrogenase. Assume that the experiment will be repeated many times and that environmental factors are kept constant. 4 marks

d Scientists have proposed that furfural is a competitive inhibitor of the enzyme alcohol dehydrogenase. Explain how furfural could act as a competitive inhibitor of the enzyme alcohol dehydrogenase. 2 marks

Question 10 (8 marks) ⬤◐○

A student wanted to test the hypothesis that plants dipped into a growth hormone will grow taller and faster than those without the hormone. Two seedlings of equal age were used from each of a rose, a hydrangea and a rosemary.

One seedling root was dipped into the growth hormone and the other was dipped in water. Both were planted in wet sand, and identical environmental conditions were maintained throughout the experiment. At the end of 5 weeks, the height difference of the plants was analysed. The results are shown below.

Plant	Height difference (cm)	
	Hormone treatment	Water treatment
Rose	6.4	5.5
Hydrangea	3.5	3
Rosemary	3.6	1.1

a What is the trend obtained from the data? 2 marks

b Why did the student use three different plants? 1 mark

c   i Which plant had the largest response to the hormone? Justify your response.   1 mark

_____

ii What is **one** way that the student could make the results more valid?   1 mark

_____

d   i Why were identical conditions maintained throughout the experiment?   1 mark

_____

_____

ii What are **two** variables that should be kept identical in the experiment?   1 mark

_____

_____

e What purpose did the cuttings dipped in water serve in this experiment?   1 mark

_____

_____

STUDENT NUMBER

Letter

BIOLOGY

Written examination

Reading time: 15 minutes

Writing time: 2 hours

QUESTION AND ANSWER BOOK
STRUCTURE OF BOOK

Section	Number of questions	Number of questions to be answered	Marks
A	40	40	40
B	11	11	80
		Total	120

- Students are permitted to bring into the examination room: pens, pencils, highlighters, erasers, sharpeners and rulers.
- Students are NOT permitted to bring into the examination room: blank sheets of paper and/or correction fluid/tape.
- No calculator is allowed in this examination.

Materials supplied
- Question and answer book of 38 pages
- Answer sheet for multiple-choice questions Instructions
- Write your student number in the space provided above on this page.
- Check that your name and student number as printed on your answer sheet for multiple-choice questions are correct, and sign your name in the space provided to verify this.
- Unless otherwise indicated, the diagrams in this book are not drawn to scale.
- All written responses must be in English.

At the end of the examination
- Place the answer sheet for multiple-choice questions inside the front cover of this book.

Students are NOT permitted to bring mobile phones and/or any other unauthorised electronic devices into the examination room.

VCE BIOLOGY
Practice Written Examination Units 3 & 4
Multiple-Choice Answer Sheet

STUDENT NAME:	*Your name will be printed here*
INSTRUCTIONS:	USE PENCIL ONLY

SIGN BELOW IF YOUR NAME AND NUMBER ARE PRINTED CORRECTLY

SIGNATURE

If your name or number on this sheet is incorrect, notify the Supervisor.
Use a PENCIL for ALL entries. For each question shade the box which indicates your answer. All answers must be completed like THIS example: A B̶ C D
Marks will NOT be deducted for incorrect answers.
NO MARK will be given if more than ONE answer is completed for any question.
If you make a mistake, ERASE the incorrect answer — DO NOT cross it out.

STUDENT NUMBER

9	9	1	2	3	4	5	6	A
0	0	0	0	0	0	0	0	A
1	1	1	1	1	1	1	1	E
2								F
3		Your student number						G
4		will be recorded here						J
5		for your to check						L
6	6	6	6	6	6	6	6	R
7	7	7	7	7	7	7	7	T
8	8	8	8	8	8	8	8	W
9	9	9	9	9	9	9	9	X

SUPERVISOR USE ONLY

USE PENCIL ONLY

Shade the "ABSENT" box if the student was absent from the examination.

ABSENT

SUPERVISOR'S INITIALS

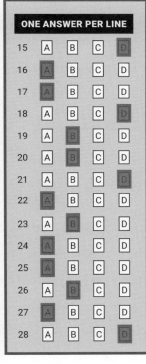

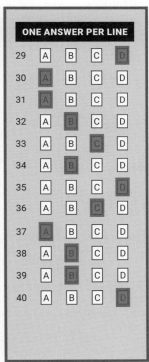

Section A: Multiple-choice questions

Instructions for Section A

- Circle correct option for multiple-choice questions.
- Choose the response that is correct or that best answers the question.
- A correct answer scores 1; an incorrect answer scores 0.
- Marks will not be deducted for incorrect answers.
- No marks will be given if more than one answer is completed for any question.
- Unless otherwise indicated, the diagrams in this book are not drawn to scale.

Question 1

The proteome in prokaryotes compared to the genome is

A larger because of post-transcriptional modification.

B equal to the genome of the chromosome.

C smaller than the genome.

D equal to the genome of any genetic material in the prokaryotes.

Question 2

A cell undergoes a large amount of protein synthesis. Which of the following functions aligns with the correct organelle involved in this process?

	Organelle	Function
A	Ribosome	Involved in the process of transcription, converting mRNA into an amino acid sequence
B	Chloroplast	Production of organic substances from inorganic compounds using solar energy
C	Mitochondrion	mRNA production from DNA during transcription via RNA polymerase
D	Golgi apparatus	Modification and folding of synthesised polypeptides

Question 3

Myoglobin is a biological macromolecule that assists in the transport of oxygen in muscle. It is composed of 153 amino acids from 10 000 bases in the DNA.

This difference in numbers can be accounted for through

A the amino acids being coded for by multiple codons.

B RNA processing with exon retention and intron removal.

C multiple bases coding for a single amino acid.

D transcriptional factors and the binding of RNA polymerase.

Question 4

Catalase is an enzyme composed of four polypeptide chains, each with more than 500 amino acids. The functional protein structure is classified as

A primary.

B secondary.

C quaternary.

D tertiary.

Question 5

The *trp* operon found in bacteria synthesises the amino acid tryptophan required in protein synthesis. The system can be repressed through

A absence of tryptophan in the bacterial cell.

B a repressor protein binding to the promoter region of the operon.

C the presence of tryptophan in the bacterial surroundings.

D synthesis of the *trp* E, D, C, B and A genes.

Question 6

The following steps are each a part of the process of bacterial transformation:

I   Bacteria are heat shocked to increase plasmid uptake.

II   Bacteria are placed on an agar plate with an antibiotic to determine transformation.

III DNA of interest and plasmid are incubated with DNA ligase.

IV DNA of interest and plasmid are incubated with an endonuclease.

V   Recombinant plasmid is formed.

Which is the correct series of events?

A IV → V → III → II → I   **B** IV → III → V → I → II   **C** III → IV → V → I → II   **D** I → II → IV → V → III

Question 7

In the CRISPR-Cas9 system, the non-repetitive sequences found in the bacterial genome are

A different Cas9 endonuclease enzyme genes.

B homologous with foreign DNA sequences from plasmids in bacteriophages.

C operator regions for RNA polymerase binding.

D viral RNA, which is produced to inhibit other virus's entry into the cell.

Question 8

A crime lab isolated the DNA from blood samples from a murder scene to assist in the identification of suspects. The samples were amplified and exposed to endonuclease before being added to a buffer and ran through the gel. The lab then ran the samples against a DNA ladder in lane 1. The results are shown below.

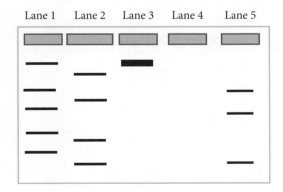

From the results above, it can be concluded that

A the DNA in lane 4 hasn't been dyed.

B lanes 2 and 3 have the longest fragments of DNA.

C lane 3 and 4 have DNA fragments which have not been exposed to endonuclease.

D lane 2 contains more genetic material than lane 5.

Question 9

Which of the following includes both a genetically modified organism and a transgenic organism?

	GMO	Transgenic organism
A	Fluorescent gene from jellyfish into fish	Increased gene expression of growth gene in tomatoes
B	Corn modified by a bacterial insecticide to increase resistance	Tomato gene suppression of pectin breakdown
C	Knock out of genes that alter the fatty acids in soybeans	Rice modified with daffodil genes to increase beta carotene
D	Knock out of a gene to alter the colour of papaya	Addition of transcription factors to increase cell division in corn

Question 10

Which of the following are the correct inputs and outputs for photosynthesis and cellular respiration?

	Photosynthesis		Cellular respiration	
	Input	Output	Input	Output
A	O_2	CO_2	H_2O	O_2
B	CO_2	O_2	O_2	H_2O
C	$C_6H_{12}O_6$	CO_2	H_2O	CO_2
D	H_2O	ATP	$C_6H_{12}O_6$	NADPH

Question 11

Which of the following affects the rate of ATP production in aerobically respiring plant cells?

A Oxygen concentration

B Carbon dioxide concentration

C Water availability

D Light availability

Question 12

Rubisco is a key enzyme involved in the production of glucose in all plants. A particular plant is exposed to extreme temperatures for a long time. Which of the following graphs represents the enzyme's function with an increasing temperature?

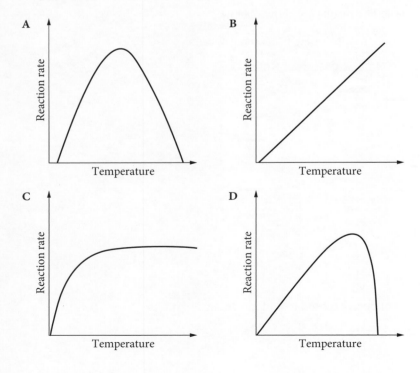

The following information relates to Questions 13 and 14.

Four groups of students investigated the effect of temperature on the rate of anaerobic respiration over 3 hours. Each group added 10 grams of yeast into a sealed beaker with glucose syrup and measured the change of pH of the solution. The results are outlined in the table.

Group	Temp. (°C)	pH			
		Trial 1	Trial 2	Trial 3	Average
1	10	6	6.2	6.4	6.20
2	20	5.3	5.5	3	4.60
3	30	4.4	4.5	4.4	4.43
4	40	6.3	6.7	6.6	6.53

Question 13

Which group experienced an error in their experiment, and which experienced the highest precision?

	Random error	Highest precision
A	Group 3	Group 2
B	Group 1	Group 4
C	Group 4	Group 2
D	Group 2	Group 3

Question 14

The students in Group 3 continued their experiment to determine how acidic the solution could become. All of the reactions they observed plateaued at pH 4.3.

A possible reason for this is

A the yeast had obtained enough ATP to function.

B the yeast underwent a conformational shape change and was unable to undergo cellular respiration.

C the pH was out of the tolerance range for the enzymes involved in respiration.

D a systematic error of miscalibration of the pH probe had occurred.

Question 15

Which of the statements correctly describes what happens when a C3 plant is exposed to high temperatures and low levels of CO_2?

A Photosynthesis decreases and photorespiration increases.

B Cellular respiration increases and photosynthesis decreases.

C Stomata close to increase the rate of gas exchange.

D It converts the carbon dioxide into a compound that can be used at night.

Question 16

The Krebs cycle is a cyclic set of nine reactions catalysed by several enzymes. One of the enzymes is citrate synthetase, which combines oxaloacetate and acetyl CoA into citrate.

If an individual has a mutation in the citrate synthetase gene altering the third amino acid to a STOP codon, the result of the mutation would be

A conversion of acetyl CoA back into pyruvate to increase glycolysis.

B an increase in the electron transport chain to increase the ATP production for the cell.

C a depletion of citrate in the mitochondrial matrix.

D an increase in the NADH and $FADH_2$ formed in glycolysis.

Question 17

The protein secretory pathway is made up of the endoplasmic reticulum, Golgi bodies and plasma membrane. Which alternative below correctly shows what happens to the proteins at each of the sites?

	Endoplasmic reticulum	Golgi bodies	Plasma membrane
A	Released by exocytosis	Produced	Modified by enzymes and packaged into secretory vessels
B	Produced	Modified by enzymes and packaged into secretory vessels	Released by exocytosis
C	Modified by enzymes and packaged into secretory vessels	Released by exocytosis	Produced
D	Produced	Released by exocytosis	Modified by enzymes and packaged into secretory vessels

The following information relates to Questions 18–20.

Bacteria often contain plasmids – small segments of DNA – as well as their circular chromosome. Plasmids pass from one bacterium to another through horizontal gene transfer. They often carry resistance genes for antibiotics through the production of proteins that block the action of the antibiotics.

Some scientists wanted to investigate whether resistance to the antibiotics Ampicillin and Bactrim could be passed between bacteria in this way. They cultured bacteria known to be sensitive (not resistant) to both antibiotics with plasmids extracted from bacteria resistant to both antibiotics. The resistance genes for the antibiotics are located on separate plasmids. It is possible for the bacteria to possess no plasmids, one of the plasmids or both plasmids.

Bacteria from the original cultures and the new culture were then transferred to nutrient agar plates and allowed to grow for several days. The table below shows the agar plates as they appeared after 4 days.

	Sensitive bacteria	Resistant bacteria	Sensitive bacteria + plasmids
Nutrient agar only	1	2	3
Agar + Ampicillin	4	5	6
Agar + Bactrim	7	8	9

Question 18

Which of the plates shows evidence that DNA can be taken up by bacterial cells?

A Plate 3 only

B Plate 6 only

C Plates 3 and 6

D Plates 6 and 9

Question 19

Bacteria that are sensitive to Bactrim and resistant to Ampicillin would be found on

A   any of plates 1, 2 or 3.

B   plate 6 only.

C   plates 3 or 6.

D   none of the plates.

Question 20

Which of the following conclusions can you draw about the bacteria growing on plate 3?

A   The bacteria have no resistance to the antibiotics Bactrim and Ampicillin.

B   Some will be resistant to both Bactrim and Ampicillin.

C   Some will be resistant to Bactrim but not to Ampicillin.

D   Some will be resistant to Ampicillin but not to Bactrim

Question 21

A patient presents at hospital with a high fever, swelling of the lymph nodes and inflammation of a recent injury sustained in a motorcycle accident. The cuts are seeping pus, swollen and warm to touch.

The patient's fever would have been caused by

A   interferon.

B   interleukin.

C   complement proteins.

D   neutrophils.

Question 22

African swine fever is a contagious viral disease in pigs that can spread rapidly, creating an epidemic.

The appropriate detection technique to identify an infection with this specific type of virus is

A   polymerase chain reaction.

B   growth on an agar plate.

C   microscopic analysis.

D   morphological identification.

Question 23

A patient presented at a doctor's clinic with virus-like symptoms. A swab was taken from the individual's nostril and sent off to the lab. The results from the diagnostic tests are shown below.

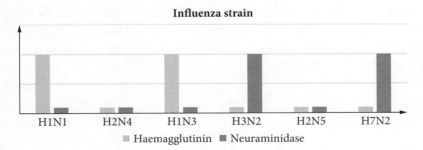

Which strain(s) of influenza is the patient suffering from?

A   H1N3 and H1N1

B   H2N5

C   H1N2

D   H2N1 and H1N5

The following information relates to Questions 24 and 25.

Joel was scratched and bitten by a stray dog while backpacking through India. To ensure he did not contract the rabies virus, he received two injections immediately after exposure. The first injection was the rabies vaccine and was provided four times over a month, and the second was the rabies immunoglobulin, which was injected at the site of injury immediately after Joel presented at the hospital.

Question 24

The immunoglobulin injection provides

A natural and passive immunity.

B natural and active immunity.

C artificial and passive immunity.

D artificial and active immunity.

Question 25

Joel continued travelling and encountered the *Macaca mulatta* monkey a couple of months later. The species has become accustomed to living alongside humans in urban areas of India. While feeding these monkeys, Joel was bitten again.

Which of the following describes the correct response to the second exposure to rabies?

A Joel received both types of injection again to prevent rabies symptoms developing.

B The humoral response was initiated due to memory being formed upon the first exposure.

C Joel's immune system has no memory of the virus and required the immunoglobulins immediately.

D Joel ingested antiseptics to clear the infection.

Question 26

Monoclonal antibodies are produced to treat many diseases and conditions. Multiple sclerosis (MS) is an autoimmune disease that causes the demyelination of nerves in the central nervous system. One treatment to slow the progression of this disease is the use of monoclonal antibodies. Which of the following would be effective for the treatment of MS?

A Addition of a radioactive drug onto the heavy chain of the antibodies injected to initiate apoptosis of the myelin cells

B Attachment via the heavy chain to the MHC II receptors on the naïve B cells in the lymph nodes

C Bind to the blood brain barrier and increase chemotaxis to the area to increase leukocyte activation and remove the damaged myelin cells

D Bind to cytotoxic T cell receptors, inhibiting recognition of myelin proteins on MHC I markers

The following information relates to Questions 27 and 28.

Mycobacterium tuberculosis is an airborne pathogen responsible for about 2 million deaths annually. The bacterium enters alveolar macrophages to replicate before escaping into the lung tissue and infecting other cells. Chronic inflammation caused by *Mycobacterium tuberculosis* can form granulomas, which consist of necrotic cells and leukocyte clusters in the lungs.

Question 27

The granuloma would form because of

A B and T cell activation and proliferation in the lymph nodes.

B chemotaxis of innate immune cells to the site of infection from release of cytokine.

C lack of an innate immune response in the infected area, resulting in apoptosis of damaged cells.

D pus formation from filtration of natural killer cells and apoptosis.

Question 28

The adaptive immune response to *Mycobacterium* can be activated by

A neutrophils breaking down the bacteria and migrating to the lymph node to present on the MHC I marker to naïve B and T cells.

B antigen-presenting cells becoming infected with the bacteria, breaking down the pathogen and presenting on its MHC I marker to T helper cells.

C dendritic cells presenting the antigen of the bacteria on the MHC II markers to T helper cells and MHC I markers to naïve T cells.

D macrophages becoming infected with the bacteria, migrating to the lymph node and lysing to expose the antigen to naïve B cells for clonal expansion to occur.

Question 29

The lymphatic system is responsible for the filtration of extracellular fluid to remove foreign material and toxins. The system is composed of a series of vessels and nodes. In some cases, lymph nodes are removed because of cancer or infection leading to lymphoedema (swelling to part of the body).

The lymph nodes

A act as a pump to circulate extracellular fluid throughout the body.

B are surrounded by valves to ensure bidirectional movement of extracellular fluid around the body.

C are composed of lymphoid tissue with clusters of lymphocytes residing within the lymphatic system.

D facilitate the reabsorption of extracellular fluid back into the circulatory system after filtration.

Question 30

The two species of *Howea* palm endemic to Lowd Howe Island established after the island formed 6.9 mya. These two palms originated from the same species, but populations experienced different flowering time and soil preference linked with height above sea level. This type of speciation can be described as

A convergent evolution.

B sympatric speciation.

C adaptive radiation.

D allopatric speciation.

The following diagram relates to Questions 31–33.

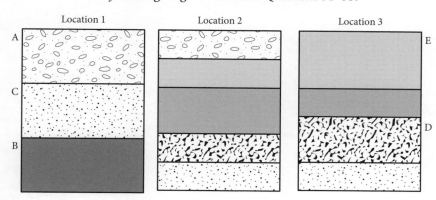

Question 31

The fossil in stratum D is widespread, abundant, and only present for a short period of time. This can be classified as a

A   transitional fossil.

B   index fossil.

C   indirect fossil.

D   trace fossil.

Question 32

A group of palaeontologists discovered a new specimen, located in rock stratum E. The palaeontologists determined that the fossil was less than 60 000 years old and performed carbon dating; carbon-14 has a half-life of 5700 years. The fossil had 12.5% of carbon-14 remaining.

The age of the fossil would be closest to

A   12 500 years.

B   7500 years.

C   17 000 years.

D   45 000 years.

Question 33

Locations 1, 2 and 3 have been isolated from different regions around the world. Based on the information in the diagram above, it can be concluded that

A   rock stratum A is the oldest of the strata.

B   rock stratum C is the oldest rock stratum present in the locations.

C   rock stratum D in location 3 is the same age as stratum B in location 1.

D   rock stratum F is older than strata A and E but younger than stratum B.

Question 34

The order pinniped includes fur seals, walruses and seals. The oldest fossil from this order dates to 30.6 million years ago. Below is the phylogenetic tree formed from molecular homology in a conserved gene within the species.

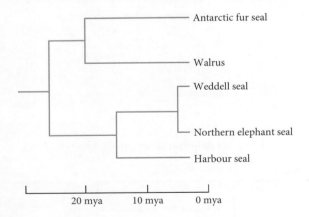

The two species with the most amino acid differences are the

A   Antarctic fur seal and walrus.

B   harbour seal and Antarctic fur seal.

C   Weddell and northern elephant seal.

D   harbour seal and northern elephant seal.

Question 35

Australian sea lions are isolated into populations that breed on rocks and sandy beaches on the coast of Australia. Each colony has a unique 17-month breeding cycle with pupping season occurring in alternate summers and winters or alternate spring and autumns. There is no evidence of genetic drift or seasonal movements. One population is exposed to a selection pressure in which the smaller seals are better suited.

Which of the following is likely to occur over time?

A Speciation due to the different selection pressures favouring different alleles

B Individuals becoming extinct due to the limited genetic variation in all populations

C Sympatric speciation from the populations in the same location

D The populations remain the same species due to genetic drift

Question 36

The differences between amino acid sequences in cytochrome c protein in several species are shown in the table.

	Human	Monkey	Whale	Chicken	Yeast	Snake
Human	0	3	23	27	37	30
Monkey		0	22	24	39	37
Whale			0	4	29	32
Chicken				0	27	22
Yeast					0	5
Snake						0

Which of the following can be concluded from the information?

A The least related species to the monkey is the snake.

B The number of differences in the DNA will be the same as the differences in the amino acid sequences.

C The yeast and chicken are more closely related than the monkey and the snake.

D The human is most closely related to the whale.

Question 37 ©VCAA VCAA 2009 (2) SB Q6 (adapted) ●●●

The images below show two *Homo* species skulls that have been discovered in separate locations and different rock strata.

Skull set 1

Skull set 2

Which of the following statements can be concluded from the above image?

A Skull set 1 is from an early *Australopithecus* due to the location of the foremen magnum.

B Skull set 2 represents a *Homo erectus* skull; skull set 1 represents a *Homo sapiens* skull.

C Skull sets 1 and 2 are the same species but different sexes.

D Skull set 2 represents an early *Homo sapiens* skull due to the prominent eyebrow ridges, sloped face and large muscle attachments on the side of the head.

Question 38

Which of the following statements describes how primates differ from other mammals?

A Primates have a hair covering over their body to increase thermoregulation.

B Primates have five digits with flat nail beds, whereas other mammals have hooves or claws on their digits.

C Primates are warm blooded compared to mammals, which can be endotherms or ectotherms.

D Primates give birth to live young and secrete milk from mammary glands.

Question 39

During the last 500 000 years, three species of the *Homo* genus have inhabited the land. *Homo sapiens* are the only species not extinct. *Homo neanderthalenis* lived from 400 000 years ago until around 40 000 years ago and *Homo denisova* evolved from their common ancestor *Homo heidelbergensis* at about a similar time, although the exact dates are difficult to determine with little available fossil evidence of this species. During the time in which these three species coevolved, potential interbreeding occurred. This is demonstrated in the diagram below.

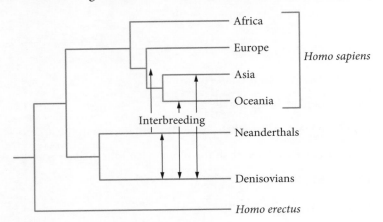

From the information above it can be concluded that

A African *H. sapiens* have the greatest diversity of genetic material.

B Neanderthals only interbred with the Denisovans and Asian *H. sapiens*.

C *H. erectus* is the common ancestor to the *H. sapiens*, Neanderthals and Denisovans.

D *H. oceania* contains more DNA from Neanderthals and Denisovans than European *H. sapiens*.

Question 40

Which of the following statements correctly describes a major difference between the genus *Australopithecus* and the *Homo* genus?

A The *Homo* genus were all bipedal, whereas *Australopithecus* were quadrupedal.

B The *Homo* genus had a larger cranial capacity and reduced prognathism compared to the *Australopithecus* genus.

C The *Homo* genus all travelled outside of Africa, whereas the *Australopithecus* genus remained in the one region.

D The *Homo* genus showed no sexual dimorphism and reduced arm length compared with the *Australopithecus* genus.

Section B: Short-answer questions

Instructions for Section B

- Answer all questions in the spaces provided. Write using blue or black pen.
- Unless otherwise indicated, the diagrams in this book are not drawn to scale.

Question 1 (10 marks)

Proteins are involved in every process in an organism's functioning.

a For each of the levels of protein structure, explain where and what type of bonds are formed to produce the protein's unique shape.

4 marks

Structure of protein (levels)	Bonds and interactions
1 Primary	
2 Secondary	
3 Tertiary	
4 Quaternary	

b Many proteins do not exhibit a quaternary structure. Explain why this is, and explain the difference between a tertiary and a quaternary structure.

2 marks

c Explain the process by which the primary structure of a protein is synthesised from the mature mRNA molecule.

4 marks

Question 2 (4 marks)

An unknown pathogen was presented at a diagnostics lab. The technician performed a number of tests with the pathogen to identify the correct treatment. The lab ran the following tests:

- growth on an agar plate; no growth determined it was non-cellular
- gene analysis, which identified a protein coat and PCR with primers to determine the exact strain of pathogen.

a Explain the steps and temperatures involved in PCR.

3 marks

b Considering the tests performed, what is the most likely type of pathogen and what would be the treatment?

1 mark

Question 3 (10 marks)

Two groups of students were investigating the rate of photosynthesis in different plants over a 24-hour period.

Group A submerged 50 grams of *Elodea* in test tubes that were stoppered and had tubes that attached to an oxygen probe. The oxygen released was measured in ppm. The students placed each test tube under light of different colours: white, blue, green, red and no light. The temperature of the water, the light intensity and the time were all controlled.

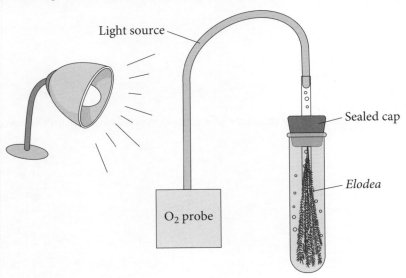

Light source

Sealed cap

Elodea

O_2 probe

a Why does temperature need to be controlled in this experiment?

2 marks

_____

_____

_____

_____

_____

_____

_____

b A student hypothesised that the green wavelengths would enable the greatest rate of photosynthesis. Is their hypothesis accurate?

2 marks

_____

_____

_____

_____

_____

c What improvement could be made to the experiment to ensure it is more reliable?

1 mark

_____

Group B identified a plant as the Australian *Calandrinia*, which possesses both C3 and CAM properties depending on the environment. In this experiment, the students placed the plants in chambers and measured the CO_2 concentration of the chamber over 24 hours.

d Explain the difference between C3 and CAM plants.

2 marks

_____

_____

_____

_____

_____

The students placed the plants under different amounts of water stress to analyse the effect on photosynthesis rates.

e What are the expected results of a plant under high stress with minimal water versus a plant under low stress with unlimited water? 3 marks

_____

_____

_____

_____

_____

_____

_____

Question 4 (8 marks)

In 2004, a rupture along the fault lines between the Burmese and Indian tectonic plates caused an earthquake with a magnitude of 9.1 in the Indian Ocean, affecting more than 14 countries. The tsunami that followed had an epicentre off the west coast of northern Sumatra, Indonesia, and caused waves of up to 30 m. These events killed about 200 000 people across the 14 affected countries.

The devastating event had continued long-lasting health effects on survivors. Many bacterial and viral infections developed due to contaminated water supply and bodies of water remaining stagnant for long periods of time. Two major diseases following this event were cholera (bacterial infection) from the contaminated water supply and dengue fever (viral infection) from mosquitoes.

a What are **two** first lines of defence that must be breached for these microbes to enter the body? 1 mark

_____

b What innate immune cell targets dengue fever in an infected cell, and how does it do this? 2 marks

_____

_____

_____

c Explain which steps in the adaptive immune response would be most effective against the cholera bacteria. 3 marks

_____

_____

_____

_____

_____

_____

d Explain how complement proteins facilitate the removal of either the virus or bacteria.   1 mark

e What role does the mosquito play in the contraction of dengue fever?   1 mark

Question 5 (6 marks)

People with type 1 diabetes are unable to synthesise enough insulin to maintain a stable blood glucose level. In the past, individuals have received insulin isolated from pig pancreases; today, with the use of gene technology, the insulin can be synthesised in the lab from bacteria.

Below is a diagram of the insulin production process.

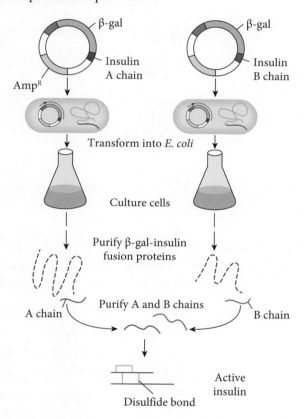

a Explain why two separate plasmids are grown in separate bacteria in separate cultures are used to produce insulin.   1 mark

b Before the insulin gene can be inserted into the plasmid all the introns need to be removed. Explain why.   2 marks

_____

_____

_____

_____

c Explain the significance of the β-gal gene.   1 marks

_____

_____

d An individual placed the transformed bacteria on two agar plates. One agar plate contained the antibiotic amp and the other the antibiotic tcl. Some bacteria grew on each of these plates. Which are the desired bacteria, and why?   2 marks

_____

_____

_____

_____

_____

_____

Question 6 (6 marks)

Australia is home to 14 of the 21 species of cockatoo that have been discovered around the world. These cockatoos have a range of habitats, colouring and food sources. Below is the phylogenetic tree constructed using molecular homology. Note that not all species have been included on this diagram to prevent confusion.

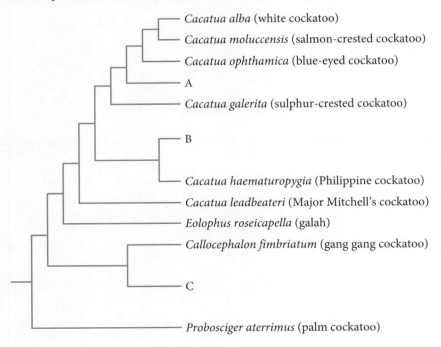

a A new type of cockatoo was discovered in the genus *Cacatua*. It is more closely related to the Major Mitchell's cockatoo than it is to the salmon-crested cockatoo. Where on the tree would you place this new species, out of clade A, B or C?

1 mark

b Explain the benefit of using mtDNA in genetic analysis over nuclear DNA in forming phylogenetic trees.   1 mark

c The transformation of the Australian landscape and climate to more arid conditions of 10–20 million years ago resulted in a change in food source for the cockatoos and isolation of different populations. Explain the effect this could have on the different species.

2 marks

c A new fossil has been located in the outback of Australia, which is in a younger rock stratum than an index fossil dated at 60 000 years. What absolute dating technique could be used to determine the age of the fossil?   2 marks

Question 7 (4 marks)

A jawbone fossil from 160 000 years ago discovered by a Tibetan monk in 1980 in a cave at the base of the Tibetan Plateau has once again altered the understanding of human evolution and the enigmatic population *Homo denisovans*. DNA from the fossil was extracted and analysed. The gene *EPAS1* found in Sherpa people is likely to have been inherited from these Denisovans and assists in their breathing at high altitudes. One theory is the Denisovans are an extinct sister group of the Neanderthals with an ancestor in common with *Homo sapiens*, called *Homo heidelbergensis*. It is thought that as they migrated out of Africa 400 000 years ago – the divergence into the *Homo sapiens*, Denisovans and Neanderthals occurred. The Neanderthals became extinct 40 000 years ago with the reasoning still disputed. The genome of the Denisovans appears to have low genetic diversity and became extinct 50 000 years ago.

a Explain why the *EPAS1* gene is not found in all *Homo sapiens*.   1 mark

b Using the information above, draw a phylogenetic tree from *Homo heidelbergensis* to present day. Include *Homo sapiens*.   2 marks

c The Denisovans interbred with populations of *Homo sapiens*. Explain why some palaeontologists may dispute the classification of these species.   1 mark

Question 8 (6 marks)

Tree nuts include almonds, Brazil nuts, cashews, chestnuts, hazelnuts, macadamias, pine nuts, pistachios and walnuts. A person who experiences inflammation after consuming some kinds of tree nuts accidently consumed a walnut in a cake.

a Compare the person's response to the allergen on the first and subsequent exposures.   2 marks

b What are **two** signs of inflammation?   1 mark

c Name and explain the action of **two** chemicals involved in inflammation.   2 marks

d What is the role of the lymph nodes in an allergic response?

1 mark

_____

_____

_____

Question 9 (8 marks)

Australia started to become isolated from other continents between 55 and 10 mya. Since the isolation of the continent, many unique animals have evolved, such as the echidna and its closest living relative, the platypus. A metazoic fossil of a humerus was discovered in sediment rock in Victoria. The fossil resembles an echidna but is incomplete and therefore its relatedness is inconclusive.

a Explain **two** requirements for fossilisation to occur.

2 marks

_____

_____

_____

b Originally, evidence suggested that the platypus and echidna diverged from 112.5 mya, but more recently evidence suggest that this occurred 19–45 mya from a common ancestor. Explain how this assumption could have been altered.

2 marks

_____

_____

_____

The Central and South American anteater and the Australian echidna both independently evolved slender, elongated and highly extensible tongues to increase the number of ants per catch, and for easy entry to the ant nests. Both species have small valvelike nostrils that prevent the prey from entering their nose.

c Name the type of evolution that has occurred.

1 mark

_____

d Explain how the separate populations evolved these traits.

3 marks

_____

_____

_____

_____

_____

Question 10 (6 marks)

A bacterium is a single-celled organism with limited access to the monomers required for protein synthesis and energy production. To increase survival, the bacterium contains a number of mechanisms.

a   Explain **one** advantage of an operon in bacteria.   1 mark

_____

_____

_____

b   In the space below, draw a labelled diagram of the *trp* operon with a bound repressor.   2 marks

c   There are two mechanisms that regulate the expression of the structural genes of the *trp* operon when the level of tryptophan in the cell is high. Name and compare these two mechanisms.   2 marks

_____

_____

_____

_____

_____

Question 11 (6 marks)

A new renewable energy production process includes the fermentation of biomass such as starch and other plant matter by bacteria to produce a high-energy molecule.

a   What is the product resulting from this fermentation product?   1 mark

_____

b   How many ATP are produced by the breakdown of biomass?   1 mark

_____

c Explain why the fermentation process requires a sealed chamber with controlled conditions.

2 marks

_____

_____

_____

_____

_____

In some countries, the burning of biomass fuel in poorly ventilated areas increases the risk of health conditions such as acute lower respiratory infections. The smoke emitted by some of these fuels increases the risk of pneumonia, which is caused by the bacteria _Streptococcus pneumoniae_.

d In the space below, draw a graph to show how the number of _Streptococcus pneumoniae_ cells increase in the extra cellular fluid during an infection of an individual.

2 marks

Question 12 (6 marks)

Australia is home to 21 of the 25 most venomous snakes in the world. Due to the varying potency of each species' venom, it is important to correctly identify the type of snake that has bitten an individual to provide the effective treatment. A diagram of the testing kit is shown below.

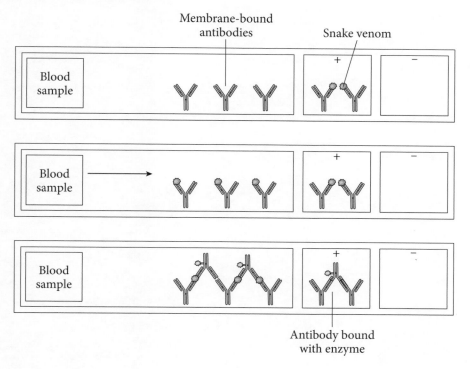

a   Monoclonal antibodies are used in tests such as the anti-venom test above. Explain how the cells
    required for these antibodies are formed.   2 marks

_____

_____

_____

_____

b   Once the cells are obtained, they are fused with a myeloma cell. Name the cell produced and explain
    why this process occurs.   1 mark

_____

_____

c   Once the type of snake is identified, the patient can receive the anti-venom. Explain what type of
    immunity this provides.   2 marks

_____

_____

_____

d   Why are a positive and a negative control both used on these tests?   1 mark

_____

Letter

STUDENT NUMBER

BIOLOGY
Written examination 2

Reading time: 15 minutes

Writing time: 2 hours

QUESTION AND ANSWER BOOK
STRUCTURE OF BOOK

Section	Number of questions	Number of questions to be answered	Marks
A	40	40	40
B	11	11	80
		Total	120

- Students are permitted to bring into the examination room: pens, pencils, highlighters, erasers, sharpeners and rulers.
- Students are NOT permitted to bring into the examination room: blank sheets of paper and/or correction fluid/tape.
- No calculator is allowed in this examination.

Materials supplied
- Question and answer book of 38 pages
- Answer sheet for multiple-choice questions Instructions
- Write your student number in the space provided above on this page.
- Check that your name and student number as printed on your answer sheet for multiple-choice questions are correct, and sign your name in the space provided to verify this.
- Unless otherwise indicated, the diagrams in this book are not drawn to scale.
- All written responses must be in English.

At the end of the examination
- Place the answer sheet for multiple-choice questions inside the front cover of this book.

Students are NOT permitted to bring mobile phones and/or any other unauthorised electronic devices into the examination room.

VCE BIOLOGY
Practice Written Examination Units 3 & 4
Multiple-Choice Answer Sheet

STUDENT NAME:	*Your name will be printed here*
INSTRUCTIONS:	USE PENCIL ONLY

SIGN BELOW IF YOUR NAME AND NUMBER ARE PRINTED CORRECTLY

SIGNATURE

If your name or number on this sheet is incorrect, notify the Supervisor.
Use a PENCIL for ALL entries. For each question shade the box which indicates your answer. All answers must be completed like THIS example: A ⬛ C D
Marks will NOT be deducted for incorrect answers.
NO MARK will be given if more than ONE answer is completed for any question.
If you make a mistake, ERASE the incorrect answer — DO NOT cross it out.

STUDENT NUMBER

9	9	1	2	3	4	5	6	A
0	0	0	0	0	0	0	0	A
1	1	1	1	1	1	1	1	E
2								F
3		Your student number						G
4		will be recorded here						J
5		for you to check						L
6	6	6	6	6	6	6	6	R
7	7	7	7	7	7	7	7	T
8	8	8	8	8	8	8	8	W
9	9	9	9	9	9	9	9	X

SUPERVISOR USE ONLY

USE PENCIL ONLY

Shade the "ABSENT" box if the student was absent from the examination.

ABSENT

SUPERVISOR'S INITIALS

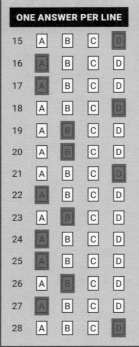

ONE ANSWER PER LINE
1   A B C **D**
2   A B **C** D
3   A B **C** D
4   A B C **D**
5   A **B** C D
6   A **B** C D
7   A B **C** D
8   A **B** C D
9   A B **C** D
10   A **B** C D
11   **A** B C D
12   **A** B C D
13   A **B** C D
14   A B **C** D

ONE ANSWER PER LINE
15   A B C **D**
16   **A** B C D
17   **A** B C D
18   A B C **D**
19   A **B** C D
20   A **B** C D
21   A B C **D**
22   **A** B C D
23   A **B** C D
24   **A** B C D
25   **A** B C D
26   A **B** C D
27   **A** B C D
28   A B C **D**

ONE ANSWER PER LINE
29   A B C **D**
30   **A** B C D
31   **A** B C D
32   A **B** C D
33   A B **C** D
34   A **B** C D
35   A B C **D**
36   A B **C** D
37   **A** B C D
38   A **B** C D
39   A **B** C D
40   A B C **D**

Section A: Multiple-choice questions

Instructions for Section A

- Answer all questions in pencil on the answer sheet provided for multiple-choice questions.
- Choose the response that is correct or that best answers the question.
- A correct answer scores 1; an incorrect answer scores 0.
- Marks will not be deducted for incorrect answers.
- No marks will be given if more than one answer is completed for any question.
- Unless otherwise indicated, the diagrams in this book are not drawn to scale.

The following information relates to Questions 1 and 2.

Collagen and myoglobin are two proteins found in the human body. A diagram of each protein is shown below.

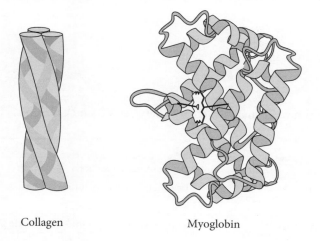

Collagen   Myoglobin

Question 1

Based on the information provided in the diagram, it is reasonable to conclude that

A collagen is an example of a protein with a tertiary structure.

B myoglobin is a fibrous protein.

C collagen is an example of a protein with a secondary structure.

D myoglobin is an example of a protein with a quaternary structure.

Question 2

Collagen and myoglobin are made up of

A glucose.

B carbohydrates.

C amino acids.

D glycerol.

Question 3

An organism's proteome is all of the

A proteins that can be expressed by the genome of an individual.

B genes found in an individual.

C amino acids that make up the proteins found within an individual.

D nucleotides that make up the genes found within an individual.

Question 4

What is the name of the enzyme used to synthesise new strands of DNA in the nucleus of a human?

A DNA polymerase

B RNA polymerase

C Taq polymerase

D DNA ligase

Question 5

Which of the following statements about enzymes is incorrect?

A Enzymes are proteins that act as catalysts

B Enzymes are specific

C Enzymes provide activation energy for reactions

D Enzymes can be used many times in the same reaction

Question 6

Identify the correct row that outlines the main differences between C3 and C4 plants.

	C3	C4
A	Only use 'light' reactions	Only use 'dark' reactions
B	Use light energy for reactions	Use chemical energy for reactions
C	Synthesise only carbohydrates	Synthesise only proteins
D	Synthesise glucose in the Calvin cycle	Synthesise glucose in the Calvin cycle plus an additional C4 cycle

Question 7

The picture below is a micrograph of a cell organelle, labelled X.

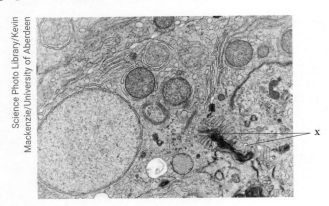

Science Photo Library/Kevin Mackenzie/University of Aberdeen

Organelle X is responsible for

A modifying proteins and packaging them for export from the cell.

B synthesising proteins.

C controlling molecules that are able to enter and exit a cell.

D generating energy in the form of ATP.

Question 8

After a person receives an organ replacement from a donor, they must be given immuno-suppressant drugs to prevent the immune system from reacting to the organ.

The immune cells most likely to be responsible for reacting to the transplanted organ are

A T helper cells.

B T cytotoxic cells.

C B cells.

D mast cells.

Question 9

Kuru is a rare and fatal disease that affects the nervous system. The symptoms of kuru include difficulty walking, chewing and swallowing, as well as a loss of coordination and muscle twitching.

Kuru is contracted by eating the brain of another organism who has also been affected by the disease.

The disease is caused by large clumps of material forming within neurons in the brain. These clumps of material are very difficult to destroy and are resistant to most chemical disinfectants. The only effective means of treatment is to heat the infective agent to high temperatures for several hours.

Based on the information stated above, it is likely that kuru is caused by

A a viral infection.

B protists.

C bacterial infection of the brain tissue being consumed.

D prions already found in the brain tissue being consumed.

Question 10

Antibodies destroy invading pathogens by

A agglutinating the pathogens to make it easier for macrophages to engulf them.

B assisting in lysis of the invading pathogens.

C neutralising toxins that are produced by invading pathogens.

D all of the above.

Question 11

CRISPR-Cas9 is a system that has evolved in bacteria to protect themselves from invading bacteriophage. Bacterial Cas enzymes cut the viral DNA into smaller pieces and the Cas9 enzyme cuts and stores a segment of the viral DNA in the bacterial protospacer. What type of enzymes are Cas?

A Protease

B Polymerase

C Ligase

D Endonuclease

Question 12

To make a fully functioning strand of mRNA once it has been transcribed from DNA

A introns must be removed, and exons spliced together.

B translation must occur.

C exons must be removed, and introns spliced together.

D coding sections of DNA must be removed.

The following information relates to Questions 13–15.

The diagram below shows the process used in generating a recombinant plasmid.

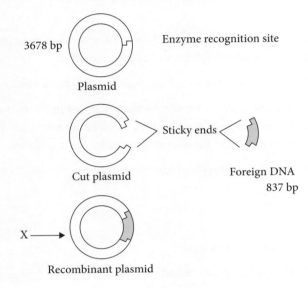

Question 13

The group of enzymes used to cut the plasmid at the recognition site would be

A restriction enzymes that recognise the sequence on the plasmid only.

B reverse transcriptase that recognises the same sequences on the plasmid and the foreign DNA.

C restriction enzymes that recognise the same sequences on the plasmid and the foreign DNA.

D DNA ligase because this can recognise all relevant sequences.

Question 14

Once the recombinant plasmid has been made, it is inserted back into a host bacterial cell ready for cloning.

In order to assist the bacteria in taking up the recombinant plasmids, the bacteria are bathed in a solution of

A calcium chloride and cultured at 37°C for 1 week.

B sodium chloride and cultured at 37°C for 1 week.

C calcium chloride and heat shocked to 42°C.

D sodium chloride and heat shocked to 42°C.

Question 15

If the recombinant plasmid is cut at point X, the number of fragments, and the size of the fragments, would be

A 2 fragments, 3678 bp and 837 bp.

B 1 fragment, 3678 bp.

C 1 fragment, 4515 bp.

D 1 fragment, 837 bp.

Question 16

Polydactyly is a disorder in which a person is born with extra fingers and toes. It is a symptom of Ellis–van Creveld syndrome. In the 1700s, a small number of immigrants to the USA set up the Amish community. Ellis–van Creveld syndrome affects 7% of the Amish community, compared with just 0.1% in the rest of the US population.

Over time, the gene pool of the Amish community would become

A larger, because there would have been greater mutations in the Amish community so more variants of the gene would be present.

B smaller, because there would be fewer variants of different genes introduced into the community.

C the same, because the Amish population would reflect the original population at all times.

D larger, because there would be more interbreeding and therefore more mutations.

Question 17

The northern spotted owl and the Mexican spotted owl once belonged to the same species of bird. Over time, the birds were separated by the Rocky Mountains in western USA and now cannot interbreed.

The original species of bird was able to develop into the northern and Mexican spotted owl species because

A their alleles become more and more different due to random mutations.

B they each adapted to their new habitat.

C the birds choose to breed with more similar birds in their new habitat.

D the birds are unable to find their mates and therefore cannot breed.

The following information relates to Questions 18 and 19.

The diagram below represents four sets of rock strata from different archaeological sites.

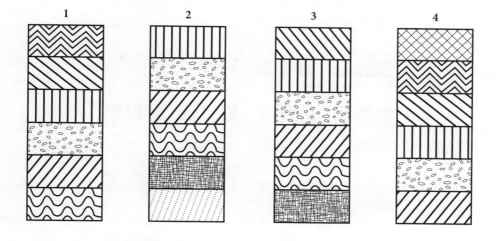

Question 18

Which of the archaeological dig sites contains the youngest rock layer?

A 3

B 2

C 4

D 1

Question 19

There are two types of dating methods for fossils discovered in the layers found at the archaeological dig sites – relative dating and radiometric (absolute) dating. Radiometric dating is most useful for

A determining the relationships between fossil species.

B determining the age of a fossilised organism.

C analysing the genetic sequence of a fossilised organism.

D determining the structure of a fossil.

Question 20

It is theorised that many early species such as single-celled organisms may never be discovered, primarily because they will not have been fossilised. The most likely cause for these species not being fossilised is

A they were terrestrial animals, which are much more unlikely to form fossils.

B the fossils are likely to have been eroded over the many millions of years since they were alive.

C the land was still forming and therefore was not capable of creating fossils at this time.

D the organisms from the early species lacked bony parts to fossilise.

The following information relates to Questions 21 and 22.

The diagram below shows several types of gene mutations labelled 1, 2 and 3. The top sequence is the original DNA sequence before any mutations have occurred.

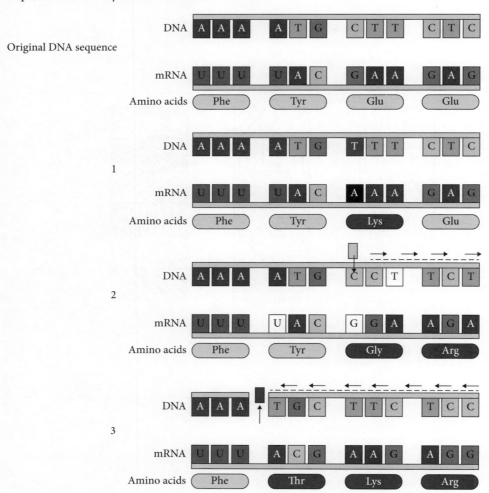

Question 21

Mutation 1 is a

A deletion mutation.

B substitution mutation.

C silent mutation.

D nonsense mutation.

Question 22

Mutations 2 and 3 have similar effects on the genetic code and resultant amino acid sequence. The two mutations would result in

A the whole gene being deleted from the genome.

B the same protein being produced as the original gene.

C a frameshift mutation that causes the remaining triplet codes to change and therefore all amino acids in the sequence will also change, creating a different or non-functional protein.

D RNA polymerase attaching to the DNA and transcribing the DNA triplets to generate an mRNA strand.

Question 23

The table shows the amino acid found at positions 1–10 in the protein cytochrome c of various different animals.

	Amino acid position									
	1	**2**	**3**	**4**	**5**	**6**	**7**	**8**	**9**	**10**
Chicken	Ala	Leu	Glu	Met	Pro	Gly	Arg	Cys	Phe	Tyr
Shark	Thr	Leu	Glu	Leu	Pro	Gly	Cys	Glu	Phe	Ala
Iguana	Ala	Leu	Glu	Met	Pro	Gly	Arg	Cys	Phe	Cys
Gorilla	Ala	Leu	Glu	Gly	Pro	Gly	Thr	Asp	Phe	Tyr
Bullfrog	Ala	Leu	Glu	Met	Pro	Gly	Cys	Glu	Phe	Cys

From the information in the table, it is safe to conclude that the

A bullfrog is most closely related to the iguana.

B gorilla has the most distant common ancestor.

C shark and iguana have a recent common ancestor.

D chicken and iguana are the most closely related species.

Question 24

A hominoid is

A an organism within the hominin group.

B a member of the family of organisms that includes humans, gibbons and great apes.

C an organism with the same skull shape and structure as humans (*Homo sapiens*).

D a group of apes that have a common ancestor with Old World monkeys.

Question 25

A Belgian blue cow is a species of cattle that has been bred for increased meat production. A picture of a Belgian blue is shown.

This breed of cow would not be found naturally in the wild. The breeding of a Belgian blue is an example of

A natural selection.

C a genetically modified organism.

B artificial selection.

D survival of the fittest.

Question 26

Which of the following statements is true about apoptosis?

A It can be caused by damaged DNA.

B Apoptosis is not regulated by the cell and is dangerous for organisms.

C An apoptotic cell releases toxins into the extracellular environment.

D It is triggered by phagocytosis of a dying cell.

Question 27

The main purpose of aerobic respiration is the

A conversion of light energy into chemical energy.

C storage of energy as starch.

B production of lactic acid.

D production of ATP from the breakdown of glucose.

Question 28

The diagram below shows a gel electrophoresis plate. Five samples of DNA have been loaded into the wells on the gel plate and show fragments of DNA of varying sizes. The wells have been labelled V–Z.

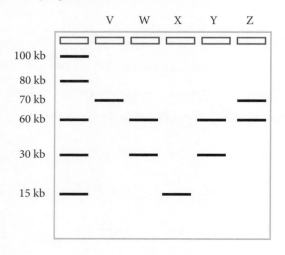

The sizes of the fragments found in the five different DNA samples are:

- Sample 1 – 15 kb
- Sample 2 – 70 kb
- Sample 3 – 60 kb, 30 kb
- Sample 4 – 60 kb, 30 kb
- Sample 5 – 70 kb, 60 kb.

Which sample of DNA matches which well on the gel electrophoresis plate?

	Well on the gel plate				
	V	**W**	**X**	**Y**	**Z**
A	Sample 2	Sample 1	Sample 4	Sample 3	Sample 5
B	Sample 2	Sample 4	Sample 2	Sample 5	Sample 1
C	Sample 2	Sample 3	Sample 1	Sample 5	Sample 4
D	Sample 2	Sample 4	Sample 1	Sample 3	Sample 5

Question 29

Two species of palm tree, *Howea belmorean* and *Howea forsteriana*, are found on Lord Howe Island, off the coast of Australia. A picture of the palm trees is shown below.

Kahuroa, Public domain, via Wikimedia Commons

The palm trees are endemic to Lord Howe Island (they are only found on this island) and the two species diverged once the island formed many thousands of years ago. The most likely reason for the formation of these two species is

A allopatric speciation.

B artificial selection.

C sympatric speciation.

D genetic modification.

Question 30

A teacher wanted to determine how well their Biology class was able to use an electric balance to measure the mass of a glass beaker. The beaker had a manufactured mass of 1.0000 g. The teacher split the class into two groups of five students – group A and group B. Each group used a different electric balance. The results of the measurements are shown below.

Mass of glass beaker (g)	
Group A	**Group B**
0.8888	1.3110
0.9959	1.3109
1.1182	1.3111
0.9938	1.3110
1.0033	1.3110
Average: 1.0000	Average: 1.3110

Based on the information provided in the table, it is reasonable to conclude that

A   group A is more precise and less accurate than group B.

B   group B is more reliable and less precise than group A.

C   group B is less precise and less reliable than group A.

D   group A is more accurate and less precise than group B.

Question 31

Cyanide is a poison that prevents ATP production by aerobic respiration, leading to eventual death. Cyanide binds to an allosteric site on the enzyme cytochrome oxidase, preventing it from catalysing its reaction.

Based on this information, it is reasonable to conclude that

A   cyanide is a non-competitive inhibitor of cytochrome oxidase.

B   cyanide is a competitive inhibitor of cytochrome oxidase.

C   cyanide is a poison that kills mitochondria.

D   cyanide activates apoptosis in the cell.

Question 32

Bt cotton is a variety of genetically modified cotton. It is engineered to manufacture a protein that acts as a pesticide against the bollworm insect. Bt cotton was made by genetically altering the cotton genome to express a microbial protein from the bacterium *Bacillus thuringiensis*.

Based on this information, Bt cotton can be considered

A   transgenic but not genetically modified.

B   transgenic and genetically modified.

C   genetically modified but not transgenic.

D   neither transgenic nor genetically modified.

Question 33

The following image shows an electron micrograph of a chloroplast.

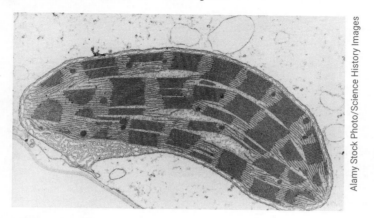

Determine which statement is correct about where the reactions of photosynthesis occur within a chloroplast.

	Thylakoid	Stroma
A	Light-independent reaction	Light-dependent reaction
B	Calvin cycle	Glycolysis
C	Glycolysis	Krebs cycle
D	Light-dependent reaction	Light-independent reaction

Question 34

Monoclonal antibodies are produced in response to a specific disease. What is the first step when producing these antibodies?

A Injecting a mouse with an antigen

B Injecting a mouse with an antibody

C Removing the spleen

D Fusing antibodies with myeloma cells

Question 35

Which of the following describes the role of Rubisco in photosynthesis?

A It is involved in the final stage of glucose formation.

B It incorporates carbon dioxide into the Calvin cycle to enable the formation of glucose.

C It catalyses the conversion of NADP to NADPH.

D It forms ATP from ADP and inorganic phosphate

Use the following information to answer Questions 36–38.

The bubonic plague was responsible for the death of one-third of the world's population in 1350. The disease most likely started in Asia and spread through several countries via the Silk Road trade route, eventually reaching Europe, where it wiped out much of the population of countries such as France and the United Kingdom.

The disease is caused by the bacterium *Yersinia pestis* and results in symptoms such as chills, muscle cramps, seizures and loss of extremities (such as fingers and toes), and eventually in death.

Question 36

The bubonic plague outbreak of 1350 can best be described as

A   an epidemic.

B   an epicentre.

C   a pandemic.

D   endemic.

Question 37

A bacterium is an example of a

A   self-antigen of cellular origin.

B   non-self antigen of non-cellular origin.

C   self-antigen of non-cellular origin.

D   non-self antigen of cellular origin.

Question 38

A first line or physical defence in humans against the bubonic plague could be

A   intact skin.

B   antibody production.

C   phagocytosis of viral proteins.

D   production of cytokines to stimulate leucocytes.

Question 39

A human gene is 1000 base pairs long. In this section of double-stranded DNA, there are 310 nucleotides containing the base cytosine.

How many nucleotides in this gene contain the base adenine?

A   310

B   1380

C   210

D   690

Question 40

Which of the following describes the type of immunity conferred when a mother breastfeeds her baby?

A   Artificial active immunity

B   Artificial passive immunity

C   Natural active immunity

D   Natural passive immunity

Section B: Short-answer questions

Instructions for Section B

- Answer all questions in the spaces provided. Write using blue or black pen.
- Unless otherwise indicated, the diagrams in this book are not drawn to scale.

Question 1 (7 marks)

a   Explain the function of each of the three forms of RNA.

3 marks

_____

_____

_____

_____

_____

The *trp* operon is a group of genes found in *E. coli* that encode enzymes required to synthesise the amino acid tryptophan.

b   Explain how the *trp* operon prevents the expression of the enzymes involved in the synthesis of tryptophan.

4 marks

_____

_____

_____

_____

_____

_____

_____

Question 2 (13 marks)

SARS-CoV-2 is a strain of virus that swept through the world in 2020 and infected millions of people.

a   Name **one** cell of the adaptive immune system that would be involved in helping to destroy the virus if it infected an individual and explain its mode of action in this response.

2 marks

Many media outlets ran stories during the pandemic that said humans were unable to generate long-term immunity to the virus. The evidence they used for this was that individuals who had been infected and recently recovered from the virus had decreasing numbers of antibodies in their blood.

b   Using your understanding of the adaptive immune system and its response to viral infections, explain whether antibody numbers is a useful indicator of long-term immunity in an individual.

3 marks

c   Name and describe **two** measures that individuals could take to minimise the spread of the virus.

2 marks

A vaccine has been developed to help prevent the spread of COVID-19. This vaccine uses a strand of synthetic mRNA that is taken up by the cells. The mRNA codes for a protein from the COVID-19 virus.

d   Using your understanding of protein production and vaccines, explain how this new vaccine would work to help protect an individual from developing COVID-19.

4 marks

e Identify the type of immunity conferred when an individual receives a vaccination.
Justify your answer.   2 marks

_____

_____

_____

Question 3 (13 marks)

Some students wanted to investigate the effect of light intensity on photosynthesis. They cut 40 discs from leaves of the same plant and added them to a beaker filled with sodium bicarbonate (this provides carbon dioxide for the leaf discs). The leaf discs sank to the bottom of the beaker at the start of the experiment. Each student set up a lamp and placed it at varying distances from the leaf discs; the closer the lamp to the beaker, the more intense the light. They recorded the number of discs that had floated to the surface every minute for 15 minutes.

A diagram of the set-up is shown below.

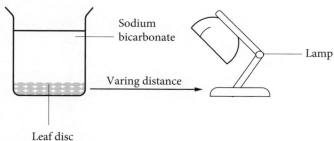

a Name the independent and dependent variables for this experiment.   2 marks

Dependent variable: _____

_____

Independent variable: _____

_____

b State the hypothesis the students are trying to investigate.   2 marks

_____

_____

_____

c State **two** controlled variables for this experiment.   2 marks

_____

_____

_____

A table and graph of the results are shown below.

Time exposed to light (min)	Number of discs floating to the surface at different distances from lamp (cm)					
	5	10	15	20	25	30
0	0	0	0	0	0	0
1	4	2	0	0	0	0
2	3	4	3	0	1	0
3	5	4	3	2	1	1
4	6	4	3	2	1	1
5	6	7	4	2	2	1
6	7	9	4	4	2	1
7	9	9	5	4	2	6
8	10	11	5	4	2	6
9	14	14	7	7	3	6
10	16	14	9	7	3	6
11	19	14	9	7	3	6
12	20	18	11	8	3	7
13	25	20	11	9	3	7
14	28	24	13	9	3	8
15	30	24	13	9	3	8

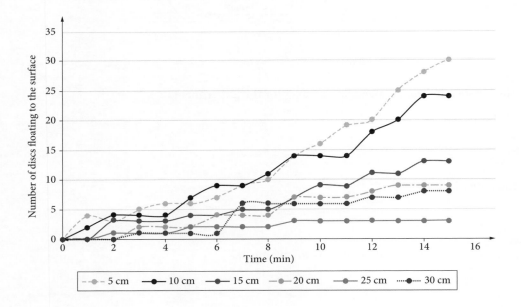

d Describe the relationship between the number of discs floating and light intensity (determined by distance from the lamp), as shown on the graph.   2 marks

_____

_____

_____

_____

e Explain the changes that occurred within the leaf tissue that allowed the leaf discs to rise to the surface.   3 marks

_____

_____

_____

_____

_____

_____

f   i Identify **one** set of results that does not seem to match the trend.   1 mark

_____

_____

_____

ii Suggest a reason for the anomalous results identified in the question above.   1 mark

_____

_____

_____

Question 4 (4 marks)

Warfarin is a chemical that has been used as a poison for more than 50 years to reduce rat populations. The drug inhibits blood clotting in rats and has been successful in controlling rat infestations all over the world. However, since 2009 some rats have shown resistance to the effects of the drug and are therefore able to survive.

Describe and explain how new colonies of warfarin-resistant rats have arisen.   4 marks

_____

_____

_____

_____

_____

_____

_____

Question 5 (17 marks)

a State the location of the following processes.   2 marks

   i Glycolysis: _____

   ii Krebs cycle: _____

d Complete the following table about aerobic respiration. Only one coenzyme is required for each
   stage of the reaction.   3 marks

Stage	Coenzyme(s) required	Purpose of coenzyme(s)
Glycolysis		
Krebs cycle		
Electron transport chain		

c What is meant by 'coenzyme'?   1 mark

_____

_____

_____

Vitamin B molecules are coenzymes involved in aerobic respiration and are essential for the reactions to proceed.
A person lacking in B vitamins can become extremely tired and fatigued and lack energy.

d Explain why people lacking in B vitamins become tired and lack energy.   3 marks

_____

_____

_____

_____

_____

e How many ATP molecules are produced in the electron transport chain?   1 mark

_____

_____

The graph below shows data collected by a group of students who investigated the effect of temperature on aerobic respiration. The students added glucose to test tubes containing yeast and placed the test tubes into incubators of water set at different temperatures. They then measured the carbon dioxide output over 10 minutes. They decided to use a test tube of yeast and water without glucose as a control.

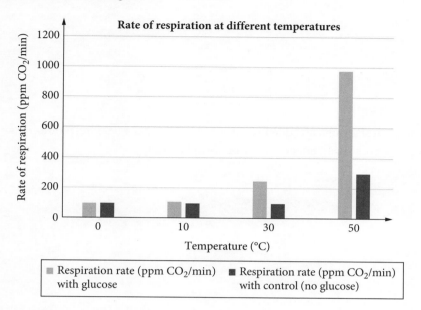

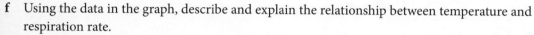

f Using the data in the graph, describe and explain the relationship between temperature and respiration rate.

4 marks

_____

_____

_____

_____

_____

_____

The same group of students then decided to test whether the type of sugar provided to the yeast affected the rate of respiration. The students added different sugars to test tubes of yeast and measured carbon dioxide output. The results are shown below.

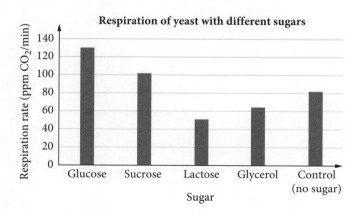

g Identify the sugar that resulted in yeast with the lowest rate of respiration. Suggest a reason for this result. 2 marks

h After 40 minutes, the control group (no added sugar) showed a higher rate of respiration than some of the yeast test tubes containing sugar. Suggest a reason for this result. 1 mark

Question 6 (7 marks)

Canola is a widely grown crop in Australia. It is used in the production of vegetable oils and spreads such as margarine.

Weeds compete with canola crops for resources, including space and soil nutrients, which compromises crop quality and yield.

Using traditional herbicides to control the weed problem can affect the crop growth. However, genetically modified canola crops have been produced that can tolerate traditional herbicides and not be damaged themselves.

The soil bacterium *Agrobacterium thuringiensis* is typically used as a vector in the genetic transformation of canola. After bacterial plasmids are modified to carry a herbicide resistant gene, the transformed bacteria are introduced to tissue cultures from the canola plant. The bacteria infect the cells and introduce the new gene.

Shutterstock.com/Daniel Prudek

a Outline the role of each of the following in the genetic modification of canola.

  i *Agrobacterium thuringiensis* 1 mark

  ii Endonucleases 1 mark

iii Ligase   1 mark

b Explain why it is important to apply the same endonuclease to cut the desired gene from the surrounding DNA and to cut the plasmid open.

2 marks

c Outline **one** advantage and **one** disadvantage of using GM canola.   2 marks

Question 7 (6 marks)

Carrots contain a compound called beta-carotene, which gives them their orange colour. In humans, beta-carotene is converted to vitamin A in the presence of an enzyme. Vitamin A helps the function of cells in the eyes, as well as reducing cholesterol in the blood.

The following reaction is required to convert beta-carotene into vitamin A:

$$\text{Beta-carotene} \xrightarrow{\text{Beta-carotene oxygenase 1}} \text{Vitamin A}$$

a Identify the substrate, product and enzyme in the reaction above.   3 marks

Substrate: _____

Product: _____

Enzyme: _____

A student wanted to investigate whether the beta-carotene oxygenase 1 protein is denatured by heat when cooking carrots.

b Explain what happens to proteins such as beta-carotene oxygenase 1 when they become denatured.   3 marks

Question 8 (5 marks)

The two skulls shown below were discovered between 1961 and 1971. When new hominin fossils are discovered, scientists must determine which species they belong to by using their understanding and knowledge of trends in the features of skulls in human evolution. When these fossils were discovered, scientists were unsure whether the skulls belonged to the species *Homo sapiens, H. neanderthalensis* or *H. heidelbergensis.*

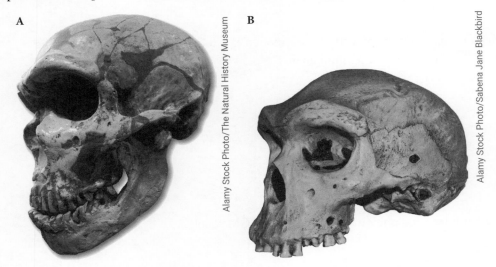

A

B

Alamy Stock Photo/The Natural History Museum

Alamy Stock Photo/Sabena Jane Blackbird

a Using your knowledge of the trends in the features of skulls within the *Homo* genus, determine which species the skulls belong to, from those listed in the information above. Justify your answer.   2 marks

Skull	Species	Justification
A		
B		

Homo neanderthalensis were once thought to be an ancestor of *Homo sapiens*. New evidence has contradicted this idea and placed Neanderthals as 'cousins' *of Homo sapiens* on many phylogenetic trees.

b Describe the evidence that has contradicted the idea that Neanderthals were ancestors of modern humans.   3 marks

_____

_____

_____

Question 9 (8 marks)

CRISPR is a gene-editing tool that was discovered and developed in Japan in 2017. CRISPRs are specialised sequences of DNA that, along with specific Cas enzymes and other guide molecules, are capable of cutting and editing strands of DNA. They are originally found in certain bacteria and archaea and are used by these organisms to defend against invading viruses. Scientists have been able to use this biochemical technology to edit genes in other organisms.

a Using your understanding of the CRISPR gene-editing tool, explain how it can be used to modify or edit a faulty gene in an organism.

3 marks

_____

_____

_____

_____

_____

b Name and explain **one** property of DNA that allows scientists to use a genetic tool from **one** organism, such as bacteria, in another organism, such as humans.

2 marks

_____

_____

_____

_____

A group of scientists wanted to insert a specific gene into an organism, but first they needed to make multiple copies of the gene. To do this, they used the process of PCR. One of the scientists failed to heat the solution to the temperature required for the first step to complete.

c Identify the temperature required to complete the first stage of PCR and explain the effect the mistake would have on the process of gene amplification.

3 marks

_____

_____

_____

_____

_____

SOLUTIONS

Solutions to Test 1: The relationship between nucleic acids and proteins [U3 Topic 1.1.1]

Multiple-choice solutions

Question 1 [A+ Study Notes p.11]

D 21%

Adenine (29%) pairs with thymine (29%) and the remaining 42% of DNA consists of guanine (21%) pairing with cytosine (21%). **A** is incorrect because 29% would be thymine as it is complementary to adenine. **B** is incorrect because 42% is the total remaining DNA % required for both guanine and cytosine. **C** is incorrect because 29% × 2 and 42% × 2 do not add up to 100%.

Question 2 [A+ Study Notes pp.10–13]

C DNA: Two nucleotide polymers of nucleotides joined together by hydrogen bonds. RNA: Polymer of nucleotides joined by phosphodiester bonds in the backbone.

Hydrogen bonds form between complementary base pairs between the two strands of DNA, forming the double helix. The backbone of both DNA and RNA is composed of phosphate and a 5-carbon sugar, which forms phosphodiester bonds. **A** is incorrect because the sugars are the wrong way around. DNA has a deoxyribose sugar and RNA has a ribose sugar. **B** is incorrect because both DNA and RNA have four bases. DNA has thymine, adenine, guanine and cytosine. RNA has adenine, cytosine, guanine and uracil. **D** is incorrect because both DNA and RNA are polymers of nucleotides formed through condensation reaction. Hydrolysis forms monomers from polymers.

Question 3 [A+ Study Notes pp.13–15]

C V → X → W → Z → Y

mRNA is produced from DNA during translation. mRNA then leaves the nucleus and enters the cytoplasm where it forms a complex with the ribosome, and tRNAs bring specific amino acids that are bonded via peptide bonds. **A** is incorrect because tRNA is not present in transcription but is present during translation after the mRNA reaches the ribosome. **B** is incorrect because mRNA must be produced before the correct tRNA with the anticodon complementary to the mRNA can enter the ribosome. **D** is incorrect because after the mRNA leaves the nucleus, it needs to form a complex with the ribosome before amino acids are bound together.

Question 4 [A+ Study Notes pp.10–11]

B Bonds are formed between chemical A of one monomer and chemical B of the next monomer.

Phosphates (A) and 5-carbon sugars (B) alternate in the backbone of nucleic acids forming phosphodiester bonds. **A** is incorrect because C is a nitrogenous base, which forms hydrogen bonds with the nitrogenous base on the other DNA strand. **C** is incorrect because the bond between a nitrogenous base (C) and a 5-carbon sugar (B) is within a nucleotide monomer not between. **D** is incorrect because the 5-carbon sugar (B) alternates with the phosphate (A) in the backbone.

Question 5 [A+ Study Notes pp.18-20]

A The folding time decreases as the number of amino acids decreases.

The fewer amino acids within the protein, the quicker the folding time. At 1000 amino acids, it takes 600 seconds for the folding, compared to 500 amino acids taking 200 seconds. **B** is incorrect because the faster the folding is complete, the smaller the protein. **C** is incorrect because the folding depends on the number of amino acids, not the type of amino acid. **D** is incorrect because the graph is linear and there is an increasing trend in the graph, showing the more the amino acids, the longer the time.

Question 6 A+ Study Notes pp.15–18

C All bacterial operons are located on a large circular chromosome within the cell.

As prokaryotes have a single circular chromosome, the operon is located on the chromosome. **A** is incorrect because there is no mention of gene activation in stem cells in the question stem, and stem cells would not need or be able to produce all proteins in their genome. **B** is incorrect because all cells could not produce all proteins. The cell would become toxic; therefore, the environment would influence the gene expression. **D** is incorrect because the question stem refers to the regulator genes being located on a different chromosome, and any mutation on the chromosomes would affect the function of these proteins within the cell.

Question 7 A+ Study Notes pp.14–15

B Some amino acids may be encoded by more than one codon.

There are 64 codon sequences and only 20 amino acids. **A** is incorrect because DNA is universal and therefore the same in all organisms. **C** is incorrect because, although a nucleotide is part of a single codon, this is not classified as degenerate. **D** is incorrect because although three codons are required for one amino acid, it is not referred to as the degenerate code.

Question 8 A+ Study Notes pp.16–18

D The presence of tryptophan represses enzyme synthesis.

It is more energy efficient for the cell to use the tryptophan available than to synthesise its own. **A** is incorrect because the presence of tryptophan would inhibit the expression of these genes. **B** is incorrect because the *trp* operon is a repressible system that is inhibited in the presence of *trp* in the bacteria's environment. **C** is incorrect because tryptophan is an amino acid required for protein production.

Question 9 A+ Study Notes pp.16–18

A A repressible system that is active if *trp* is not present.

The bacteria require tryptophan for protein synthesis because it is one of the 20 amino acids. Therefore, the operon is active until *trp* is present in the bacteria's surroundings where it uses this instead of synthesising its own. **B** is incorrect because it is always active until *trp* is present. An inducible system is inactive until a molecule is present. **C** is incorrect because the *trp* operon does not metabolise lactose for the bacteria. **D** is incorrect because multiple operons can be expressed simultaneously in the bacteria.

Question 10 A+ Study Notes pp.16–18

B UGA codes for a stop codon which stops the ribosome and enables a hairpin loop to form which pulls the mRNA away from the DNA thereby terminating transcription.

When the ribosome stops at the stop codon it overlaps sections 1 and 2 thereby blocking sections 1-3 from forming hairpin loops but not sections 3 & 4. **A** is incorrect as the ribosome does not overlap three sections of the mRNA. A hairpin loop can still form. **C** and **D** are incorrect as the codon UGA is a stop codon, it does not code for tryptophan.

Question 11 A+ Study Notes pp.13–14

C There to be different editing of the pre-mRNA.

Post-transcriptional modification removes introns and shuffles exons. **A** is incorrect because RNA polymerase is the only enzyme to synthesise RNA from DNA. **B** is incorrect because RNA polymerase can only read the mRNA 3' to 5' and synthesise 5' to 3'. **D** is incorrect because mutations are random and by chance.

Question 12 A+ Study Notes p.11

D An RNA molecule found in the cytoplasm of a eukaryotic cell.

> Produced through transcription and contains the nitrogenous base uracil. **A** is incorrect because uracil is not found in any DNA, regardless of the species. **B** is incorrect because DNA does not contain uracil. **C** is incorrect because RNA codes for proteins composed of amino acids, not phospholipids.

Question 13 A+ Study Notes p.16

B Produce chemicals that control the action of other genes.

> These genes can produce proteins that increase or inhibit structural gene expression. **A** is incorrect because genes can be structural or regulatory depending on their role in the cell. **C** is incorrect because they are involved in protein production, not lipid synthesis – this is the SER. **D** is incorrect because splicing of introns occurs in post-transcriptional modification and regulatory proteins are involved in transcription.

Question 14 A+ Study Notes pp.18–20

A Is determined by its sequence of amino acids.

> The properties of the amino acids and their interactions with each other determine the unique structure for each protein. **B** is incorrect because the protein does not change shape in response to surrounding molecules. **C** is incorrect because the protein becomes functional at the tertiary level where it forms the three-dimensional shape. **D** is incorrect because the active site and shape is dependent on the interactions between the amino acids and the location in the primary polypeptide chain.

Question 15 A+ Study Notes pp.12–15

A CGTACGTTTTATTTG

> The mRNA strand is GCAUGCAAAAUAAAC and correctly codes for the amino acid sequence Ala–Cys–Lys–Ile–Asn. **B** is incorrect because GCAACGAAAAUAAAC codes for the sequence Ala–Unknown–. **C** is incorrect because GCAUGAAAAAUGAAC Ala–Unknown–. **D** is incorrect because GCUUGUAAGAUAAAA Ala–Cys–Lys–Ile–Lys.

Question 16 A+ Study Notes p.14; 22

B Polymer being formed at molecule X is a polypeptide chain formed through an anabolic reaction.

> **A** is incorrect because the tRNA molecules have an anticodon, not a codon, on their base. **C** is incorrect because hydrolysis is the breakage of bonds (polymer to monomer). **D** is incorrect because mRNA (the 5' to 3' molecule) contain codons, not anticodons.

Question 17 A+ Study Notes pp.18–20

D Quaternary protein structure.

> Quaternary structure is two or more polypeptide chains bonded together. **A** is incorrect because primary protein structure is the polypeptide chain and not functional. **B** is incorrect because the secondary structure has not formed the interactions between the variable groups. **C** is incorrect because if it were a singular polypeptide chain, it could be functional at the tertiary level, but because it is four polypeptide chains bonded together, the individual chains do not function solo.

Question 18 A+ Study Notes pp.23–24

C ribosome → endoplasmic reticulum → Golgi body → vesicle

> The protein is translated at the ribosome, transported in the endoplasmic reticulum to the Golgi body, where it is folded and modified before being packaged into a vesicle for export from the cell through exocytosis. **A** is incorrect because the mitochondria are not directly involved in protein export. Mitochondria produce the ATP required for exocytosis (as the vesicle fuses with the membrane). **B** is incorrect because protein is produced at the ribosome and transported through the endoplasmic reticulum to reach the Golgi body for folding and modification. **D** is incorrect because the Golgi apparatus comes after the endoplasmic reticulum.

Question 19 A+ Study Notes p.21

C No, different genes are expressed in different cells depending on the function of the cell.

> The proteome is the whole set of proteins produced by a cell, tissue or organism. Each cell contains the same complement of genes but only some are expressed depending on the cell function. **A** is incorrect as not all the genes in every cell are expressed. **B** is incorrect because a skin cell and a white blood cell have different functions and would require different proteins. **D** is incorrect because a skin cell and a white blood cell in the same person would have the same set of chromosomes in each cell.

Question 20 A+ Study Notes pp.18–20

A A change in the tertiary structure of a protein may result in the protein becoming biologically inactive.

> If a protein is subject to high temperature or a pH outside its pH tolerance range, the interaction between the variable groups alters, changing the protein shape, which may inhibit its function. **B** is incorrect because some proteins do not require multiple polypeptide chains to function. **C** is incorrect because the sequence of amino acids may differ in the two proteins, which means different side groups and interactions. **D** is incorrect because denaturation breaks the bonds involved in the tertiary structure.

Short-answer solutions

Question 1 A+ Study Notes pp.10–11

a DNA (deoxyribonucleic acid)

> This is because there are two strands bonded in an antiparallel manner and thymine is present.

b 1 Hydrogen bond
   2 Adenine (nitrogenous base)
   3 Deoxyribose sugar

   4 Phosphate
   5 Cytosine

c

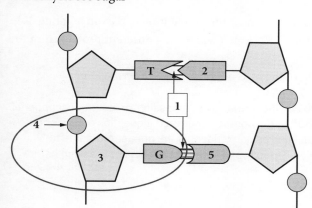

d A condensation polymerisation reaction produces a polymer from nucleotide monomers. (1 mark)

   Phosphodiester bonds form between phosphate and sugar. Hydrogen bonds form between nitrogenous bases to form the double helix. (1 mark)

Question 2 A+ Study Notes pp.11–15

a Deoxyribonucleic acid, because it contains thymine, rather than uracil, which is in RNA.

b The molecule would be located in the nucleus and mitochondria.

c mRNA   AUG UGU CCG CUA UCC GGC
   tRNA   UAC ACA GGC GAU AGG CCG
   amino acids   Met–Cys–Pro–Leu–Ser–Gly

> Always use the mRNA (codons) to code for amino acids, not the tRNA, unless the question provides you with an anticodon chart.

d The amino acid would be the same in the bacterial cells as in the human cell. DNA code is universal with the same nitrogenous bases. The same codon codes for the same amino acid in all species.

Question 3 A+ Study Notes pp.13–15

a Translation   (1 mark)

Bond J – peptide bond formed through condensation polymerisation. Group H – anticodon on the tRNA. Structure K – ribosomal complex.   (1 mark)

The ribosome binds to or reads the mRNA, and tRNA brings in specific amino acids, or the tRNA anticodon is complementary to the mRNA codon. The amino acids are joined by condensation polymerisation to form a peptide bond.   (2 marks)

> Your answer must state what occurred at the ribosome, not just restate translation. Descriptions of corresponding instead of complementary tRNA anticodons and mRNA codons would not be accurate enough to receive a mark. You are also required to make a valid point about the production of a protein.

b i Product G is mature mRNA – uracil instead of thymine, ribose sugar instead of deoxyribose.

DNA has coding and non-coding regions whereas mature mRNA has had the introns removed and exons retained.

> To achieve full marks, your answer needs to include a comparative term, such as 'whereas' or 'compared to' to assist in the linking of these concepts. A statement such as 'DNA has thymine, whereas RNA does not' is not adequate. Your answer must include what RNA has instead.

ii RNA processing or post-transcriptional modification enables a larger proteome than genome.   (1 mark)

One of (1 mark):
- Discuss two of: intron splicing (non-coding regions), intron retention and exon juggling (coding regions) to form a different mature mRNA that will form a different protein from the same pre-mRNA strand.
- Methyl cap and poly A tail for protection of the mRNA; guidance to the ribosome.

> 'Outline' is another term for 'explain'. VCAA can also use the terms 'RNA processing' or 'post-transcriptional modification' when discussing this stage. Students often get confused between a question asking about modifications that occur during this stage, and a question asking to identify the greater proteome to genome – gene expression. Always read the question carefully and know exactly what it is asking before you start.

c Eukaryotic (no mark)

One of (1 mark):
- Post-transcriptional modification occurring where the pre-mRNA is modified to mature mRNA – introns removed whereas prokaryotes mRNA undergoes no modification.
- The nucleus is present, separating transcription and translation within the cell, whereas prokaryotic cells can complete both processes simultaneously.

> It is important to know that you will not receive a mark for a 50/50 answer. The mark is received through the justification of your answer.

Question 4 A+ Study Notes pp.13–15

a i Gly–Ala–Val–Pro

ii CGT CGC CAG GGA

b i Point mutation, where one nucleotide is changed from G to C. This is an example of missense mutation.

(1 mark)

Amino acid 1 is changed from Gly to Ala.

(1 mark)

ii The change of amino acid could alter the secondary and tertiary structure of the protein due to the different properties and interactions between the amino acids. This could reduce the protein function or completely inhibit it from functioning.

c Yes; due to the degenerate code, there are multiple codons coding for the same amino acid.   (1 mark)

For example, proline can be coded for by CCU, CCC and CCA on the mRNA.   (1 mark)

Question 5 A+ Study Notes pp.16–18

a A structural gene codes for a protein that becomes part of the structure or function of an organism.

b An operon is a cluster of genes with similar function under the control of a single promoter and operator region.

c Regulatory genes form regulatory proteins, which control the expression of structural genes (1 mark), increasing transcription (activators) or inhibiting the transcription (repressors). Also referred to as transcriptional factors (1 mark).

> Your answer needs to state that regulatory genes are functional through the protein and not the gene itself.

d In the presence of tryptophan the repressor protein binds to some of the amino acid causing a conformational shape change of the repressor protein complementary to the operator region, enabling it to bind.

(1 mark)

RNA polymerase is unable to bind to the promoter region, and transcription of the structural genes downstream (*trp E, D, C, B* and *A*) is inhibited.

(1 mark)

e The repressor gene could code for a different amino acid sequence forming a different protein shape. Not complementary to the operator region of *trp* operon and therefore unable to repress the system.

(1 mark)

One of (1 mark):

- A potential side effect is the build-up of enzymes and tryptophan in the cell.
- Depletion of ATP due to the amount of transcription and translation occurring.

f The functional gene on the plasmid would transcribe and translate into a functional repressor protein. This repressor protein would bind with tryptophan when present in the cell, and form a complementary shape to the operator enabling its repression.

> It does not matter where the gene for a protein is located, because proteins can move to different parts of the cell and function.

g Repression involves a repressor protein binding to an operator region whereas attenuation does not involve a repression protein

(1 mark)

Repression blocks the initiation of transcription whereas in attenuation transcription is started but is terminated before the ribosome reaches the five structural genes

(1 mark)

> To achieve full marks, your answer needs to include a comparative term, such as 'whereas' or 'compared to'.

Question 6 A+ Study Notes pp.18–20

a Polymerise: Joining of tubulin dimers (the monomer) to create the protofilament (the polymer) (1 mark)

b Primary structure: The sequence of amino acids (1 mark)

Secondary structure: The coiled or pleated structure. (1 mark)

c Tertiary structure – A 3D structure composed of secondary structures (1 mark)

Quaternary structure – Two or more polypeptide chains joined together. (1 mark)

> A significant number of students got all parts of this question wrong. Rather than naming polypeptide chains for the quaternary structure, many students incorrectly mentioned proteins or tertiary structures being joined together.

Question 7 A+ Study Notes pp.15; 18–20

a One of (2 marks):
- Primary structure – polypeptide chain has a different sequence of amino acids bonded together.
- Tertiary structure – different interactions between the variable groups forming a different shape of the haemoglobin molecule.

> Your answer should refer specifically to the molecules discussed in the stem of the question and should be able to explain each level of protein hierarchy with where the bonds are formed.

b

	Bonds and interactions
mRNA	AUG GUG CAC CUG ACU CCU GAG GAG
Amino acid sequence	Met–Val–His–Leu–Thr–Pro–Glu–Glu

> The key here is that the strand above is the coding sequence, not the template.

c CAC is altered to GAG coding for Glu instead of His. (1 mark)

The amino acid may have different chemical properties and interact with other amino acids in the polypeptide chain differently, altering the secondary and tertiary shape of the haemoglobin protein, which would then alter its function. (1 mark)

Question 8 A+ Study Notes pp.23–24

a Exocytosis (1 mark)

Bulk transport, which requires ATP. Fusing of a vesicle to the plasma membrane to release the protein into the extracellular fluid (1 mark)

b

	Name	Function
i	Ribosome	Site of protein synthesis. Polypeptide chain formed from amino acids by a condensation polymerisation reaction
ii	(Rough) Endoplasmic reticulum	Folds and transports the protein from the ribosome to the Golgi apparatus
iii	Golgi body/apparatus	Modifies and packages the protein into a vesicle for export out of the cell
iv	Vesicle	Carries the protein from the Golgi body to the membrane where it fuses to release the protein into the extracellular fluid (exocytosis)

> An exam question may ask for two organelles involved in the process. It is important for you to know the function of all organelles involved in the protein excretion from a cell.

c   The other organelle is mitochondrion.   (1 mark)

One of (1 mark):
- ATP is required for the process of exocytosis (the release of the insulin from the vesicle to the extracellular fluid).
- ATP for the process of translation – forming a polypeptide chain (polymer) from amino acids (monomer).

Question 9 A+ Study Notes p.13

a   Transcription

b   Transcriptional factors are required in eukaryotic cells for the RNA polymerase to bind. They can also increase or inhibit the expression of genes depending on the cell's requirements.

c   Two of:
- DNA has deoxyribose sugars whereas RNA has ribose sugars.
- DNA has a double-stranded helix whereas RNA has a single-stranded polymer.
- DNA contains thymine whereas RNA contains uracil.
- DNA is located in the nucleus, mitochondria and chloroplast, whereas RNA is present in the nucleus and cytosol.

Part c asks you to explain how DNA differs from mRNA. To achieve full marks, your answer needs to include a comparative term, such as 'whereas' or 'compared to'.

Question 10 A+ Study Notes pp.13–14

a   The promoter region is upstream of the gene where the RNA polymerase and transcriptional factors bind.

b   The proteome is bigger due to post transcriptional modification – intron retention, exon juggling forms different mature mRNA strands from the same pre mRNA sequence.   (1 mark)

Promoter	Intron 1	Exon 1	Intron 2	Exon 2	Intron 3	Exon 3

                                                                                        (1 mark)

Your diagram must show two or more different mature mRNA strands from the original DNA sequence; for example, exon 2, exon 1, exon 3 or exon 1 intron 1, exon 2, exon 3.

c   DNA unwinds, exposing the template strand of DNA. RNA polymerase binds to the promoter region upstream of the gene of interest on the template strand (3'–5').   (1 mark)

RNA polymerase catalyses the synthesis of the pre mRNA (5'–3') from the template strand.   (1 mark)

Pre mRNA is produced by complementary base pairing.   (1 mark)

Solutions to Test 2: DNA manipulation techniques and applications U3 Topic 1.1.2

Multiple-choice solutions

Question 1 A+ Study Notes pp.33–34

C Corn modified with an insecticide gene from bacteria to produce a toxin to prevent infection.

The corn is a transgenic organism with a gene from a different species and is building up resistance to infection through toxin resistance. **A** is incorrect because although it is a transgenic organism, improving nutrition is not building up resistance to disease in the plant. **B** is incorrect because the tomato plant is not a transgenic organism; its own genome has been modified. **D** is incorrect because the cotton plant is not a transgenic organism; its own genome has been modified.

Question 2 A+ Study Notes pp.28–29

C Denaturing 95°C, annealing 55°C, extension 72°C

The DNA must be denatured to separate the strands by breaking hydrogen bonds, the temperature is lowered to allow the primers to anneal to the single strands of DNA. The temperature is then increased to the optimal temperature of the Taq polymerase to enable the synthesis of the DNA strand. **A** is incorrect because, although the steps are correct, they are not in the correct order and the temperatures do not align. **B** is incorrect because the DNA must be denatured (separated) first to allow the primers to anneal. The steps are in the wrong order, but the temperatures are correct. **D** is incorrect because, although the steps are in the right order, the temperatures do not align.

Question 3 A+ Study Notes pp.24–25

C HindII and PstI

HindII cuts twice on the plasmid and PstI cuts in the selectable marker. Neither could be used. **A** is incorrect because both these enzymes only cut once. **B** is incorrect because, although EcoRI could be used as it cuts the plasmid once, PstI cuts in the selectable marker. **D** is incorrect because BamII could be used in genetic engineering but not PstI.

Question 4 A+ Study Notes p.29

B 850 bp and 2792 bp

The fragment between BamII and EcoRI is 850 bp. To work out the length of the other fragment, start at the origin of replication and go to the BamII cut (2000 bp) from the EcoRI to the origin of replication 4642−2850 bp. Add them together. **A** is incorrect because you do not start the length of the DNA from the origin of replication. **C** and **D** are incorrect because only two fragments are formed.

Question 5 A+ Study Notes pp.29–31

C S2 could be the father but S1 could not.

S2 bands align with the child's bands that do not align with the mother's. All of the child's bands align with either the mother's or the father's. **A** is incorrect because the child's bands that align with S1 also align with the mother and there are bands unaccounted for in the child. **B** is incorrect because the child's bands that align with S1 also align with the mother, and there are bands in the child's sample that are unaccounted for. **D** is incorrect because the banding patterns align with the mother and S2. If neither man was the father, bands from the child would not match the mother or the suspects.

Question 6 A+ Study Notes p.25

A Two fragments of double-stranded DNA, each with a sticky end.

The linear strand of DNA would be cut into two fragments and the enzyme cuts at different locations. **B** is incorrect because the enzyme only cuts the fragment once, not twice. **C** is incorrect because only AluI and HaeIII would produce blunt ends. **D** is incorrect because the enzyme only cuts once and is not AluI and HaeIII.

Question 7 A+ Study Notes p.25

C AluI and HindIII only.

Both AluI and HindIII restriction sites are present in the strand provided. **D** is incorrect because, although AluI and HindIII are correct, there is no restriction site for HaeIII. **A** is incorrect because EcoRI restriction enzyme site is not present in the sequence. **B** is incorrect because there is more than just HindIII present.

Question 8 A+ Study Notes pp.26–27

C Spacer sequences.

The bacteria isolate genes from the virus to protect them from future exposures through recognition. **A** is incorrect because the Cas gene segment is the segment that remains relatively constant in all bacteria and produces the protein for the CRISPR function. **B** is incorrect because repeat sequences are placed between spacers as the bacteria save the viral genomes for subsequent exposures. These are non-coding regions. **D** is incorrect because the promoter is where the RNA polymerase binds to initiate transcription.

Question 9 A+ Study Notes p.26

D Invading bacteriophage DNA contains small segments called protospacer adjacent motif which identify it as non-self.

Bacterial DNA does not contain PAMs, only bacteriophage DNA does thereby allowing the cell to identify the bacteriophage DNA as non-self. **A** is incorrect as the bacteriophage DNA is not kept in a separate compartment in the cell. **B** is incorrect as it would be identified as non-self. **C** is incorrect as bacterial DNA does not contain protospacer adjacent motif.

Question 10 A+ Study Notes pp.26–27

D Direct the Cas9 to the specific gene sequence.

The guide RNA recognises a specific sequence in the genome to guide Cas9 for cleaving the DNA. **A** is incorrect because RNA would not be stable in the DNA double helix. **B** is incorrect because the Cas9 is the enzyme that cuts the DNA at the sequence the gRNA has guided it to. **C** is incorrect because the Cas9 is the part that can cut a gene.

Question 11 A+ Study Notes p.28

D A portion of its DNA sequence.

To allow a primer to be created for the sequence and PCR to proceed. **A** is incorrect because the whole sequence is not required for the primers to anneal. **B** is incorrect because the amount of adenine and thymine would not alter the PCR method. **C** is incorrect because the sample once amplified can be run through a gel to assist identification.

Question 12 A+ Study Notes p.15

B Gene and the upstream promoter.

The promoter is required for RNA polymerase to bind and transcribe the gene into mRNA. **A** is incorrect because the gene could not be transcribed without the flanking regions. **C** is incorrect because the RNA polymerase is not able to bind. **D** is incorrect because the gene does not contain a promoter, and any piece of DNA inserted into a cell can be transcribed without integration into the existing genome.

Question 13 `A+ Study Notes` `pp.26–27`

D Removal of the plasmid transfer gene.

The treatment referred to is the application of modified bacteria to the soil around plants in order to destroy *Agrobacterium tumefaciens*. It is important that the plasmid transfer gene is removed (option D) so that *A. radiobacter* is not able to transfer the antibiotic resistance gene to *A. tumefaciens*. **A** is incorrect because the removal of all plasmids would remove the antibiotic that is lethal to *A. tumefaciens*. **B** is incorrect because the removal of the antibiotic gene would make the bacteria ineffective at killing the *A. tumefaciens*. **C** is incorrect because if the antibiotic resistance gene is removed from *A. radiobacter*, it would be susceptible to the antibiotic on the plasmid and therefore die.

Question 14 `A+ Study Notes` `p.29`

B A should be negative, and B should be positive.

DNA has a negative charge in the backbone of the double helix. It is repelled from the negative terminal and attracted to the positive terminal. **A** is incorrect because DNA is attracted to the positive terminal and therefore will run to the left of the wells. **C** is incorrect because the DNA is attracted to the positive terminal and therefore would not be present on the gel because it would run off to the left of the wells. **D** is incorrect because although the DNA is repelled from the negative charge, it is not attracted to a neutral charge.

Question 15 `A+ Study Notes` `pp.29–30`

D Lane 4 only.

The DNA fragment has migrated the furthest down the gel to terminal B (away from the wells). **A** is incorrect because lane 3 does not have a fragment that has migrated furthest down the gel towards B. **B** is incorrect because while lanes 1 and 2 have two relatively short fragments near the B terminal of the gel, there is a fragment in a different well that has migrated further. **C** is incorrect because lane 2 contains the longest fragment – it has migrated the shortest distance from the well.

Question 16 `A+ Study Notes` `pp.29–30`

B 1 and 5.

The bands of the child match in all locations with either parent 1 or parent 5. **A** is incorrect because there are bands in the child that do not match with any of the bands in lanes 1 or 2. **C** is incorrect because there are bands in the child that do not match with any of the bands in lanes 4 or 5. **D** is incorrect because there are bands in the child that do not match with any of the bands in lanes 2 or 5.

Question 17 `A+ Study Notes` `pp.31–33`

A Restriction enzyme and DNA ligase.

Restriction enzymes cut at specific recognition sequences and DNA ligase re-forms the phosphodiester bonds between complementary base pairs. **B** is incorrect because restriction enzymes cut the plasmid at the specific recognition site but DNA polymerase replicates DNA strands not reforming bonds between genes. **C** is incorrect because DNA ligase reforms the phosphodiester bonds after gene insertion and DNA polymerase is involved in DNA replication. **D** is incorrect because DNA polymerase is involved in DNA replication. DNA ligase is correct for inserting genes.

Question 18 `A+ Study Notes` `pp.31–33`

B Can replicate non-bacterial sequences of DNA in a short time.

Recombinant plasmids can be inserted to replicate the DNA before isolation. This gene can then be isolated, or the bacteria can be induced to produce proteins such as insulin. **A** is incorrect because the restriction enzymes are used prior to insertion of the plasmid into bacteria. **C** is incorrect because bacteria replicate via binary fission. **D** is incorrect because bacteria are prokaryotic and do not contain a nucleus.

Question 19 A+ Study Notes pp.31–33

C Plate Y

The bacteria contain the recombinant plasmid and the arabinose to allow GFP protein expression. **A** is incorrect because the untransformed bacteria do not contain the plasmid with the GFP protein. **B** is incorrect because untransformed bacteria do not contain the plasmid with the GFP protein or the antibiotic resistance so would die on plate X. **D** is incorrect because, although the bacteria is transformed, it does not contain the arabinose for GFP expression.

Question 20 A+ Study Notes pp.31–33

D Plate X shows that ampicillin was effective in killing the untransformed bacteria.

The lack of the plasmid killed off the bacteria as they did not contain the gene for resistance. **A** is incorrect because the bacteria had not taken up the plasmid. **B** is incorrect because the bacteria were placed on a nutrient agar plate to act as a control group. **C** is incorrect because the nutrient agar plate was plate W. Plate X showed the bacterial sensitivity to ampicillin when the plasmid was not present.

Short-answer solutions

Question 1 A+ Study Notes pp.29–31

a Lane 1 is a standard/ladder. This contains fragments of DNA with known lengths to determine the length of the unknown DNA samples in the other wells. Additionally, it provides a baseline to ensure the gel has been on long enough to separate the DNA fragments but before the DNA reaches the end of the gel.

b Size of the DNA fragment: the larger the fragment, the longer it takes to migrate.

Agarose density/concentration/viscosity: the more concentrated the agarose, the longer it takes the DNA to migrate; the smaller the pores in the gel for DNA movement. (1 mark)

Voltage: the greater the electric current, the quicker the movement of the DNA. (1 mark)

Students received 1 mark for stating the two properties and 1 mark for explaining how it affects the movement. Students should not refer to the charge of the DNA or the time in the gel.

c The gel has been left on with the current for too long. (1 mark)

The standard in lane 1 should have four fragments and only three are present. Lane 3's shortest fragment is at the end of the gel. These results are inconclusive, and the gel would need to be rerun. (1 mark)

d The investigator had not added the restriction enzyme(s) into the DNA sample before running on the gel. Therefore, the sample would have been a long fragment and moved slowly through the agarose.

Question 2 A+ Study Notes pp.31–33

a A vector is a vehicle used to transmit a piece of genetic material from one host to another.

b The functional gene of interest (gene without the mutation) is isolated from the genome of a healthy individual. This gene is cut with a restriction enzyme at a specific recognition site on both ends outside of the gene to produce sticky ends. At the same time, a plasmid with one specific recognition site for the same restriction enzyme is cut, producing complementary sticky ends to the gene of interest. The plasmid and gene of interest are combined. (1 mark)

DNA ligase is added to the solution to re-form the phosphodiester bonds, forming a recombinant plasmid. (1 mark)

c   One of:
- A plasmid is more stable than a linear piece of DNA because nucleotides are not lost off the ends and the sequence is less likely to mutate.
- The DNA may insert into a section of the nuclear chromosomes, disrupting other gene function.

Question 3   A+ Study Notes   pp.33–34

a   As GMO crops increased from 2008 to 2020 within the community, mortality rates decreased from 800 to 100 deaths per 100 000 people.   (1 mark)

From 2016 to 2020, the deaths remained at 100 per 100 000 people despite a continued increase in GMO crops.   (1 mark)

Questions like this always require specific quotes and use of data. Look for any change in trends; is the data always decreasing, or increasing? Does it plateau?

b   Any one of: natural disaster, change in weather conditions, infectious disease

c   Transgenic organism has had their genetic material altered. (gene from a different species inserted into the organism)

A transgenic organism is a type of genetically modified organism (organism that has had its genetic material altered)

d   Selectable markers – antibiotic resistance

Gel electrophoresis – plasmid length before and after the insertion of the gene

e   One of:
- The rice strain from the local areas would be best suited to the environment in which it is growing.
- The farmers are used to growing the local strains and know how to care for the crop.

Question 4   A+ Study Notes   pp.28–31

a   **i**   Polymerase chain reaction.

ii   The DNA is heated to 95°C to denature the strands by breaking hydrogen bonds, then cooled to 55°C to allow the primers to anneal to the start and end of the DNA sample on both single strands of DNA. The temperature is then increased to 72°C (the optimal temperature of the Taq polymerase) to allow the synthesis of the new DNA strand.

b   Individual A does not have sickle cell anaemia. Their DNA has been cut by the restriction enzyme, resulting in two shorter strands of DNA rather than one large segment of DNA as seen in B.

c

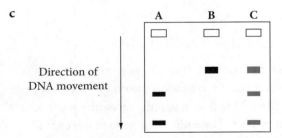

Direction of DNA movement

Your diagram needs to show the separation of DNA, the primers and the outcome. To achieve the whole mark, you must also show one round of the process (the three stages).

Question 5 A+ Study Notes pp.33–34

a A genetically modified organism is any organism that has had its DNA altered through genetic engineering techniques (1 mark), whereas a transgenic organism is one whose genome has been altered through the addition of DNA from another species (1 mark).

b

	Example	Organism modified	Transgenic or genetically modified	Reason
i	A rat with rabbit haemoglobin genes	Rat	Transgenic	The new gene comes from a rabbit to a rat.
ii	A human treated with insulin produced by *E. coli* bacteria	*E. coli*	Transgenic	The new gene comes from humans to *E. coli*.
iii	A rice plant that has had a rice gene knocked in to produce double the amount of rice protein	Rice	GMO	The new gene is a rice gene.

Question 6 A+ Study Notes pp.26–27

a Single guide RNA with the sequence of the faulty gene guides the Cas9 enzyme to the region of the non-functional cystic fibrosis gene. (1 mark)

The Cas9 enzyme cuts the DNA at this sequence and the functional cystic fibrosis gene can be inserted with the use of other manipulation tools. (1 mark)

b One of:
- Cas9 can cut at the non-desired gene and inhibit another gene from functioning.
- The inserted gene could be inserted into the wrong region or with addition of bases.
- There may be another similar sequence that the guide RNA recognises.

c The promoter region enables the binding of RNA polymerase for transcription.

d A: Repeat – identical sequences between variable regions to space out the viral genome segments (2 marks)

B: Spacer – variable regions formed after an infection by a virus (2 marks)

Question 7 A+ Study Notes pp.30–32

a quaternary (1 mark)

b Bacterial DNA does not contain introns (1 mark). Leaving introns in the genes will cause the genes not to be expressed. (1 mark)

c *lacZ* gene – Grow the bacterium on an agar plate. White colonies demonstrate the bacteria that have taken up the plasmid without the gene of interest (1 mark) whereas blue colonies demonstrate the transformed bacterium with the gene of interest. (1 mark)

d Ethical – religious or cultural objections due to the unnatural process or 'playing God', altering an organism's genetic composition; unforeseen consequences, concern about exploiting bacteria for own cause, influence on natural selection (1 mark)

Social – more accessible, cheaper, equitable, fewer side effects or allergic responses due to the protein being human based, constant reliable source, no animal welfare concerns, more jobs in the biotechnology sector. (1 mark)

Question 8   A+ Study Notes   pp.24–26

a   Cuts at specific recognition sequences through the phosphodiester bonds

To answer this question, you must be able to read a restriction enzyme cut and identify whether a blunt or a sticky end is produced.

b   i   TTA GAA TTC CCC GGG ACA **GENE OF INTEREST** CGA ATT CGT AAG CTT AAA AAG

AAT CTT AAG GCC CCC TGT **GENE OF INTEREST** GCT TAA GCA TTC GAA TTT TTC

Your answer needs to show three different designs of line for the enzymes. Students need to be able to identify areas in DNA that restriction enzymes will recognise – in some cases these sequences are back to front to add a level of difficulty.

ii

Restriction enzyme combination	Number of fragments
EcoRI and HindIII	4
HindIII	2

c   SmaI because it produces blunt ends

Blunt ends make re-forming bonds in genetic engineering more difficult because the DNA is less likely to combine.

d   EcoRI (1 mark) because it cuts the gene of interest twice outside of the gene coding area, whereas the other restriction enzymes only cut the strand once – the gene of interest is not isolated (1 mark).

As a general rule, the gene of interest requires two cuts on either side to isolate it from a linear piece of DNA; these restriction sites should occur in the flanking regions (upstream and downstream of the gene).

Question 9   A+ Study Notes   pp.29–31

a   The polymerase enzyme catalyses the production of a new strand of DNA or is involved in making multiple copies of DNA or amplification of DNA (1 mark), and DNA polymerase replicates the DNA by (one of the following) (1 mark):
- extending from the primer
- complementary base pairing
- using the original DNA as a template.

Some students incorrectly identified the enzyme and discussed the role of another enzyme. Many other responses gave one part of the expected answer. To avoid doing this in your own responses, it is important to remember to use the number of marks allocated to the question as an indication of the depth required in the answer.

b   Short tandem repeats are short non-coding sequences and differ in the number of repeats. They are more variable than genes as a mutation or duplication does not affect the organisms' functioning.

c   No; one STR is not adequate to make an identification.   (1 mark)

8–10 STRs are required for identification and a profile. Some individuals could have the same number of repeats at a particular region.   (1 mark)

d   Standard/DNA ladder.   (1 mark)

Known DNA fragment lengths that the unknown fragments can be compared to/ensure the gel has been run for the right amount of time.   (1 mark)

e A DNA fragment will move according to its charge. DNA is negatively charged and moves to the positive pole. (1 mark)

Molecular weight (size): smaller DNA fragments move further or faster than larger fragments. (1 mark)

f The DNA would migrate faster through the gel because of the decrease in viscosity.

Question 10 A+ Study Notes pp.26–27

a

Component	Function
Single guide RNA	Complementary to the target sequence; guides the Cas9 enzyme to the specific segment required for modification
Cas9 enzyme	Type of endonuclease that cuts the DNA at the sequence specified by the single guide RNA. Cuts through the phosphodiester bonds in the backbone of the DNA

b Identify the sequences of the form of SWEET genes that the bacteria activate. (1 mark)

Form a guide RNA for the sequence identified. (1 mark)

Insert the guide RNA Cas9 complex into the cells. (1 mark)

One of (1 mark):

- The complex cuts the SWEET genes and knocks them out, inhibiting the ability of the bacteria from activating the non-functional genes; therefore, no blight.
- Adding a form of the gene (knock in – more difficult and requires additional tools) that is not activated through the bacterial transcriptional factors; therefore, no blight.

Solutions to Test 3: Regulation of biochemical pathways in photosynthesis and cellular respiration [U3 Topic 1.2.1]

Multiple-choice solutions

Question 1 [A+ Study Notes] [p.60]

C Amino acids.

These are the monomers of a polypeptide chain and are proteins. Enzymes are types of proteins. **A** is incorrect because glucose molecules (monomers) combine to form complex carbohydrates. **B** is incorrect because lipids form fats. **D** is incorrect because nucleic acids (DNA and RNA) are composed of nucleotides and carry genetic information.

Question 2 [A+ Study Notes] [pp.61–62]

C Concentration of enzyme.

The maximum concentration of product is reached in a shorter period of time due to more active sites binding with the substrate. **A** and B are incorrect because the change in the substrate (or reactant) would change the concentration of product that is produced. The more substrate (or reactant), the greater the concentration of product. **D** is incorrect because a graph of pH would show a maximum concentration of product at an optimal pH with a decrease in product either side as the enzyme denatures.

Question 3 [A+ Study Notes] [pp.61–62]

B After 14 minutes, all of the substrate was consumed.

The amount of product remains constant, suggesting that no more reaction is occurring. **A** is incorrect because enzymes act as catalysts, which reduce activation energy but are not consumed in a reaction. **C** is incorrect because the enzyme's active site is specific to the substrate, not the product. **D** is incorrect because the enzyme facilitates the reaction from substrate to product.

Question 4 [A+ Study Notes] [pp.69–70]

B Have high concentrations of lactic acid in its muscle cells.

The lack of oxygen would cause the turtle to undergo anaerobic respiration. **A** is incorrect because when the turtle is submerged, there is no access to oxygen from the air so O_2 levels would be low. **C** is incorrect because high activity would require large amounts of ATP, which is hard to produce in anaerobic respiration. **D** is incorrect because the muscles cannot store a large amount of ATP reserve. ADP + P_i would be in limited amounts.

Question 5 [A+ Study Notes] [p.60]

C the reactant in this reaction is A.

A is the molecule that enters the reaction to be acted upon by enzymes. **A** is incorrect because B is the product of enzyme 1. **B** is incorrect because if enzyme 3 could not function then it would not produce Product D. **D** is incorrect because the reaction would still occur slowly, enzymes only speed up the rate of the reaction.

Question 6 [A+ Study Notes] [pp.67–68]

B A molecule of glucose in aerobic respiration.

A large amount of energy is released from the glucose molecule (30–32ATP) throughout the three stages of aerobic respiration. **A** is incorrect because the energy content would increase from the reactants to the glucose product. **C** is incorrect because only a small amount of energy is lost from the glucose in the formation of pyruvate/lactic acid or ethanol during anaerobic respiration. **D** is incorrect because protein synthesis requires an input of energy for the bond formation between amino acids.

Question 7 A+ Study Notes pp.61–62

B Has the same amino acid sequence as before.

The amino acid sequence formed in translation (primary structure) has not altered, but the interactions between these amino acids have. **A** is incorrect because if the shape were the same as the original, it would still be complementary to the substrate and act as a catalyst for the reaction. **C** is incorrect because when a protein is exposed to high temperatures, the conformational shape change is irreversible. Proteins can only return to functioning after they are exposed to cold temperatures. **D** is incorrect because the number of amino acids (primary structure) has not altered, just the interactions between the amino acids in the tertiary structure.

Question 8 A+ Study Notes pp.67–69

D Cellular respiration.

NADH and $FADH_2$ are produced during the Krebs cycle and release their electrons in the electron transport chain where 32–34ATP is produced. **A** is incorrect because NADPH is the electron carrier used in photosynthesis. **B** and **C** are incorrect because there is no electron carrier involved in these processes.

Question 9 A+ Study Notes p.62

C the product is released from the active site.

In both models the reaction occurs at the active site and the product is released. **A** is incorrect as only the lock-and-key model proposes this. **B** is incorrect as only the induced-fit model proposes this. **D** is incorrect as the enzyme is not altered by the reaction and can be reused.

Question 10 A+ Study Notes pp.61–62

D The heat-tolerant bacteria are unlikely to survive in water at 100°C.

100°C is above the heat-tolerant bacteria's enzyme tolerance zone and therefore the enzymes would have denatured and no reactions could take place. **A** is incorrect because enzymes only denature above their optimal temperature. Below their optimal temperature, the rate of collisions decreases owing to the molecules having less kinetic energy. **B** is incorrect because the enzymes only function between 40°C and 90°C. **C** is incorrect because if an enzyme is heated above its optimal temperature, it will be denatured, resulting in irreversible conformational shape change. Cooling does not change the shape back.

Question 11 A+ Study Notes pp.62–63

A Have a chemical structure complementary and specific to enzyme's active site.

The molecule must be specific in shape to the active site to temporarily block the substrate from binding. **B** is incorrect because the structure could be different as long as it has a region that is complementary to the enzymes active site. **C** is incorrect because the amino acid structure is not altered with the binding of a competitive or non-competitive inhibitor. **D** is incorrect because the molecules only enter into the active site temporarily to block the substrate from binding.

Question 12 A+ Study Notes pp.63–65

C Sunlight

Sunlight forms the first ATP during the light-dependent reaction, which allows the formation of glucose. Glucose is later broken down in cellular respiration to produce more ATP. **A** is incorrect because ADP combines to form the high-energy ATP compound. **B** is incorrect because glucose cannot be formed without ATP during the Calvin cycle. **D** is incorrect because the phosphate is the molecule that binds with the ADP to form ATP.

Question 13 `A+ Study Notes` `pp.63–66`

A   A is with no inhibitor, B is with a competitive inhibitor and C is with a non-competitive inhibitor.

The fastest reaction is without any inhibitor (A); as substrate increases, the competitive inhibitor (B) becomes less effective; and with a non-competitive inhibitor, the rate is always lower because it is binding to a region other than the active site (C). **B** is incorrect because above optimal temperature would cause the enzymes to denature and the rate of reaction would be low to none. **C** is incorrect because all enzymes must be specific and complementary to the substrate to facilitate the reaction. **D** is incorrect because a plant in shade will always have a lower rate of photosynthesis owing to lower light intensity and the rate of photosynthesis would be 0 in the dark.

Question 14 `A+ Study Notes` `pp.61–62`

B   Substrate concentration.

As the substrate concentration increased in each experiment, the rate continued to increase. **A** is incorrect because temperature remained constant throughout the experiment. **C** is incorrect because enzyme concentration would cause the graph to plateau as all active sites would be occupied. This is called saturation. **D** is incorrect because product saturation would not limit a rate of reaction unless it became an inhibitor for the enzyme.

Question 15 `A+ Study Notes` `pp.22–23; 60`

D   Energy is released in reaction P only.

When reading the graphs it was important for students to identify the energy levels of the reactants and the products. In reaction M, energy has been required for the reaction to proceed and is endothermic. In reaction P, the energy level of the products is lower, so energy has been released and it is therefore exothermic. **A** is incorrect because the products have less energy than reactants – release of energy to the surroundings. **B** is incorrect because activation energy is from the products to top in M and from energy of reactants to top in reaction P. **C** is incorrect because reaction M is an endothermic reaction and Reaction P is an exothermic reaction.

Question 16 `A+ Study Notes` `pp.22–23; 60`

C   Is the energy required to start the reaction.

Activation energy is the energy required for bonds to be broken or re-formed. **A** is incorrect because activation energy is reduced in the presence of an enzyme because the enzyme acts as a catalyst. **B** is incorrect because activation energy is not changed with temperature. **D** is incorrect because activation energy is required for any endothermic or exothermic reaction.

Question 17 `A+ Study Notes` `pp.61–62`

C   At pH 3 and a temperature of 37°C, the active site of enzyme W binds well with its substrate.

A is incorrect because below their optimal temperature the reactions rate of enzymes decreases but the enzymes not denature until above the optimal temperature. **B** is incorrect because human enzymes function optimally at body temperature (35–37°C). **D** is incorrect because enzyme X functions optimally at pH 7.

Question 18 `A+ Study Notes` `pp.63–65`

B   Requires an overall input of energy.

Energy is required through light or movement of ions at ATP synthase to produce ATP from ADP + P_i. **A** is incorrect because the anabolic reaction is forming a larger molecule from two smaller molecules. **C** is incorrect because ATP can be formed under aerobic or anaerobic conditions. **D** is incorrect because ATP is formed in photosynthesis and in the cytosol of the cell (glycolysis).

Question 19 A+ Study Notes pp.63–69

D Uses water as a reactant in the first stage. Forms water as a product in the final stage.

> Water is split by light energy in the light-dependent reaction of photosynthesis. Water is formed with the electrons and oxygen at the end of the electron transport chain. **A** is incorrect because photosynthesis requires energy, so is endergonic, and cellular respiration is exergonic. **B** is incorrect because glycolysis occurs in the cytosol of the cell in aerobic respiration. **C** is incorrect because electron transport is involved in the light-dependent reaction of photosynthesis and the third stage of respiration.

Question 20 A+ Study Notes pp.61–62

D

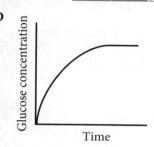

> The rate would increase and then plateau as all of the substrate was consumed/reacted. **A** is incorrect because the glucose concentration would only continually increase if there was an unlimited supply of maltose substrate. **B** is incorrect because glucose is the product, so the rate would not drop over time. **C** is incorrect because the glucose concentration would only increase if there was an unlimited supply of maltose substrate.

Short-answer questions

Question 1 A+ Study Notes pp.61–62

a

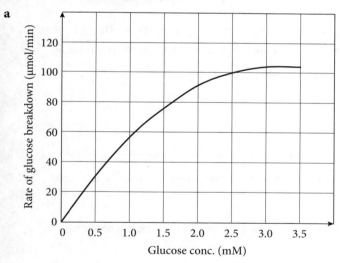

b **i** Enzyme and water

   ii Tube 1 was the control group, which lacks the independent variable and acts as a baseline with which to compare the experimental groups.

> It contained the same quantity of the enzyme as the other tubes, the same volume and was kept at the same temperature. However, it did not contain glucose, which was the variable being investigated.

c One of:
- At high glucose concentrations, the rate of reaction is limited by the availability of the enzyme.
- The enzyme concentration is the limiting factor at high glucose concentrations.

d **i** Temperature or pH

   ii Temperature: The rate of chemical reactions generally increases with increasing temperature. For enzyme-catalysed reactions, this rate increases to a point and then rapidly declines at higher temperatures as enzymes are denatured.

    pH: Enzymes work most efficiently at an optimum pH. There may be some activity at varying pH levels, but enzymes are denatured if the pH is too far from the optimum.

Question 2   A+ Study Notes   pp.67–68

a   Glucose + oxygen $\longrightarrow$ carbon dioxide + water OR $C_6H_{12}O_6 + 6O_2 \longrightarrow 6H_2O + 6CO_2$

> A significant number of students got this question wrong during the exam. Remember to always read the stem of the question to ensure you are providing the correct type of equation – word or chemical. If the chemical equation is provided, it must be balanced. Energy/ATP was not required to gain the mark.

b   One of (2 marks):
- The electron transport chain would be unable to provide larger amounts of ATP.
- There would be insufficient energy available to maintain life.

> Most students answered this question incorrectly. An answer such as 'There would be no energy available to the cell' was incorrect because glycolysis, anaerobic respiration and the Kreb's cycle would produce some ATP. Writing 'this would result in death of the cell' was simply repeating the stem of the question.

c   The cyanide needs to be specific and complementary to the enzymes involved in the reactions.   (1 mark)

    If the molecules do not have a complementary shape, the cyanide will have no effect on the species.   (1 mark)

Question 3   A+ Study Notes   pp.66–67

a   A biochemical pathway is a series of reactions catalysed by enzymes that produce a final product through intermediate products, which become the reactants in the next step of the pathway.

> Some products can then become inhibitors for enzymes earlier in the pathway to control the rate of the reactions.

b   Cytosol of the cell

c   Glucose-6-phosphate would build up and there would be a depletion of fructose-6-phosphate. (1 mark)

    Pyruvate would be depleted within the cell, preventing aerobic and anaerobic respiration from occurring.   (1 mark)

> Your answer should identify the implications for the specific reaction and for the whole process of respiration, i.e. the effect on cell function.

d **i** Non-competitive inhibition

> Non-competitive inhibition occurs when a molecule binds onto the allosteric site of the enzyme to alter the enzyme's active site, preventing the substrate from binding.

   ii

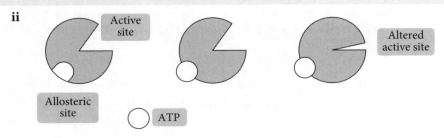

iii One of:
  • Through gene regulation – transcriptional factors preventing the transcription of the mRNA for the enzymes.
  • Through wrapping of DNA around histones, preventing RNA polymerase from binding and transcribing.
  • Break down of mRNA in the cytosol of the cell – no translation.

You would receive 1 mark for stating the inhibitor, and 1 mark for explaining the regulation.

e Provides a usable energy source for the cell and the reactions that take place.

Question 4 A+ Study Notes pp.13–15; 18–20; 61–62

a Amino acids

Your answer should demonstrate that you are able to recognise monomers and polymers of nucleic acids and proteins. A monomer is the smallest functional unit, whereas a polymer is many of the monomers joined together.

b Two of:
  • Ribosome: Translation – formation of the primary structure through adjacent amino acids forming peptide bonds. Eventually forming a polypeptide chain.
  • Endoplasmic reticulum: Transport of the polypeptide chain and the folding of the enzyme.
  • Golgi apparatus: Modification and folding of the protein, packaging into a vesicle.
  • Vesicle: Transports the protein to the membrane where it fuses with the membrane during exocytosis to release the enzyme into the extracellular fluid.

Ensure the functions of the two organelles mentioned in your answer are different. Referring to the ribosome for synthesis of the protein and the rough endoplasmic reticulum for the synthesis of proteins is the same concept and would not receive the full marks.

c One of:
  • Enzyme denatured during the transfer from the mouse to the human – irreversible conformational shape change.
  • No substrate complementary and specific to the enzyme in the human cell.

Question 5 A+ Study Notes pp.67–68

a Energy is required (1 mark) to add a free inorganic phosphate onto ADP   (1 mark)

b Glycolysis 2ATP, Krebs cycle 2ATP, electron transport chain 26 or 28ATP.   (2 marks)

You would receive 1 mark for naming each stage and 1 mark for the ATP amounts.

c Oxygen is required to bind to the electrons and hydrogen ions to form water in the electron transport chain. The more oxygen, the greater the rate of the Krebs cycle and the electron transport chain.

d More ATP in the muscle cell than in the skin cell because it is a more active cell.   (1 mark)

  There would be more mitochondria in the muscle cell than in the skin cell.   (1 mark)

Question 6 A+ Study Notes pp.61–62

a Concentration, amount of hydrogen peroxide, amount of substrate   (1 mark)

b 10°C. Low kinetic energy in the molecules. Less enzyme substrate collisions. Lower rate of reaction.   (1 mark)

  50°C. Enzyme denatures. Irreversible conformational shape change. Reaction rate would decline to zero.   (1 mark) c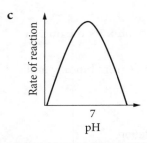

1 mark for the right shape. 1 mark for the labels and pH 7.

Question 7 A+ Study Notes pp.61–62; 69

a

	CO$_2$ level (increase, decrease or constant)	O$_2$ level (increase, decrease or constant)	Glucose level (increase, decrease or constant)
Rate of photosynthesis is greater than rate of cellular respiration	Decrease	Increase	Increase
Rate of photosynthesis equal to rate of cellular respiration	Constant	Constant	Constant
Plant exposed to 0 light intensity	Increase	Decrease	Decrease

b The reactants of one reaction are the products of another reaction. (1 mark)

CO$_2$ is the output of cellular respiration and input for photosynthesis. O$_2$ is the output of photosynthesis and an input for cellular respiration. (1 mark)

c As temperature increases to the enzyme's optimal temperature, the average kinetic energy of the molecules increases. Therefore, the number of substrate–enzyme collisions increases, increasing the rate of reactions. (1 mark)

Above their optimum temperature, enzymes denature, resulting in an irreversible change in conformational shape. The reaction rate would lower. (1 mark)

Photosynthesis and cellular respiration are catalysed by enzymes. Students commonly overlooked the influence of enzymes in these processes when discussing temperature effect on photosynthesis in previous VCAA exams.

Question 8 A+ Study Notes pp. 61–63

a Acts as a catalyst to lower the activation energy of reactions and therefore increase the rate of reactions.

b Two of:
  • pH – Any variation in pH can slow the rate of reaction as the enzymes begin to denature.
  • Temperature – Enzymes have an optimal temperature. Low temperature can reduce the rate of reaction, high temperature can denature the enzymes.
  • Substrate concentration – Without substrate the reactions cannot occur.

Students often overlook the function of enzymes in photosynthesis and cellular respiration. You must be clear when discussing temperature, which causes denaturation (high) and slows collisions (low).

c Competitive and non-competitive inhibitors. (1 mark)

A competitive inhibitor enters the active site of the enzyme and competes with the substrate, whereas anon-competitive inhibitor binds onto the allosteric site and alters the active site of the enzyme, preventing the substrate from binding. (1 mark)

To achieve full marks, your answer needs to include a comparative term, such as 'whereas' or 'compared to'.

d Independent variable, dependent variable and controlled variable. (1 mark)

Experimental group and control group (1 mark)

Experimental set-up (1 mark)

Results to support competitive vs non-competitive (1 mark)

Example: Independent variable – changing concentrations of substrate. Dependent variable – rate of CO_2 produced. Controlled variables – Temperature, pH, same plant species.

Control group: no pesticide. Experimental group: 100 mL glucose, 200 mL glucose, 300 mL glucose

Experimental set-up: 100 g muscle cells in each group in a sealed container, constant environment. Expose all experimental groups to 50 mL pesticide.

Test the CO_2 concentration over a 24-hour period. Observe the rate of CO_2.

Results: If more glucose (substrate) results in more CO_2, the reaction is increasing, and the inhibitor is competitive. If the rate remains at the same rate, the inhibitor is non-competitive.

> You should be aware that non-competitive inhibitors will always have a lower rate of reaction, whereas competitive inhibitors are less effective as the concentration of substrate increases. The marking guide for experimental design changes depending on the VCAA assessors. A hypothesis can be included if you want to include one, but does not receive marks.

Question 9 A+ Study Notes pp.68–69

a

Molecule	Name
Process 1	Photosynthesis
Molecules A and B	Glucose and oxygen
Molecule C	ATP or H_2O

b NADH and $FADH_2$ (coenzymes) are formed during the Krebs cycle through loading of electrons. (1 mark)

These coenzymes release their electrons into the cristae of the mitochondria, which move through and facilitate 26 or 28ATP forming. (1 mark)

> Process 2 is cellular respiration.

c Light energy (1 mark)

Light energy splits water molecules in the electron transport chain to release the electrons into the thylakoid membrane. (1 mark)

d Oxygen inhaled and diffused through the phospholipid bilayer from the blood/extracellular fluid. (1 mark)

Glucose digested from food, which enters into the cell through facilitated diffusion. (1 mark)

Question 10 A+ Study Notes pp.66–67; 194–195

a Same amount of water

Same species of tomato

Same age and size of tomato plant

> Avoid giving vague answers or equipment, such as test tubes' as the controlled variables. The controlled variables should be specific to the experiment and units should be used where appropriate.

b Experiment 1 – Increase the number of experimental groups.

Experiment 3 – There was no control group. The experiment could be improved by including a chamber that lacked all light as the control group, as well as providing more specific light intensity.

None of the experiments are repeated, which is important to form an average and prevent outliers.

c The higher the temperature, the greater the amount of oxygen produced through an increase in photosynthesis.

d The rate of photosynthesis increases as the light intensity increases. (1 mark)

As the light intensity increases, the amount of light hitting the chlorophyll pigments increased. The more water molecules split, the greater the rate of ATP and NADPH produced, meaning that more glucose was formed. (1 mark)

e As the pH of the environment changes, the rate of photosynthesis would alter. (1 mark)

If the pH is not in the enzyme's range of tolerance, the enzyme undergoes an irreversible conformational shape change and denatures. (1 mark)

Solutions to Test 4: Photosynthesis as an example of a biochemical pathway U3 Topic 2.2.2

Multiple-choice solutions

Question 1 A+ Study Notes pp.60–61

D $NADP^+ + 2e^- + 2H^+ \rightarrow NADPH + H^+$

Two electrons and two positive charges balance out the NADPH, leaving a single H^+ ion. **A** is incorrect because $NADP^+$ is the oxidised form. **B** is incorrect because NADH is a different type of coenzyme used during cellular respiration, whereas NADPH is used during photosynthesis. **C** is incorrect because the reaction is not balanced.

Question 2 A+ Study Notes pp.61–62; 64

C Amount of oxygen in the cell.

Oxygen is the output of the light-dependent reaction. **A** is incorrect because the Calvin cycle is catalysed by enzymes. The lower the temperature, the lower the number of substrate–enzyme collisions. **B** is incorrect because carbon dioxide is the input for the Calvin cycle. The less substrate there is, the slower the rate of reaction. **D** is incorrect because light initiates the splitting of H_2O in the light-dependent stage. The less light there is, the fewer electrons are released into the membrane and less ATP and NADPH produced for the Calvin cycle.

Question 3 A+ Study Notes pp.63–64

C O_2 is a product of the reaction that occurs at A.

A shows the site of the light-dependent stage of photosynthesis. The input is water, NADP+, ADP and Pi and light, the products are oxygen NADPH and ATP. **A** is incorrect because B is the site of the light-independent stage. **B** is incorrect as CO_2 is an input of this reaction. **D** is incorrect as ATP is a product of this reaction.

Question 4 A+ Study Notes p.64

B ATP

ATP is produced during the light-dependent reaction and used in the cyclic set of reactions in the light-independent stage of photosynthesis. **A** is incorrect because CO_2 diffuses in through the stomata of the plant and is the input for the Calvin cycle. **C** is incorrect because H_2O is the input for the light-dependent reaction. **D** is incorrect because O_2 is the by-product/waste product, which diffuses out of the cell through the stomata from the light-dependent reaction (produced when water splits).

Question 5 A+ Study Notes p.64

B It relies on the products of the light-dependent reactions.

The ATP and NADPH produced. **A** is incorrect because the light-independent reactions do not directly depend on the light to undergo the reactions. **C** is incorrect because collisions between enzyme and substrate would slow but not stop. **D** is incorrect because water enters the plants from the roots, and leaves through the stomata.

Question 6 A+ Study Notes p.64

A CO_2.

CO_2 is the input for the Calvin cycle, a series of reactions that produces glucose ($C_6H_{12}O_6$). **B** is incorrect because ATP does not contain oxygen. It provides energy for the Calvin cycle. **C** is incorrect because H_2O is the input for the light-dependent reaction and is broken into O_2, which is released from the plant, and H^+ ions, which bind to $NADP^+$. **D** is incorrect because O_3 is not involved in the photosynthesis reaction.

Question 7 `A+ Study Notes` `p.69`

C Oxygen produced by photosynthesis is equal to the oxygen used by aerobic respiration.

$6O_2$ are produced in photosynthesis and $6O_2$ are used in cellular respiration. **A** is incorrect because photosynthesis would be occurring with the light intensity at 10 AU. If no photosynthesis was occurring, there would be an overall uptake of oxygen. **B** is incorrect because if aerobic respiration was not occurring, there would be an overall release of oxygen. **D** is incorrect because the inputs of one are outputs of the other.

Question 8 `A+ Study Notes` `p.64`

B Carbon dioxide concentration.

Carbon dioxide is an input for the light-independent stage of photosynthesis. **A** is incorrect because photosynthesis is occurring at a greater rate than cellular respiration and glucose is an output of photosynthesis. **C** is incorrect because the rate is continually increasing with intensity even though the rate plateaus at around 40 AU of light. Carbon dioxide is only the limiting factor prior to it plateauing. **D** is incorrect because photosynthesis is occurring at a greater rate than cellular respiration and oxygen is an output of the light-dependent stage of photosynthesis.

Question 9 `A+ Study Notes` `pp.65–66`

B CO_2 is absorbed at night and converted into a storage molecule for use during the day.

CO_2 is converted into a 4-carbon compound during the night to allow stomata to remain closed during the day, resulting in lower level of water loss. **A** is incorrect because CAM plants are located in hot arid conditions so would lose too much water with stomata open in the middle of the day. **C** is incorrect because light energy cannot be stored within a plant. **D** is incorrect because the 3-carbon compound is used in the C3 plants and cannot be stored.

Question 10 `A+ Study Notes` `pp.65–66`

C Binding to the active site of Rubisco enzyme.

Oxygen can become a substrate for Rubisco and undergo photorespiration. **A** is incorrect because water is split through the light energy being absorbed by the chlorophyll pigments in the thylakoid membrane and oxygen is produced. **B** is incorrect because NADPH does not contain any oxygen and is formed on the opposite side of the membrane to the oxygen formation. **D** is incorrect because denaturation occurs due to high temperatures (above optimal) and extreme pH levels.

Question 11 `A+ Study Notes` `p.69`

C Input of CO_2 and release of O_2 during photosynthesis.

CO_2 is an input for the light-independent stage of photosynthesis and O_2 is the output. **A** is incorrect because oxygen is the output of photosynthesis and is released into the air. Carbon dioxide is an input for photosynthesis. **B** is incorrect because although water is lost from stomata and CO_2 is an input, O_2 is not an input; it is an output. **D** is incorrect because water is absorbed by the roots and lost through the stomata.

Question 12 `A+ Study Notes` `pp.65–66`

C Hot and dry conditions.

Hot and dry conditions result in low water availability for the plant and lead to loss of water by transpiration when the stomata open due to the heat and sun. **A** is incorrect because damp conditions would result in the plant having adequate water and loss from the stomata would not have a large impact on survival. **B** is incorrect because transpiration would be lower under cold conditions than under hot conditions. **D** is incorrect because humid conditions would reduce the transpiration rate of water loss because there is a smaller difference in water concentration between the plant and the air.

Question 13   A+ Study Notes pp.65–66

B   Separating the reactions between the mesophyll and bundle sheath cells.

By separating the reactions, C4 plants can reduce the effect of oxygen on the enzyme Rubisco and the rate of photorespiration. **A** is incorrect because their stomata would be open during the evening and night to allow gas exchange and would not open at peak temperature. **C** is incorrect because C3 plants form the 3-carbon compound. C4 and CAM plants produce a 4-carbon compound from CO_2 when it enters the plant. **D** is incorrect because photorespiration is when Rubisco uses oxygen as a substrate and produces a molecule that cannot be used to make glucose.

Question 14   A+ Study Notes p.65

A   Plant 2 will be more likely to survive in a dry environment than plant 1.

B is incorrect because the water loss is the same – the lines cross at the same point. **C** is incorrect because the water loss is greater in plant 2, indicating greater stomatal aperture than in plant 1. **D** is incorrect because photosynthesis requires CO_2 as an input for the light-dependent stage. If the stomata were open for this CO_2 to enter, there would be water lost from the plant.

Question 15   A+ Study Notes pp.63–65

D   Carbon dioxide is the reactant for the light-independent reaction, which occurs in the stroma.

Carbon dioxide is the substrate for the cyclic set of reactions in the Calvin cycle. The enzymes for this reaction are located in the stroma (fluid) of the chloroplast. **A** is incorrect because water is the input for the light-dependent reaction where it is split to release electrons into the thylakoid membrane and forms oxygen. **B** is incorrect because oxygen is the output of the light-dependent reaction on the thylakoid membrane. **C** is incorrect because glucose is formed in the light-independent reaction that occurs in the stroma of the chloroplast.

Question 16   A+ Study Notes p.65

D   The amount of carbon dioxide entering the leaf decreased.

The wilting of the leaf would have caused the stomata to close to reserve levels of water and reduce the level of transpiration; therefore, CO_2 could not enter to be a reactant for photosynthesis. **A** is incorrect because if enzymes denatured the cell would not be able to recover and produce ATP or more enzymes before toxic substances had built up. **B** is incorrect because the pigments are not susceptible to heat. **C** is incorrect because as the plant was not moved from the original location, the light intensity was not altered.

Question 17   A+ Study Notes p.64

A   12; 12

As there are 12 units of NADP+ and ADP and Pi produced as outputs there must be 12 units of each as inputs. **B**, **C** and **D** are all incorrect as there must be 12 units of each.

Question 18   A+ Study Notes p.64

B   In the grana during the light-dependent reaction.

The water is split when light hits the chlorophyll pigments in the membranes of the grana. **A** is incorrect because the light-independent reaction has CO_2 as an input. **C** is incorrect because the thylakoid membrane is where the light-dependent reaction occurs. **D** is incorrect because the outer membrane of the chloroplast does not contain the chlorophyll pigments for the light energy to split the water. It is only located in the inner thylakoid membrane.

Question 19 A+ Study Notes pp.63–65

D The rate of the light-independent reaction in the stroma increases with the increase in CO_2 level.

> CO_2 is an input for the Calvin cycle, the more substrate the greater the enzyme substrate collisions. **A** is incorrect because the light-independent reaction occurs in the stroma of the chloroplast. **B** is incorrect because the more water that is lost from the stomata, the less is available for the light-dependent reaction, slowing the overall rate of photosynthesis. **C** is incorrect because a change of pH away from the optimal will slow the rate of photosynthesis; the greater the change in pH, the lower the rate as enzymes involved in the Calvin cycle begin to denature.

Question 20 A+ Study Notes p.66

C Other wavelengths from the visible light spectrum and reflect green wavelengths.

> Blue and red light have the highest rate of absorption. Plants appear green because they reflect this wavelength and do not absorb it. **A** is incorrect because the plant will reflect the wavelength that is not absorbed. **B** is incorrect because different wavelengths are absorbed more easily by the chlorophyll pigments. Blue and red are the optimal wavelengths for chlorophyll pigments. **D** is incorrect because the plant could not photosynthesise without absorbing some wavelengths.

Short-answer solutions

Question 1 A+ Study Notes pp.63–65

a

Label	Name	Role
A	H_2O (12)	Input for the light-dependent reaction Provides electrons and hydrogen ions
F	$ADP + P_i$ and $NADP^+$	Inputs for the light-dependent reaction. Provides energy and H^+ ions and electrons for the Calvin cycle, Stage 2
G	Light energy	Provides the energy to split the H_2O

b Calvin cycle   (1 mark)

Cyclic set of reactions catalysed by enzymes and facilitated by ATP (energy source) and NADPH (electron/hydrogen source) to form $C_6H_{12}O_6$ (glucose) from $6CO_2$.   (2 marks)

> You must discuss all inputs and outputs in your answer. Always use chemical formulas with the amounts.

c Low temperature will affect the rate of substrate enzyme collisions in the Calvin cycle. The lower the temperature, the lower the average kinetic energy of the molecules, (1 mark) slowing down the rate at which ATP and NADPH are broken down and the $ADP + P_i$ and $NADP^+$ are returned for the light-dependent reaction. This slows the overall rate of photosynthesis (1 mark).

> This question is asking specifically about the rate of the reactions. You must demonstrate a link with the substrates.

Question 2 A+ Study Notes pp.65–66

a Rubisco enzyme

b One of:
- C3 plants fix the carbon dioxide directly into a 3-carbon compound to use for the Calvin cycle, whereas C4 and CAM plants produce a 4-carbon compound from the carbon dioxide that enters the plant, which has a greater affinity to Rubisco than oxygen.
- The C4 and CAM plants are located in hot, dry environments where stomata need to remain closed to prevent water loss, whereas C3 plants do not face this pressure.

> To achieve full marks, your answer needs to include a comparative term, such as 'whereas' or 'compared to'.

c Oxygen functions as a competitive inhibitor to the Rubisco enzyme and can react during photorespiration with Rubisco. (1 mark)

An increase in the substrate CO_2 would decrease the effect of the oxygen on the enzyme, increasing the rate of photosynthesis and decreasing the rate of photorespiration. (1 mark)

d CAM plants are better adapted to a hot temperature because they are able to keep their stomata closed during the day to prevent water loss while still undergoing photosynthesis as the carbon dioxide is transformed into a 4-carbon compound, which has a high affinity to the Rubisco enzyme than oxygen.

e Photorespiration is where Rubisco uses oxygen and ATP to produce a molecule from the fixed carbon dioxide into a compound that cannot be used in the Calvin cycle, whereas photosynthesis is where Rubisco binds with carbon dioxide and the Calvin cycle produces glucose.

> To achieve full marks, your answer needs to include a comparative term, such as 'whereas' or 'compared to'.

Question 3 A+ Study Notes pp.61–64

a Photosynthesis

b Carbon dioxide + water $\xrightarrow{\text{light energy, chlorophyll}}$ Sugar + oxygen

> You must include light energy and chlorophyll above the arrow to receive a mark.

c The student's conclusion is not supported. (1 mark)

Although the rate of cellular respiration would be greater than the rate of photosynthesis due to light intensity acting as a limiting factor, photosynthesis would still be occurring. (1 mark)

d Between 50 and 80 units, the rate of photosynthesis and cellular respiration remained constant because of another limiting factor, such as CO_2 input.

> It could be any limiting factor, except for light intensity.

Question 4 A+ Study Notes pp.64–67

a **i A** Thylakoid membrane or grana; light-dependent reaction

  ii B Stroma; light-independent, Calvin cycle

b **i** First stage: $6O_2$, ATP and NADPH

  ii Second stage: $6H_2O$, $C_6H_{12}O_6$ ADP +P_i and $NADP^+$

Question 5 A+ Study Notes pp.64–67

a

Experiment number	Carbon dioxide	Oxygen
1	Decrease	Increase
2	Increase	Decrease
3	Decrease	Increase

> There would only be a very small amount of photosynthesis occurring, but, as the stem of the question states, the chlorophyll absorbs all wavelengths.

b

Experiment number	Glucose level	Reasoning
1	Increase	Photosynthesis (glucose an output) would be occurring at a greater rate than cellular respiration (glucose an input)
2	Decrease	Cellular respiration would be occurring at a greater rate than photosynthesis

c The Calvin cycle is a series of reactions catalysed by enzymes. A change in pH beyond the enzyme's range of tolerance can reduce the rate of photosynthesis. (1 mark)

The enzyme's active site would begin to denature, causing a conformational shape change and become unable to bind to the substrate. (1 mark)

Question 6 [A+ Study Notes pp.67–69]

a Glucose or $C_6H_{12}O_6$

b Glycolysis: Glucose converted to pyruvate, 2ATP produced. (1 mark)

Krebs cycle: Pyruvate converted to carbon dioxide, 2ATP produced and electron carriers. (1 mark)

Electron transport chain: Electron carriers release electrons into the membrane. Hydrogen combines with oxygen to produce water, 26 or 28ATP produced. (1 mark)

> The question required a brief description of three stages of cellular respiration. Answers could also include the inputs or outputs of each stage, where each stage occurred or the amount of ATP produced. Many students described different stages of photosynthesis and did not gain any marks. Some students wrote large amounts of information for each part; this was unnecessary and wasted valuable examination time. Some students gave contradictory information.

c The oxygen produced in photosynthesis can be used in cellular respiration.

> Few students were able to answer this question correctly.

Question 7 [A+ Study Notes pp.64; 66]

a As the absorption increases, the rate of photosynthesis increases. At wavelengths with less absorption, the rate of photosynthesis also decreases.

> You must refer to both the high and low rates. Always refer to specific data points and the units used.

b Light energy is absorbed by the chlorophyll and carotenoid pigments embedded within the thylakoid membrane. (1 mark) This provides energy to split water (1 mark).

c The assumption is not correct. Even though blue light shows the greatest rate of photosynthesis compared with other colours of light, blue light is greatest is at 425 nm and not 450 nm. (1 mark)

White light would provide a greater rate of photosynthesis because it contains all wavelengths of light. (1 mark)

Question 8 [A+ Study Notes p.64]

a Oxygen is increased because photosynthesis is occurring at a high rate. Oxygen is an output of the light-dependent stage of photosynthesis.

b i

ii No photosynthesis could occur in the dark (1 mark) and any oxygen present would be used as an input for cellular respiration (1 mark) – decreasing the partial pressure between X and Y.

Question 9 [A+ Study Notes pp.194–195]

a To act as a control group (negative control), act as a baseline that lacks the independent variable (light intensity)

> Always write 'control group' or 'controlled variable'. 'Control' is not clear enough and will not receive marks.

b Light is absorbed by chlorophyll pigments and provides energy to split water into H^+ ions, $\frac{1}{2}O_2$ and release an electron into the membrane. (1 mark)

Low light intensity slows the rate of photosynthesis as the ATP and NADPH in the light-dependent reaction is reduced and this is the energy source that is required for the light-independent reaction. (1 mark)

c Any set of results that show the greater the light intensity, the greater the number of discs floating to the top in a shorter amount of time. All discs in the dark must remain at the bottom. (1 mark)

Must use units (time). (1 mark)

d Weakness: One of (1 mark):
- small sample size in each light intensity
- not repeated
- distance from light not determined.

Improvement: One of (1 mark):
- include more beakers at each light intensity.
- run the experiment multiple times and form an average to decrease effect of outliers.
- control all variables other than light intensity – measure distance from light, controlled temperature.

> Students often correctly identify a weakness but miss the second part of these types of questions. Your answer must explain the improvement and not just state the weakness.

Question 10 [A+ Study Notes p.67]

a The enzymes denaturing in the light-independent reaction will eventually inhibit the light-dependent reaction from occurring because of the lack of ADP + P_i and $NADP^+$ to act as reactants for the light-dependent reaction. (1 mark)

The electrons will saturate the thylakoid membrane and prevent any further water splitting. (1 mark)

b Above the optimal temperature range, the enzyme will begin to denature and undergo an irreversible conformational shape change, altering the active site and preventing it from binding to the substrate.

c With a reduction in temperature, the enzymes that have denatured will still be unable to function. (1 mark)

The chloroplast will have to synthesise new proteins before photosynthesis can occur. (1 mark)

Solutions to Test 5: Cellular respiration as an example of a biochemical pathway U3 Topic 2.2.3

Multiple-choice solutions

Question 1 A+ Study Notes p.69

B O_2

Oxygen is the input for the electron transport chain where electrons and hydrogen ions combine to form H_2O and 30 or 32ATP. In the absence of O_2, anaerobic respiration occurs. **A** is incorrect because it is an output of the Krebs cycle. **C** is incorrect because water is the output of the electron transport chain after the electrons and hydrogen ions have bound with oxygen. **D** is incorrect because NADPH is present in photosynthesis not cellular respiration. The electron carriers in respiration are NADH and $FADH_2$.

Question 2 A+ Study Notes pp.68–69

C Water.

Water can be classified as a by-product or waste product of the production of ATP. **A** is incorrect because oxygen is an input for the electron transport chain. **B** is incorrect because carbon dioxide is the waste product of the Krebs cycle. **D** is incorrect because ATP is not a waste product but is the main output for the reaction and the purpose of the reaction.

Question 3 A+ Study Notes pp.68–69

A Mitochondrion and is the site of aerobic respiration.

Only the Krebs cycle and the electron transport chain are part of aerobic respiration. **B** is incorrect because anaerobic respiration (glycolysis and fermentation) occurs in the cytosol. **C** is incorrect because the image is showing cristae in the centre of the diagram rather than stacks of membranes. **D** is incorrect because glycolysis does not occur in organelles. It occurs in the cytosol.

Question 4 A+ Study Notes p.68

A In the cytosol of cells.

The breakdown of glucose to two pyruvate molecules occurs in the cytosol under aerobic or anaerobic conditions. **B** is incorrect because acetyl CoA enters into the Krebs cycle in the matrix of the mitochondria. **C** is incorrect because the outer surface of the mitochondrial membrane regulates the movement of substances into and out of the mitochondria. **D** is incorrect because the chloroplast is responsible for photosynthesis only.

Question 5 A+ Study Notes pp.68–69

D Produce more ATP than anaerobic pathways do.

2ATP are produced in anaerobic pathways compared with 30 or 32ATP in aerobic pathways. **A** is incorrect because aerobic is in the presence of oxygen. Anaerobic is in the absence of oxygen. **B** is incorrect because the Krebs cycle and the electron transport chain occur in the mitochondria. **C** is incorrect because glucose is required for the production of pyruvate for both aerobic and anaerobic respiration.

Question 6 A+ Study Notes p.68

B Pyruvate and ATP.

Two pyruvate and 2ATP are produced during the reactions involved in glycolysis. **A** Carbon dioxide is produced in the Krebs cycle and water in the electron transport chain. **C** is incorrect because ethanol and ATP are produced during fermentation with yeast. **D** is incorrect because glycolysis is the first of three stages of cellular respiration.

Question 7 A+ Study Notes p.68

A $NAD^+ + 2e^- + H^+ \rightarrow NADH$

B is incorrect because NADPH is present during photosynthesis. NADH and $FADH_2$ are the coenzymes in cellular respiration. **C** is incorrect because NADPH is present during photosynthesis. **D** is incorrect because the reaction is not balanced with hydrogen ions in the reactants and NADH is formed in the Krebs cycle, not broken down.

Question 8 A+ Study Notes p.69

B H_2O

The electrons and hydrogen ions released into the membrane from NADH and $FADH_2$ combine with oxygen to form water as a by-product. **A** is incorrect because CO_2 is an output of the Krebs cycle (stage 2). **C** is incorrect because ATP is the main product. The question is asking for the by-product. **D** is incorrect because NADH and $FADH_2$ are the inputs for the electron transport chain, releasing electrons and hydrogen ions.

Question 9 A+ Study Notes p.68

D One fatty acid X molecule produces more ATP in aerobic respiration than one glucose molecule does.

The fatty acid produces 8 acetyl CoA, which are involved in the aerobic stages of respiration, whereas glucose is converted into 2 pyruvate – 2 acetyl CoA. **A** is incorrect because the Krebs cycle only produces 2ATP. **B** is incorrect because pyruvate is broken down to acetyl CoA in the matrix of the mitochondria. **C** is incorrect because 2ATP are produced with each glucose molecule in anaerobic respiration.

Question 10 A+ Study Notes p.68

C Carbon dioxide.

Carbon dioxide is formed during the cyclic breakdown of acetyl CoA. **A** is incorrect because electron carriers are produced for use in the final stage of the electron transport chain. **B** is incorrect because oxygen is an input for the electron transport chain. **D** is incorrect because glucose is the input for glycolysis.

Question 11 A+ Study Notes p.69

C Aerobic respiration in the mitochondria would be disrupted.

The electron transport chain would be inhibited, which would cause the build-up of loaded electron carriers and therefore stop the Krebs cycle. **A** is incorrect because increasing the rate of glycolysis would increase the amount of ATP. **B** is incorrect because ATP would not be produced in the mitochondria owing to the interference with the electron transport chain. **D** is incorrect because oxygen can diffuse through the phospholipid bilayer.

Question 12 A+ Study Notes pp.68–69

A Rotenone is not absorbed through the plasma membranes of people who have eaten poisoned fish.

The rotenone does not lead to the individuals being poisoned; therefore, rotenone either cannot enter into the cells or cannot bind to the electron transport proteins. **B** is incorrect because if the rotenone was not absorbed by the fish, then it would not be able to inhibit the ATP production. **C** is incorrect because the electron transport system is involved in cellular respiration in all species with mitochondria. **D** is incorrect because it disrupts the electron transport system, which is part of the aerobic respiration process.

Question 13 A+ Study Notes pp.69–70

B Increase glucose availability.

Increase in substrate will increase the amount of ethanol. **A** is incorrect because increasing the oxygen would induce aerobic respiration and therefore decrease ethanol production. **C** is incorrect because decreasing the temperature would lower the kinetic energy of the particles and reduce the number of substrate–enzyme collisions. **D** is incorrect because CO_2 is an output of respiration.

Question 14 A+ Study Notes p.70

D

The substrates and enzymes increase in kinetic energy, which increases collisions. After optimal temperature is reached, the enzymes begin to denature and undergo a conformational shape change. **A** is incorrect because the graph represents an increase in substrate before the enzyme becomes saturated. **B** is incorrect because the temperature reaches an optimal before dropping. **C** is incorrect because the rate of reaction would not' continuously increase unless enzyme and substrate were unlimited and temperature remained at the optimal.

Question 15 A+ Study Notes pp.68–69

A pyruvate is a 3-carbon compound

Glucose is a 6-carbon compound and is broken down into two molecules of 3-carbon pyruvate. **B** is incorrect as pyruvate can proceed into aerobic respiration to be broken down further. **C** is incorrect as there is a net production of 2 ATP. **D** is incorrect because 4 ADP and Pi are used to produce 4 ATP (Note: 2 ATP are used to initiate glycolysis so the net production of ATP is 2).

Question 16 A+ Study Notes p.68

B Increase carbon dioxide concentration.

CO_2 is an output for the Krebs cycle, increasing the CO_2 would not increase the production of NADH and $FADH_2$. **A** is incorrect because the increase in oxygen would enable more electrons to be released into the membrane in the electron transport chain from NADH and $FADH_2$. **C** is incorrect because an increase in pyruvate would increase the amount of acetyl CoA for the Krebs cycle. **D** is incorrect because the increase in unloaded coenzymes would allow more electrons being taken up during the Krebs cycle.

Question 17 A+ Study Notes pp.69–70

C Animal cells become more acidic and yeast forms ethanol and CO_2.

Lactic acid build-up results in the animal cells becoming more acidic. **A** is incorrect because pyruvate is broken down into ethanol and CO_2. **B** is incorrect because although animal cells produce lactic acid, no CO_2 is produced, and acetyl CoA is not involved in anaerobic respiration in yeast. **D** is incorrect because anaerobic respiration produces 2ATP in any species.

Question 18 A+ Study Notes p.68

C The majority of ATP is produced in the cell.

Y is where the electron transport chain occurs, which produces 26 or 28ATP. **A** is incorrect because glycolysis occurs in the cytosol of the cells. **B** is incorrect because NADH is formed in X (the matrix of the mitochondria). **D** is incorrect because pyruvate is broken down into acetyl CoA in the matrix of the cell (X).

Question 19 A+ Study Notes pp.68–69

A Occurs in the third stage of cellular respiration as the electrons and hydrogen ions from NADH and $FADH_2$ facilitate the conversion of ADP + P_i into ATP.

The release of electrons into the membrane and the movement of hydrogen ions facilitates the conversion of ADP + P_i into ATP. **B** is incorrect because ATP is produced with the breakdown of glucose into pyruvate. The NADH is reduced in the formation of lactic acid from pyruvate. **C** is incorrect because the Calvin cycle is a part of photosynthesis, which breaks down ATP into ADP + P_i. **D** is incorrect because the Calvin cycle only uses NADPH not $FADH_2$.

Question 20 A+ Study Notes p.69

A A decrease in ATP production.

> ATP production would be reduced because the electron transport chain, where 26 or 28ATP are produced, is inhibited. **B** is incorrect because ATP is the usable energy source for cells. **C** is incorrect because oxygen is an input not an output of cellular respiration. **D** is incorrect because aerobic respiration would be inhibited if the electron transport chain was interfered with.

Short-answer solutions

Question 1 A+ Study Notes p.68

a Krebs cycle

b Acetyl CoA is used to convey the carbon atoms from the amino acids, monosaccharides and fatty acids to the citric acid cycle at the mitochondria.

c The energy released in the citric acid cycle is used to produce NADH.   (1 mark)

The energy in NADH is used to produce ATP.   (1 mark)

d ATP transports chemical energy within cells for reactions to take place.

ATP breaks down to ADP releasing energy for metabolism.

e

Stage	Number of ATP molecules
Formation of acetyl CoA from glucose	2
Citric acid cycle	2
Oxidative phosphorylation	26 or 28

> 26-28ATP are produced in the electron transport chain depending on whether the NADH is from glycolysis.

Question 2 A+ Study Notes pp.69–70

a Temperature and pH

b Carbon dioxide. This is the product of fermentation by yeast.

c 38°C and pH 7

d At pH 5, at least one of the enzymes necessary for the metabolism of sucrose is denatured – irreversible conformational shape change.

e i The rate of fermentation increases with increasing temperature between 15°C and 38°C. At the highest temperature, 75°C, the rate of fermentation is very low due to denaturation of enzymes.

> The rate of reaction increased up to the optimal rate due to enzyme–substrate collisions increasing.

ii The rate of a chemical reaction increases with increasing temperature due to the average increase in kinetic energy of the substrate and enzymes.   (1 mark)

At high temperatures, the temperature is above optimal for the enzymes and they denature, leading to irreversible conformational shape change, which alters its active site.   (1 mark)

f The yeast lacks an enzyme that will convert lactose or maltose to glucose for respiration.   (1 mark)

The enzymes that convert lactose and/or maltose to glucose are inactive at pH 7.   (1 mark)

Question 3 `A+ Study Notes` `pp.9; 67–70`

a   Glucose + oxygen + 30 or 32 ADP + 30 or 32 Pi $\longrightarrow$ Carbon dioxide + water + 30 or 32ATP

You would receive 1 mark for the correct word equation and 1 mark for the correct amount of ADP/Pi and ATP.

b CO_2 is a product of both aerobic and anaerobic respiration.   (1 mark)

The conclusion cannot be supported because there is no clear evidence of which reaction is taking place to produce the CO_2.   (1 mark)

c   The formation of lactic acid would lower the pH, and the cell contents becoming more acidic. (1 mark)

If pH decreases below the tolerance range of an enzyme, denaturation can result, i.e. an irreversible conformational shape change of the protein.   (1 mark)

d   Facilitated diffusion.   (1 mark)

Carrier proteins enable the large molecule to pass through the phospholipid bilayer.   (1 mark)

e   Passive diffusion through the phospholipid bilayer.

Question 4 `A+ Study Notes` `p.194`

a   The greater/lower the dosage the lower/higher the rate of lactic acid.

Alternatively, you could formulate a specific mass (mg) of drug.

b   Independent and dependent variables   (1 mark)

Controlled variables   (1 mark)

Large sample size   (1 mark)

Experimental design (control group and experimental group.   (1 mark)

Example set up:

Independent variable –drug dosage/amount of drug (mg)

Dependent variable – lactic acid production (mL)

Controlled variables – amount of muscle cells, oxygen concentration, temperature of culture, same time and regularity of the dosage.

Set up 100 cultures with the same amount of muscle cells and with no oxygen supply (enclosed container). Separate into five groups of 20. Group 1 would be exposed to no drug (control group); Group 2 would be exposed to 5 mg of drug, Group 3 to 10 mg of drug, Group 4 to 15 mg of drug and Group 5 to 20 mg of drug (experimental groups).

Measure the amount of lactic acid produced (pH probe) in each of the chambers every hour for 24 hours.

c   One of:
- The lower the concentration of drug, the lower the pH became.
- The increase of drug from 5 to 20 mg caused the cell to become less acidic due to the decreased production of lactic acid.

Question 5 `A+ Study Notes` `pp.68–69`

a   Products: NADH and $FADH_2$.   (1 mark)

Release electrons and hydrogen ions into the cristae of the mitochondria to facilitate the formation of ATP from ADP + P_i.   (1 mark)

b The substrate malate would build up and there would be no oxaloacetate in the mitochondria. (1 mark)

The entire citric acid/Krebs cycle would be inhibited. (1 mark)

> Always refer to the specific substrates and products if provided with an image. You are not expected to know each of these enzymes to answer this question.
>
> You should be aware that the build-up of substrate malate and resulting lack of oxaloacetate in the mitochondria would inhibit the electron transport chain and aerobic respiration would be halted.

c A mutation in the DNA of the malate dehydrogenase gene will result in a different mRNA strand being formed. (1 mark)

The different sequence of codons could code for different amino acids with different properties, resulting in a different secondary and tertiary structure. (1 mark)

Change in tertiary structure would alter the active site and inhibit the substrate from binding to the enzyme. (1 mark)

d Lowering the temperature lowers the average kinetic energy of the molecules in the cell. (1 mark)

Fewer enzyme–substrate collisions in the Krebs cycle, slowing down the series of reactions and the amount of NADH and $FADH_2$ produced. Therefore, this would slow down the electron transport chain and the production of ATP. (1 mark)

e NADH, $FADH_2$ and the waste product CO_2

Question 6 [A+ Study Notes pp.68–70]

a Aerobic respiration occurs in the presence of oxygen, whereas anaerobic occurs in the absence of oxygen. (1 mark)

Benefit: Aerobic – 30 or 32ATP produced relative to 2ATP in anaerobic. Anaerobic is quicker for ATP production than aerobic. (1 mark)

> To achieve full marks, your answer needs to include a comparative term, such as 'whereas' or 'compared to'.

b Two of:
- double membrane from being engulfed into a vesicle
- circular chromosome
- own ribosomes
- lack membrane-bound organelles
- replicate via binary fission.

> Bacteria do not have a double membrane – this was a common mistake made in previous VCAA exams. 'Similar size and appearance to bacteria' is a weak answer.

c

	Name of structure	Process or role in cellular respiration
A	Cristae of mitochondria	Electron transport system Forms 26 or 28ATP through the release of electrons into the membrane
B	Outer membrane	Diffusion of oxygen/carbon dioxide OR Facilitates pyruvate into the cell
C	Matrix of mitochondria	Krebs cycle/cyclic set of reactions that break down acetyl CoA to form NADH and $FADH_2$ (CO_2 by-product)

Question 7 A+ Study Notes pp.61–62; 66–69

a Oxygen availability

> Remember that if the graph shows that the rate is continuing to increase, the variable on the *x*-axis is the limiting factor. When the graph plateaus, it indicates that another factor has become limiting.

b CO_2 concentration

c An increase in temperature from 10°C to 30°C increases the rate of cellular respiration from 2.5 to 6.5 (relative rate).   (1 mark)

An increase in temperature increases the average kinetic energy of the molecules – increasing substrate–enzyme collisions and therefore increasing rate of reaction.   (1 mark)

d Electron transport chain on the cristae of the mitochondria.   (1 mark)

NADH and $FADH_2$ release their electrons into the membrane from the Krebs cycle.   (1 mark)

Oxygen combines with the electrons and the hydrogen ions from the membrane to form water. ATP production occurs through the movement of hydrogen ions through the membrane.   (1 mark)

e The hypothesis would not be supported. An increase in temperature would increase the rate of collisions until the optimal temperature was reached.   (1 mark)

Above the optimal temperature, there would be denaturation of the enzymes involved in glycolysis and Krebs cycle.   (1 mark)

f Two of:
- multiple test tubes in each experimental group (to form an average and reduce outliers)
- repeating the experiment
- increasing the number experimental groups to observe the effect of temperature.

Question 8 A+ Study Notes pp.67–70

a CO_2 concentration would increase from 8 AU over the 180 minutes.   (1 mark)

CO_2 is produced in both aerobic and anaerobic respiration in yeast.   (1 mark)

b Both aerobic respiration with the reduction in oxygen (3 to 1) and anaerobic respiration with the increase in ethanol produced (0 to 2).

> Use the data from the table. You should always refer to data any time there is a graph or table.

c ATP production would be 30 or 32ATP at 30 minutes as oxygen would still be present.

ATP production would be two ATP per glucose molecule with no oxygen in anaerobic respiration.

d

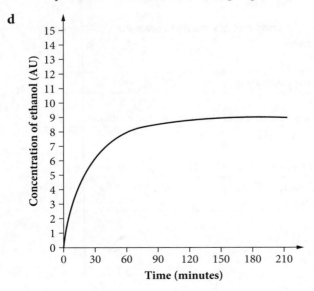

e One of:
- The amount of ethanol would continue to increase.
- Ethanol concentration would increase until levels became toxic to the yeast and the rate plateaued.

f The build-up of NADH depletes the cell of NAD^+, thus stopping glycolysis from occurring.   (1 mark)

The breakdown of pyruvate to ethanol releases the electrons and hydrogen from NADH to form NAD^+, which allows the substrate for glycolysis.   (1 mark)

> You should be aware that the production of pyruvate produces 2ATP and 2NADH.

Question 9 A+ Study Notes pp.69–70; 194

a To produce ATP when there is no oxygen available

b Ethanol, carbon dioxide and ATP

c Independent variable: presence or absence of furfural.   (1 mark)
Experiment set up with same amount of glucose.   (1 mark)
Experiment set up with same amount of alcohol dehydrogenase.   (1 mark)
Dependent variable: measure the amount of product produced, e.g. carbon dioxide.   (1 mark)

> About 50% of students answered the various components of this 2016 VCAA question incorrectly. Students who were able to understand the information and identify the key components of experimental design scored at the highest level. Many students were confused by the terms and their responses lacked clarity.

Question 10 A+ Study Notes pp.68–69

Experiment 1 – Suspension of mitochondria

Suspension added	Change in oxygen concentration (increase/decrease/no change)	Reason
Glucose	No change	Glucose is not metabolised (broken down) by mitochondria
Pyruvate	Decrease	Pyruvate is a substrate of the Krebs cycle, which occurs in the mitochondria

Experiment 2 – Cytosol of cells from which the mitochondria would have been removed

Suspension added	Change in oxygen concentration (increase/decrease/no change)	Reason
Glucose	No change	Glycolysis is anaerobic (Glucose when converted to pyruvate uses no oxygen)
Pyruvate	No change	No aerobic respiration can occur without the mitochondria

Experiment 3 – Suspension of mitochondria and cytosol of cells

Suspension added	Change in oxygen concentration (increase/decrease/no change)	Reason
Glucose	Decrease	Glucose is converted into pyruvate, which is metabolised by the mitochondria using oxygen
Pyruvate	Decrease	Pyruvate is metabolised by the mitochondria in a process that uses oxygen

> This question requires you to give a correct change and a correct reason to gain 1 mark for each situation. However, if a reason was incorrect or not given, you could still gain a mark for every two correct changes.

Solutions to Test 6: Biological applications of biochemical pathways U3 Topic 2.2.4

Multiple-choice solutions

Question 1   A+ Study Notes pp.71–72

C Fermentation of bacteria in culture mediums can produce H_2 used for a biofuel.

> Bacteria can undergo fermentation-based processes through the breakdown of organic matter to form H_2, acetic acid and carbon dioxide. **A** is incorrect because biofuels are formed through renewable organic compounds. **B** is incorrect because lactic acid is produced in anaerobic respiration in animals, not yeast. **D** is incorrect because natural gas is a non-renewable resource.

Question 2   A+ Study Notes pp.71–72

C Biofuels draw down carbon from the existing atmosphere whereas fossil fuels contain carbon dioxide from ancient atmospheres.

> Biofuels are carbon neutral as they contain carbon from the existing atmosphere whereas fossil fuels contain carbon from ancient atmospheres and this is added to the existing atmosphere when fossil fuels are burnt. **A** is incorrect as this is not the main reason for using biofuels. **B** is incorrect as fossil fuels are very costly to extract and purify. **D** is incorrect as biofuels are carbon neutral as explained.

Question 3   A+ Study Notes pp.71–72

B people reducing their carbon footprint in response to the changes in climate.

> A social factor is one which involves people and their community. **A** is incorrect as this is a political factor. **C** is incorrect as this is an economic factor. **D** is incorrect as this is a legal factor.

Question 4   A+ Study Notes p.71

D The cells and cells produced from these in the modified plant only.

> The modification was present in the somatic cells and will remain with that plant. For the modification to be passed onto future generations it must be integrated into the germline cells. **A** is incorrect because it is only present in the cells and the cells from these because it occurred in the somatic cells. **B** is incorrect because it was not modified in the gamete cells and was not present in all cells in the current plant. **C** is incorrect because it was not modified in the gamete cells and cannot be passed onto future generations.

Question 5   A+ Study Notes pp.71–72

D Limited amount of available resources for the production of biofuels.

> Biofuels are produced from organic materials; therefore, there is an unlimited supply. **A** is incorrect because as farmers increase their crop production to meet demand for food and fuels, fertiliser and pesticides will be increased, flowing into waterways. **B** is incorrect because crops such as corn and sugar cane – the common crops for biofuels – require large amounts of land to grow. **C** is incorrect because as we transition to the use of renewable fuels, traditional engines will need to be modified to account for the changing products and properties.

Question 6   A+ Study Notes pp.61–62

D temperature of 37°C and pH of 7.

> As fermentation involves enzymes you would expect the conditions inside the fermenter to be conducive to enzyme activity. **A** is incorrect because this would be too hot and a very acidic environment. **B** is incorrect as this is too hot and would denature enzymes. **C** is incorrect as this is a very alkaline environment.

Question 7 `A+ Study Notes` `p.71`

B There are fewer offsite effects and risk of mutations in the organism's genome.

As the Cas9 enzyme is not cutting the genome, there is not the risk of offsite cuts and mutations in other genes within the genome. **A** is incorrect because the inactivated Cas9 cannot cut any DNA and the guide RNA is specific to one sequence in the genome due to its 18–20 nucleotide sequence. **C** is incorrect because both are specific due to the single guide RNA. **D** is incorrect because both contain enzymes and RNA that will eventually degrade in a cell, and traditional CRISPR alters the genome, thus altering the cell until it is removed or divides.

Question 8 `A+ Study Notes` `pp.71–72`

A Seasonal growth period, energy production, labour.

If the plant or organism is only active for short periods, then the plant is not favoured for biofuel because of its slow growth. If the energy isolated from the biomass is low and the isolation/purification process is labour intensive, it would not be a considered biofuel reactant. **B** is incorrect because the chlorophyll pigments would not affect the amount of ATP produced per glucose molecule. Cell density would not affect the ATP production. **C** is incorrect because chromosomal complexity has no effect on photosynthesis and cellular respiration. **D** is incorrect because the replication style of the plant would not be of concern providing it produced enough offspring for a consistent crop.

Question 9 `A+ Study Notes` `pp.71–72`

C Ethanol and carbon dioxide.

ATP is produced as an end product of the fermentation of sugar. CO_2 and ethanol are waste products as a consequence of the reactions. **A** is incorrect because ATP is not a waste product. Heat is a by-product of fermentation. **B** is incorrect because although CO_2 is a by-product, ATP is not. **D** is incorrect because pyruvate is formed during glycolysis. This is further broken down into ethanol to release the electron from the NADH.

Question 10 `A+ Study Notes` `pp.71–72`

C Facilitate the formation of glucose through a biochemical pathway, digesting the cellulose and lignin.

The lignin and cellulose are broken down to form the starting material (reactant) of fermentation – yeast. **A** is incorrect because enzymes reduce the activation energy required to form products from reactants. **B** is incorrect because the enzymes are involved in the formation of glucose, not its breakdown. This is completed by the yeast in fermentation. **D** is incorrect because the active sites of the enzymes are specific and complementary to the lignin, cellulose and intermediate product that form glucose as a final product.

Short-answer solutions

Question 1 `A+ Study Notes` `pp.71–72`

a $2ADP + 2Pi + sugar \longrightarrow$ carbon dioxide + ethanol + 2ATP

b Ethanol (ethyl alcohol)

c One of:
- As more CO_2 is produced and dissolves into the solution, the pH becomes toxic to the yeast.
- Ethanol build-up can become toxic to the yeast cells, inhibiting them from performing their usual functions.

(Either response would receive 1 mark.)

d Starchy crop waste contains complex sugars and structures such as cellulose, lignin and hemicellulose. These require additional steps to break them down into sucrose or glucose, which yeast can use in fermentation.

(1 mark)

Corn or sugar cane can be used directly because they contain simple sugars that the yeast can ferment directly.

(1 mark)

e Oxygen atoms in the ethanol help the fuel to burn more completely and therefore reduce the harmful carbon monoxide emissions.

f (One of each) Economical: cost of production of biofuel; cost to consumer; cost of any alterations required to car to use biofuel. (1 mark) Political: legislating any subsidies to biofuel manufacturers; reducing cost of cars that run on biofuel. (1 mark)

Question 2 A+ Study Notes pp.71–72

a To ensure the bacteria survive and (one of):
- The controlled culture needs to have no access to oxygen to ensure the bacteria are undergoing anaerobic respiration and not aerobic digestion.
- Enzymes are involved in biochemical pathways and are functional at their optimal range.

b Constant supply of substrate–organic compound for fermentation to occur. (1 mark)

Controlled optimal temperature and pH to prevent enzymes denaturing and to maximise substrate–enzyme collisions. (1 mark)

c

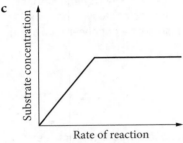

You would receive 1 mark for the graph and 1 mark for correctly labelling the x-axis 'Rate of reaction' and y-axis 'Substrate concentration'. Note that although substrate concentration continues to increase, the rate of reaction will eventually plateau because of another limiting factor; for example, the amount of bacteria or enzyme saturation in the bacteria.

d Methane

e Methane is used in biogas where it makes up 50–70%, the remaining consisting of CO_2. (1 mark)

It used in a similar way to natural gas for production of power. (1 mark)

Question 3 A+ Study Notes p.71

a Two of:
- more specific targeting of the desired traits
- keep variation of other genes within the population
- quicker outcomes such as increased crop yield present in the next generation.

b 'Transgenic organisms' refers to the insertion of DNA from a different species, whereas CRISPR is the modification of the existing genome with no additional genes from other species. (1 mark)

Advantage of CRISPR: only modifying the genome through the removal of a gene or the insertion of a different form of the gene already existing in the species will improve crop production. CRISPR is a very specific technique and since no new genes or foreign genes are introduced there is less likelihood of adverse outcomes in the crops produced. (1 mark)

Disadvantage of use of transgenic organisms: unexpected outcomes or cross-pollination; for example, may result in passing the foreign genes onto wild species. (1 mark)

You should define both 'transgenic organisms' and 'CRISPR' in your response.

c Cas9 is an endonuclease (restriction enzyme) that cuts the gene of interest at the specific sequence determined by the guide RNA.

Single guide RNA is 18–20 nucleotides that guide the Cas9 enzyme to the specific sequence within a gene with the complementary sequence.

d By removing genes, you can observe the role they play in the species; for example, the role of specific genes in the timing of fruit maturation or produce size, which would increase the crop yield.　(1 mark)

The expression of these genes can then be targeted; for example, by inhibiting the production of a certain enzyme which results in an increase in crop yield.　(1 mark)

Question 4　A+ Study Notes　pp.26–28

a Blunt ends would be produced.

b One of (1 mark):
- PCR: Denaturing (breaking hydrogen bonds between the two DNA strands), annealing (adding primers to either end of the DNA strand) and extension (Taq polymerase catalysing the synthesis of the complementary strands from the primers).

- Plasmids: Incorporating into a plasmid – recombinant plasmid – and inserting into bacteria (transformation of the bacterium).

Gel electrophoresis can then be used in both techniques to isolate the sequence.　(1 mark)

c Guide RNA specifically to the sequence of interest attached to the Cas9 enzyme.　(1 mark)

Gene of interest inserted with the complex and an insertion protein.　(1 mark)

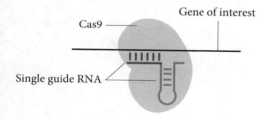

　(1 mark)

d To ensure only one sequence. The desired gene is targeted by the Cas9 enzyme, which reduces undesired side effects with other genes.

e A gene in a species that produces larger crops or is pest resistant can be isolated and used in the CRISPR system to introduce it into other plants. This process is quicker than selective breeding.

Question 5　A+ Study Notes　pp.71–72

a Biofuel is a form of renewable energy produced from biomass and can a gas or a liquid.

b Marine biofuel is produced in areas where terrestrial crops cannot grow, and does not take away from food production on land. Marine biofuel grows faster than terrestrial biofuel and has a higher photosynthetic efficiency. It can be used to produce biodiesel, bioethanol and biogas in algae refineries.　(1 mark)

By comparison, terrestrial biofuel uses leftover organic matter from crops, reduces air pollutants, and has less transport and isolation costs.　(1 mark)

To achieve full marks, your answer needs to include a comparative term, such as 'whereas', 'by comparison' or 'compared to'.

c   Genetic modification can increase the amount of substrates organisms can break down; for example, by adding enzymes to break up complex structures.   (1 mark)

By adding enzymes, the rate of biofuel production can increase, thus increasing productivity or increasing the range of substrates (waste products in addition to crops).   (1 mark)

Genetic modification can include modifying an existing genome or adding additional genes from other species.

d   Two of:
- the amount of land required to grow the crop, which would result in less land for food production
- environmental impacts such as pollution from fertilisers or pesticide
- lower energy output than with fossil fuels.

Solutions to Test 7: Responding to antigens U4 Topic 1.1.1

Multiple-choice solutions

Question 1 A+ Study Notes p.95

A Process antigens and present them on their cell surfaces to the T helper cell.

Dendritic cells and macrophages are antigen-presenting cells that activate the third line of defence. **B** is incorrect because mast cells and basophils release histamine when damaged. **C** is incorrect because dendritic cells migrate via chemotaxis from the release of chemicals from other cells. **D** is incorrect because dendritic cells migrate in response to chemical release. They endocytose pathogens.

Question 2 A+ Study Notes p.93

A Mucus lining in the respiratory tract.

The mucus catches large particles and pathogens in the fluid, slowing the movement and entry into tissue. **B** is incorrect because memory cells are produced in clonal expansion of B and T cells and are specific to an antigen. **C** is incorrect because cytotoxic T cells recognise a specific antigen in a cell's MHC I marker. **D** is incorrect because antibodies are produced with a variable region specific to an antigen.

Question 3 A+ Study Notes pp.93–94

D Hydrochloric acid in the stomach.

The low pH (acidic) solution kills off some pathogens that enter through food and the digestive tract. **A** is incorrect because interferon is a chemical produced in response to an infection in the cells and body tissue. **B** is incorrect because histamine is released when mast cells are damaged. **C** is incorrect because complement proteins are activated after the pathogens have entered the tissue.

Question 4 A+ Study Notes p.98

B Macrophage.

Macrophages and dendritic cells are antigen-presenting cells that migrate to the lymph nodes with the antigen in their MHC II marker. **A** is incorrect because the neutrophils are first at an infection but engulf the pathogens and undergo apoptosis. **C** is incorrect because mast cells are cells in tissues that release histamine when damaged. They do not circulate. **D** is incorrect because natural killer cells recognise the absence of MHC markers on self-cells.

Question 5 A+ Study Notes p.96

C The capillaries would become more permeable.

Histamine increases vascular permeability and vasodilation to allow leukocytes to enter into the tissue. A is incorrect because mast cells produce histamine; B plasma cells produce antibodies specific to an antigen. **B** is incorrect because the increase in blood flow to the area increases the temperature. **D** is incorrect because red blood cells are involved in the transport of oxygen around the body. Phagocytes engulf the foreign material.

Question 6 A+ Study Notes p.97

D Activation of the humoral and cell-mediated responses.

The adaptive immune system takes a few days to be activated and active in the body. **A** is incorrect because histamine is produced when the mast cells in the tissue are damaged at the start of the inflammation process. **B** is incorrect because chemotaxis occurs when the cells release cytokines when injured. **C** is incorrect because the macrophages release interleukin while engulfing the foreign debris to raise the body temperature.

Question 7 `A+ Study Notes` `pp.93–94`

B Natural flora in the digestive and respiratory tract.

The presence of natural flora prevents pathogenic bacteria from growing in these areas due to lack of space and nutrients. **A** is incorrect because cytotoxic T cells are specific to an antigen and are part of the third line of defence. **C** is incorrect because complement protein is part of the second line of defence (non-specific). **D** is incorrect because histamine is part of the second line of defence (non-specific).

Question 8 `A+ Study Notes` `p.99`

C An allergen.

An allergen produces an exaggerated immune response to a usually innocuous molecule. **A** is incorrect because an antigen is a molecule that is associated with a pathogen and initiates an immune response. **B** is incorrect because it is not a cellular or non-cellular pathogen that initiates disease. **D** is incorrect because the IgE antibody is the antibody that binds onto the mast cells and causes degranulation of the histamine when the allergen binds to the variable region.

Question 9 `A+ Study Notes` `p.97`

A Interleukin.

Interleukin is released from macrophages from an area of inflammation. **B** is incorrect because histamine causes vasodilation and increased vascular permeability. **C** is incorrect because interferon is released from viral infected cells to protect the surrounding cells through the increase in antiviral proteins. **D** is incorrect because neurotransmitters are released from a neuron to stimulate an action potential in a connecting neuron or muscle.

Question 10 `A+ Study Notes` `p.97`

A Interferon builds resistance in the surrounding cells through antiviral proteins.

The surrounding cells receive signals, which results in an increase in gene expression of antiviral proteins to prevent newly formed viruses from entering these cells. **B** is incorrect because interleukin is released from T helper cells for clonal expansion for B and T cells and from macrophages for fever. **C** is incorrect because chemokines are involved in chemotaxis and the migration of leukocytes to sites of infection. **D** is incorrect because histamine initiates the vasodilation and increased vascular permeability of the blood vessels in injured tissue to increase the leukocyte movement into the tissue.

Question 11 `A+ Study Notes` `pp.99–100`

C There is no immediate innate inflammatory response to stimulate the humoral or cell-mediated response.

The proteins are recognised as self; therefore, there is no adaptive immune response. **A** is incorrect because if the proteins were recognised as foreign, they would act as an antigen and initiate an immune response. **B** is incorrect because, although prion proteins can be intracellular, they are not obligate parasites. **D** is incorrect because leukocytes can recognise pathogens in all areas of the body.

Question 12 `A+ Study Notes` `p.95`

B Leukocytes migrate to site of infection from low to high concentrations of chemokine release to areas of infection.

The infection site releases chemicals to attract the leukocytes to the area to reduce the infection spreading. **A** is incorrect because the movement of the antigen-presenting cells is after the chemotaxis of leukocytes to the area of infection. **C** is incorrect because histamine is the chemical responsible for the increase in the permeability and vasodilation. **D** is incorrect because platelets are activated with the break in vessels and form a net-like structure over the wound.

Question 13 A+ Study Notes p.100

B Prion.

A is incorrect because bacteria are cellular pathogens that contain genetic material in the form of a circular chromosome. **C** is incorrect because viruses are non-cellular but contain genetic material in the form of DNA or RNA. **D** is incorrect because protozoa are cellular pathogens that are cellular and contain linear DNA.

Question 14 A+ Study Notes p.100

C Bacteria grow in the extracellular environment, whereas viruses multiply within the cell.

Bacteria grow in the extracellular environment, whereas viruses are intracellular and extracellular (enter and replicate within the cell before lysing).The bacteria would replicate by binary fission, giving exponential growth. Virus numbers plateau while they are inside cells replicating and increase in numbers when the infected cells lyse. **A** is incorrect because viruses replicate in cells and so transition between intracellular and extracellular fluid. **B** is incorrect because, although viruses mutate, this does not occur that rapidly and this wouldn't explain the change in the number of pathogens. **D** is incorrect because bacteria and viruses are recognised as foreign pathogens by the innate immune system, which then activates the adaptive immune system.

Question 15 A+ Study Notes pp.97–98

A Natural killer cell.

These cells act in a similar function to the cytotoxic T cells but are not specific to the antigen. **B** is incorrect because cytotoxic T cells recognise a specific antigen on the MHC I marker of viral-infected cells, cancer cells or transplanted cells. **C** is incorrect because neutrophils are a type of phagocyte that phagocytoses foreign pathogens. **D** is incorrect because dendritic cells phagocytose and present the antigen to the adaptive immune system.

Question 16 A+ Study Notes pp.93–94

D Sap and natural secretions.

These can contain digestive enzymes or toxins that inhibit the growth and entry of pathogens. **A** is incorrect because this is a structural modification to reduce the amount of sunlight and to make it more difficult for insects and pathogens to remain on the leaf. **B** is incorrect because, although hairs and thorns can be coated in a chemical to prevent pathogens, they are structural. **C** is incorrect because a waxy cuticle prevents water loss and is a structural barrier to make it more difficult to enter the internal leaf structure.

Question 17 A+ Study Notes pp.96–97

A Includes phagocyte migration to the site of the injury.

Phagocytes, such as macrophages and neutrophils, migrate to a high chemical concentration from damaged cells due to chemotaxis. **B** is incorrect because inflammation is immediate and involves cells from the innate immune response. **C** is incorrect because the inflammatory response occurs in the same steps regardless of the type of pathogen or foreign body that has entered the body, or breached the first line of defence. **D** is incorrect because lymphocytes such as B and T cells can be activated by the macrophages after they have encountered the foreign body, but these cells already exist before the inflammatory response.

Question 18 A+ Study Notes pp.93–94

C Use of an epidermal layer to inhibit the invasion of pathogens.

The intact skin in animals acts as a first line of defence; similarly in plants, the cuticles prevent pathogens entering the tissue. **A** is incorrect because plants do not have an adaptive immune system as they cannot circulate cells throughout. **B** is incorrect because plants do not contain a pump and circulatory in the same way animals do to move specific immune cells to an area of infection. **D** is incorrect because salicylic acid is only produced in some plants and is a specialised defence chemical.

Question 19 A+ Study Notes p.99

C William is allowed to keep his cat.

> The cat produced a negative response (no shading present). **A** is incorrect because no inflammation was present in the saline section (no shading present). **B** is incorrect because there is a positive response to histamine and it is acting as a positive control to ensure the histamine release is occurring. **D** is incorrect because the allergen from each species will have a unique shape and the immune system will have produced specific IgE to each of these antigens.

Question 20 A+ Study Notes p.98

B The recipient's cells and the donor cells would recognise different surface receptors.

> The MHC markers on the recipient's cells and the donor cells would be different, resulting in the newly formed leukocytes recognising the recipient's cells as foreign. **A** is incorrect because their proteins would be slightly different in shape depending on the alleles (form of genes) in their DNA. **C** is incorrect because only macrophages, dendritic cells and B cells contain MHC II markers. **D** is incorrect because it is the recognition of foreign that initiates the apoptosis. The inserted cells are not cancerous and will not continuously replicate.

Short-answer solutions

Question 1 A+ Study Notes pp.93–98 ●●●

a Intact skin

> You must include 'intact' in your response. VCAA would not accept an answer of just 'skin'.

b Release of histamine – vasodilation and increased vascular permeability.   (1 mark)

Neutrophils engulf foreign pathogens and apoptose.   (1 mark)

Chemotaxis – immune cells migrate to the area of infection, low to high chemical concentration.   (1 mark)

c One of: Activated proteins that bind to foreign material to neutralise (prevent entry or toxin release)/ agglutinate pathogen/bind with antibodies and lyse the pathogen.

d Pus is composed of neutrophils that have undergone apoptosis after engulfing foreign material.   (1 mark)

Pathogen debris/fluid.   (1 mark)

Question 2 A+ Study Notes p.99

a Allergen

b An allergen enters the body and is recognised as foreign by the innate immune cells. The allergen is engulfed by a macrophage or dendritic cell, which migrates to the lymph node.   (1 mark)

Humoral response is activated, which produces IgE antibodies specific to the antigen.   (1 mark)

IgE antibodies enter the bloodstream and bind to mast cells in tissue via the heavy chain resulting in the individual becoming sensitised.   (1 mark)

c The allergen binds onto the variable region of the IgE antibodies, cross-linking.   (1 mark)

The binding results in degranulation of histamine, producing an inflammatory response.   (1 mark)

Question 3 A+ Study Notes pp.93–94; 98–99

a A pathogen is any agent that causes harm or disease to an organism, whereas an antigen is a molecule that initiates an immune response.

> A pathogen can have multiple antigens.

9780170479448

b MHC I are located on all nucleated cells and continually recycle proteins from the cell, and present on the surface of the cell.
(1 mark)

MHC II are located on antigen-presenting cells (dendritic cells/macrophages) and present an antigen to the T helper cell (adaptive immune system).
(1 mark)

c

	Barrier example		
	Physical	**Chemical**	**Microbiological**
Animals	Intact skin/cilia	Stomach acid, mucosal membrane, lysosomes	Natural flora or microbiota prevent colonisation by pathogens
Plants	Bark, hair, thorns, thick waxy cuticle, spines	Sap, secretions, digestive enzymes, oils, resins	

Question 4 A+ Study Notes p.97

a The virus enters self-cells and uses the cells' organelles to synthesise new viral particles.
(1 mark)

After the new viruses have been made, the cell is lysed and the virus particles are released into the extracellular fluid.
(1 mark)

The new virus particles bind onto surrounding cells and repeat the process, resulting in an inflammatory response from the lysed cells.
(1 mark)

b Two of:
- Macrophages and dendritic cells: engulfing/phagocytosing the virus and breaking it down to present the antigen on its MHC II marker and migrating to the lymph node.
- Neutrophils: engulfing/phagocytosing the virus and undergoing apoptosis.
- Natural killer cell: recognition of viral-infected cells through the removal of the MHC I marker and lysing/initiating apoptosis.

c The mother would have active natural immunity through formed memory cells from encountering the virus in the past, whereas the child would have passive natural immunity received from the mother through antibodies in the breast milk.

d Viral-infected cell releases interferon to the surrounding cells.
(1 mark)

Interferon upregulates the expression of antiviral genes in the surrounding cells.
(1 mark)

Question 5 A+ Study Notes p.100

a **i** Ticks live in an ever-changing environment. An internal pathogen lives in an internal environment that is constant in many respects, for example temperature.

ii One of:
- A tick burrows through the outer layers of hair or fur and bites into and attaches to the underlying skin to ensure a reasonably constant temperature.
- A tick has a hard exoskeleton.

b Graph X : Bacteria

Graph Y: Virus

c Viruses need to enter living cells for their reproductive cycle. On maturity of the viruses within a cell, the cell bursts open, viruses are released and then the cycle is repeated. Every time viruses burst from cells, there is a sudden increase in the number of viruses. When the virus is inside the cells, the graph is level, running parallel to the horizontal axis.
(1 mark)

In contrast, bacteria enter a host but are not dependent on any cycle within cells for their reproduction. They show a single ongoing phase of exponential growth indicated in graph X.
(1 mark)

Many students understood that a virus must enter a cell before it can reproduce. Few could relate the shape of the graph to the increase in number of viruses after infection.

Question 6   A+ Study Notes pp.97–98

a   Interferon is not specific and is produced in response to many different viruses.

b   Interferon binds to the plasma membrane of infected cells and triggers the production of cellular enzymes that interfere with viral replication within the cell.

c   Interferon works rapidly to contain a viral infection until the slower-reacting specific immune responses can take over.

d   Viral nucleic acid directs the production of viral proteins by the host cell. A change in viral nucleic acid leads to a change in protein synthesis.   (1 mark)

   A change in one base can lead to a different amino acid. A different amino acid means that the protein may be different.   (1 mark)

Question 7   A+ Study Notes pp.93–94; 96–97

a   One of:
   • When the skin is cut or broken
   • When there is a deep burn.

b   Two of:
   • Lysozyme – an enzyme produced in tears – destroys bacteria.
   • Gastric juice produced in the stomach, destroys bacteria.
   • Saliva in the mouth dilutes microorganisms.
   • Acidity on the skin inhibits bacteria growth.
   • Sebum on the skin inhibits bacterial growth.

c   i   Swelling is due to the build-up of fluid in the area.   (1 mark)

      The splinter would have damaged cells that release histamine. Histamine causes the capillaries in the vicinity to become leaky, causing a build-up of fluid in the area.   (1 mark)

   ii   Symptom 1: Presence of pus or pain.

      Reason: Phagocytic white blood cells such as neutrophils would be phagocytosis and then undergoing apoptosis – pus formation.

      Symptom 2: Feeling hot and red.

      Reason: Histamine caused vasodilation and increased vascular permeability. The increased blood flow would make the area red and warm.

Question 8   A+ Study Notes pp.93–94

a   To protect them from pathogens.

b   Two of:
   • thick cuticle
   • secondary cell wall
   • vertical alignment of leaves
   • thorns.

c   Two of:
   • compounds affecting vectors of plant viruses
   • antifungal compounds
   • antimicrobial compounds
   • compounds that interfere with pathogen nutrition and retard their development.

d   When the plant cells detect that the pathogen has penetrated the physical and chemical barriers.

e One of:
- Repair of damaged tissue to limit spread of pathogen.
- Infected cells and those surrounding them die, preventing further spread and, in some cases, killing the pathogen.
- Isolation of the infected area, e.g. formation of galls.

Question 9 A+ Study Notes pp.93–94

a Through a break in the skin.

b i One of:
- Through horizontal gene transfer of plasmids with the antibiotic resistance genes, meaning plasmids can be transferred from one resistant bacteria to another.
- Natural variation in the ability to resist antibiotics exists in bacterial populations. When exposed to antibiotics, those that are resistant are selected for and survive; they are most fit. Non-resistant bacteria are selected against and do not survive to reproduce. The surviving resistant bacteria reproduce, passing on this resistance to their offspring. The next generation contains a larger proportion of resistant bacteria.

ii Hospitals provide an environment where there is extensive use of antibiotics and antiseptics. The wide range of these chemicals means that all but resistant bacteria are selected against, producing populations of resistant bacteria.

Hospitals also contain good breeding grounds for bacteria, such as surgical wounds and other injuries.

c Two of:
- short breeding or generation time
- very large populations and high genetic variation within populations
- ability to exchange genetic material through exchange of plasmids
- high mutation rate.

Question 10 A+ Study Notes pp.95; 99 ●●

a Eosinophil

b
- Three correct labels, such as allergen, antibodies, antibody binding site, cross-linking of allergen, histamines, etc.   (1 mark)
- Elaboration of terms/labels   (1 mark)
- A description of events occurring   (1 mark)
- A complete description of the allergic reaction. This question gave students the opportunity to demonstrate their knowledge and be suitably rewarded for the depth of their response.   (1 mark)

The following is an example of a good response: Antibodies to the allergen are produced by the humoral response. These antibodies bind to the mast cell and act as receptors to the allergen. When the allergen is again encountered, the mast cells release histamines, which cause the allergic symptoms such as swelling and itchiness.

Solutions to Test 8: Acquiring immunity U4 Topic 1.1.2

Multiple-choice solutions

Question 1 A+ Study Notes p.105

C Vaccination with a solution containing killed viruses.

Vaccination initiates the adaptive immune response to form memory cells – subsequent exposures will be larger and more rapid. **A** is incorrect because antivenom contains antibodies specific to the spider bite. These proteins are broken down by the immune system. **B** and **D** are incorrect because the antibodies provide immediate protection but are broken down over time.

Question 2 A+ Study Notes p.101

D Are the major sites of B and T cells.

Naïve mature B and T cells and memory cells reside in the lymph nodes for activation through antigen-presenting cells or free-floating antigens. **A** is incorrect because leukocytes circulate in the blood. **B** is incorrect because lymph nodes enlarge due to proliferation of cells during an infection. **C** is incorrect because red and white blood cells circulate and reside in the circulatory system.

Question 3 A+ Study Notes p.102

B

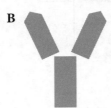

The antigen-binding site is complementary in shape on both sides to one of the antigens on the pathogen. **A** and **C** are incorrect because the antigen-binding site is the same as the antigen, not complementary. **D** is incorrect because the antigen-binding site on the antibody must be the same on both arms.

Question 4 A+ Study Notes pp.102–104

D Large quantities of antibodies specific to the measles virus remain in the circulation for a short time.

Antibodies continue to circulate in the body until they are broken down and the B plasma cells that produce these undergo apoptosis. **A** is incorrect because antibodies are specific to an antigen. The antigen on the bacteria would be different from that of the virus. **B** is incorrect because antibodies are only produced by B plasma cells. T helper cells release cytokines to activate other immune cells. **C** is incorrect because clonal selection of the naïve B cell produces a single type of antibody. Clonal selection would have occurred for the viral and bacterial antigens.

Question 5 A+ Study Notes pp.102–104

B Patient 2 will suffer from measles for a shorter time than patient 1.

Patient 2 has memory cells and therefore can produce more antibodies in a shorter amount of time, reducing the number of viral particles in the body fluid in a shorter amount of time. **A** is incorrect because there is no information to suggest either has been vaccinated. Patient 2 may have been infected with measles previously. **C** is incorrect because on day 12, patient 2 has a higher concentration of antibodies than patient 1. **D** is incorrect because on subsequent exposures, patient 1 will have memory cell, providing a larger and quicker response.

Question 6　A+ Study Notes　p.104

D　They are activated through antigen-presenting cells and MHC restriction.

Naïve T cells proliferate into cytotoxic T cells and memory cells after the antigen is presented on the MHC marker of an antigen-presenting cell. **A** is incorrect because B plasma cells produce antibodies, which bind to extracellular pathogens **B** is incorrect because T cells are unable to recognise free-floating antigens or target infected cells. **C** is incorrect because histamine causes vasodilation and increased vascular permeability. The immune response is rapid and the adaptive immune response takes days to be activated.

Question 7　A+ Study Notes　p.104

D　Cytotoxic T cells.

Foreign proteins can be presented on the MHCI markers of the infected cells. These are recognised by specific cytotoxic T cells and apoptosis is initiated. **A** is incorrect because antibodies bind to extracellular pathogens through their antigens. **B** is incorrect because T helper cells release cytokines to activate other immune cells. **C** is incorrect because complement proteins bind to extracellular pathogens.

Question 8　A+ Study Notes　p.104

C

Apoptosis of some B and T cells after infection. Memory cells reside in lymph nodes.	Second line involved in the response to assist the antibodies after neutralisation of the toxin/pathogen.

Active cells (cytotoxic and B plasma cells) undergo apoptosis after the infection has cleared. The innate immune system is activated with the presence of a pathogen and assists in the breakdown of the pathogen/toxin after the antibodies from an injection have neutralised. **A** is incorrect because on subsequent exposure to the pathogen the immune system would respond at the same rate in short-term immunity; therefore, no is memory formed. **B** is incorrect because inflammation is associated with the second line of defence. **D** is incorrect because broad-spectrum antibodies provide no memory cells (short term). Vaccination initiates an adaptive immune response (long term).

Question 9　A+ Study Notes　p.101

D　Develop from bone marrow cells.

All leukocytes and lymphocytes are formed in the bone marrow. Once these cells are mature, they reside in the lymph nodes and lymph vessels. **A** is incorrect because lymphocytes include cell mediates, humoral and natural killer cells. **B** is incorrect because T cells mature in the thymus, but B cells mature in the bone marrow. **C** is incorrect because lymphocytes reside in the lymph vessels and lymph nodes. Leukocytes circulate in the blood.

Question 10　A+ Study Notes　p.102

A　Production of antibodies by plasma cells.

Antibodies are specific to an antigen. **B** is incorrect because histamine is released if a mast cell is damaged in an area of connective tissue. **C** is incorrect because phagocytes are part of the innate immune response. They engulf any foreign material. **D** is incorrect because complement proteins bind to any foreign material. They have similar functions to antibodies but are not specific to an antigen.

Question 11　A+ Study Notes　p.105

A　An antigen.

An antigen is anything that initiates an immune response. The scientists would be isolating a protein which could initiate a response. **B** is incorrect because an allergen is a molecule that produces an exaggerated immune response and inflammation. **C** is incorrect because the antibodies are what are formed to bind to the antigen from the B plasma cells. **D** is incorrect because the complement protein is part of the innate immune response.

Question 12 `A+ Study Notes` `p.24`

B   Mitochondria.

Exocytosis of the antibodies (proteins) in vesicles requires a large amount of ATP. **A** is incorrect because each cell only contains a single nucleolus where the ribosomal subunits are produced. **C** is incorrect because centrioles are involved in cell division, not the synthesis of proteins for export. **D** is incorrect because proteins synthesised for export are produced on the ribosomes on the endoplasmic reticulum.

Question 13 `A+ Study Notes` `p.104`

C   The innate immune response would remain the same, whereas the adaptive immune response would be faster and larger than the first exposure.

The memory cells would increase the size and response of the adaptive immune response. The innate immune response will always remain the same rate regardless of how many times it has encountered a particular pathogen. **A** is incorrect because the adaptive immune response would have memory cells, which can proliferate and differentiate faster on subsequent exposures. **B** is incorrect because the innate immune response has no memory of encountering the varicella virus and therefore will respond the same as first exposure. **D** is incorrect because the innate immune response does not speed up with age.

Question 14 `A+ Study Notes` `p.105`

C   Received artificial passive immunity and would need to be reinjected with antivenom.

Artificial passive antibodies were provided when they were injected into the patient. **A** is incorrect because no memory cells are formed in response to the venom. Antivenom contains the antibodies for the venom antigen. **B** is incorrect because antibodies are proteins and therefore broken down by the patient's immune system over time. **D** is incorrect because the patient has not formed memory cells and has not encountered the antibodies naturally (not from breast milk).

Question 15 `A+ Study Notes` `pp.100–101`

C   Circulates the macrophages, neutrophils, mast cells and platelets throughout the body.

Although macrophages and neutrophils can be found in the lymph fluid, platelets are only found in the circulatory system, and mast cells are not circulating cells. **A** is incorrect because the valves ensure the fluid flows in a single direction from the body through the lymph nodes to be filtered past the naïve B and T cells. **B** is incorrect because the afferent is entering into the lymph node and the fluid leaving is the efferent. **D** is incorrect because the lymphatic system does not contain a pump like the circulatory system does, so the movement of muscles facilitate the fluid movement.

Question 16 `A+ Study Notes` `pp.101–102`

B   Presents on the MHC II marker, activating the T helper cell to release interleukins to activate naive B and T cells.

The antigen-binding cell binds onto the T helper cell, releasing the interleukin that is required, as well as naïve cell activation with the antigen for clonal expansion. **A** is incorrect because MHC I markers are on all nucleated cells, while T helper cells recognise MHC II markers. They also release interleukin, not interferon. **C** is incorrect because the T helper cell does not come into contact directly with the naïve B and T cells. Naïve B cells encounter free -floating antigens and naïve T cells recognise antigens on antigen-presenting cells MHC I marker. **D** is incorrect because the T helper and naïve T cell can only recognise antigens when presented on MHC markers – not free-floating antigens.

Question 17 `A+ Study Notes` `p.104`

B   Cytotoxic T cells.

The infected cells will present proteins on their MHC I marker, which will be recognised by the specific cytotoxic T cell, initiating apoptosis or lysing the cell. **A** is incorrect because the T helper cells are not involved directly with infection; they release cytokines to activate other immune cells. **C** is incorrect because antibodies are only able to bind to pathogens and antigens in the extracellular fluid. **D** is incorrect because the stem of the question is specifically looking at the toxin antigen.

Question 18 `A+ Study Notes` `p.105`

D Artificially acquired, active immunity.

The pathogen is encountered by the body through an injection (artificial) and the adaptive immune system is activated, forming memory cells for future exposure (active). **A** is incorrect because the fact did not catch the virus (natural) and would have formed memory cells with a vaccination (active not passive). **B** is incorrect because the infant is receiving a vaccine (artificial) not encountering the pathogen (natural). **C** is incorrect because vaccinations produce memory cells (active).

Question 19 `A+ Study Notes` `p.101`

C Reduced production of T cells.

The T cells undergo maturation and self-tolerance in the thymus. Without this organ, there would be fewer T cells present, thus a reduction in the cell-mediated response. **A** is incorrect because the lymph nodes would contain fewer naïve T cells if the thymus is unable to mature these cells. **B** is incorrect because B cells are produced and mature in the bone marrow. **D** is incorrect because mast cells are part of the innate immune system and do not mature in the thymus.

Question 20 `A+ Study Notes` `p.102`

D HIV antigen has a complementary shape specific to the HIV antibody.

The HIV antibody in the patient's serum would have a variable region on each side that is specific to the HIV antigen. **A** is incorrect because the dye is added at the end of the process and reacts with the enzyme attached to the man-made antibody. **B** is incorrect because the HIV antigen is bound to the surface of the test tube and the enzyme is attached to the antibody. The enzyme has an active site complementary to the dye, which is washed over. **C** is incorrect because the man-made antibody will bind to the HIV antibodies if they are present in the individual's sample. If they are not present, then the man-made antibody will not bind and will be washed off.

Short-answer solutions

Question 1 `A+ Study Notes` `p.105`

a The flu virus mutates at a high rate, altering the amino acid structure of the proteins, thus altering the antigens on the surface.

b The antigen is engulfed by a macrophage or dendritic cell and migrates to lymph node.   (1 mark)

The antigen is presented to a T helper cell, releasing cytokines to the naïve B cell.   (1 mark)

The antigen is encountered by a specific naïve B cell, activating it. This is an example of clonal expansion (proliferation and differentiation into B plasma cells and B memory cells).   (1 mark)

Before clonal expansion, the naïve B cell must encounter the free-floating antigen and the cytokines from the T helper cell. These processes occur simultaneously in the lymph node.

c Dead or attenuated pathogen/protein coat/antigens.

Adjuvant to increase the immune response. Adjuvants are added to vaccines to increase the immune response, by creating a more local reaction and increasing the number of antibodies produced.

d Two of:
- Decreases mortality (death)
- Decreases morbidity (poor health) within the population
- Decreases potential hosts or carriers; therefore, protects immunocompromised/elderly
- Economic value – increases productivity and the economy (fewer individuals off sick)
- Decreases healthcare costs.

You could also approach this question in terms of immunity or the economic value.

Question 2 A+ Study Notes p.102

a B plasma cells

b Amino acid

Antibodies are types of proteins.

c

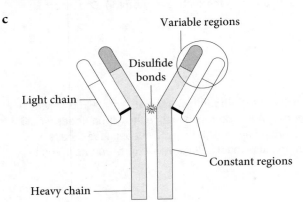

d Two of:
- Agglutination – clumping the pathogens together.
- Opsonisation – binding onto the pathogen to attract other leukocytes.
- Neutralisation – binding to the pathogen to prevent entry into cells or pathogenicity.
- Lysing in combination with complement proteins.

Question 3 A+ Study Notes pp.100–102

a Keeps the lymph flow in one direction/out of the lymph node.

b Answer could include:

Phagocytes (macrophages/dendritic cells) engulf and destroy pathogens.

Mast cells release histamines that cause vasodilation.

Natural killer cells develop in the lymph tissue.

c B cells produce antibodies, rough endoplasmic reticulum produces protein that makes up antibodies.

Some students gave inappropriate cell types, including memory cells and phagocytes. Transport and packaging of proteins was also incorrect (this is a role of the Golgi complex). Students were required to name a cell type of the adaptive immune response and give the function of rough endoplasmic reticulum: making proteins.

Question 4 A+ Study Notes p.102

a **i** Light chain

   ii Constant region

   iii Hinge region

b **i** They comprise the antigen-binding sites. They bind to foreign antigens, producing an antigen–antibody complex.

   ii Each different antibody is made with a shape that is specific to a particular antigen. Variety in antigen-binding sites enables the immune system to respond specifically to a large variety of foreign antigens.

c Two of:

- Antibodies can bind to antigens on two pathogens causing clumping or agglutination.
- The antigen–antibody complex promotes phagocytosis by making the pathogen more recognisable to phagocytes.
- The antigen–antibody complex activates complement proteins.
- The binding of antibody to an antigen, which is a toxin, neutralises the toxin.

Question 5 A+ Study Notes pp.102–103

a Humoral response.

b Stage 1: clonal selection

Stage 2: clonal expansion

c A: naïve B cell, B: free-floating antigen, C: T helper cell, D: antigen presenting cell (macrophage or dendritic cell). (1 mark for identifying each)

Naïve B cell encounters a free-floating antigen specific to their receptor. (1 mark)

Antigen-presenting cell presents the antigen to the T helper cell on the MHC II marker. (1 mark)

T helper cell releases cytokines/interleukin to the stimulated naïve B cell. (1 mark)

d Proteins – antibodies are produced at the rough endoplasmic reticulum and modified in the Golgi apparatus packaged into a vesicle. This fuses with the membrane to release the protein. (1 mark)

Exocytosis ATP required. (1 mark)

Question 6 A+ Study Notes pp.24; 102

a T helper cells release cytokines (interleukins) (1 mark), which stimulate clonal expansion of the B and T cells (1 mark).

b These people lack the third line of defence – specific immunity. (1 mark)

Therefore, the common cold can continue to spread throughout the body (opportunistic pathogen), which can eventually lead to death. (1 mark)

c Antiviral

d Histamine is released from damaged mast cells, causing vasodilation and increased vascular permeability. This would result in swelling and aching experienced in the affected area. (1 mark)

Damaged cells release chemokines, resulting in chemotaxis leading to migration of neutrophils and macrophages to the area (low to high chemical concentration), again resulting in swelling and aching. (1 mark)

Macrophages release interleukin, which stimulates the hypothalamus to raise the body temperature, resulting in fever. (1 mark)

High-achieving students will link each of the marks to the symptoms being addressed in the stem of the question. This is a good habit to develop to ensure answers are being specific to the scenario.

Question 7 A+ Study Notes pp.93; 96

a Intact skin

Your answer must refer to the skin as intact. 'Skin' alone is not accepted by VCAA.

b The immediate response would be histamine release from damaged mast cells, resulting in vasodilation and increased vascular permeability. (1 mark)

One of (1 mark):
- chemotaxis (cytokine release from damaged cells) would occur to enable the migration of neutrophils and macrophages to the site of infection (low to high chemical concentration).
- phagocytosis of foreign materials and debris by neutrophils.

c The first line of defence is non-specific consistent barriers that remain in a specific area (1 mark), whereas the second line of defence is non-specific and is able to move to the site of infection and be upregulated in an infection. The second line of defence assists in the activation of the specific line of defence (1 mark).

Use linking words such as 'while', 'whereas' or 'compared to' in your answer.

d Timing: some of the defences are instant, whereas some are activated within a few hours and the specific immune system is active within days.

Having two lines of defence forms a consistent barrier, which reduces a large amount of exposure, but when breached there are other ways to protect the body.

Question 8 A+ Study Notes pp.100–104

a Thymus

b Macrophage or dendritic cell

c MHC I – antigen presented on MHC I to the naïve T cell (clonal selection). (1 mark)

MHC II – antigen presented on MHC II to the T helper cell, which releases cytokines to stimulate the naïve B and T cells into clonal expansion after clonal selection. (1 mark)

d Cytotoxic T cell – active cells that recognise the specific antigen on viral-infected or cancer cells on the MHC I marker and initiate apoptosis or lyse with perforin. (1 mark)

T memory cells – reside in the lymph nodes for faster and larger responses on subsequent exposure to the same antigen. (1 mark)

e The transplant will have different proteins and MHC markers which will be recognised as foreign. (1 mark)

These antigens will be presented to the adaptive cells and the cell-mediated response will be activated. Cells will be removed by the cytotoxic T cells. (1 mark)

f The immune response for a kidney transplant is due to the non-self-antigens being recognised and removed in a similar way to pathogens. (1 mark)

Self-tolerance is where the naïve B or T cells which react to auto antigens are not removed in the maturation process. (1 mark)

These cells recognise the self-antigens in a similar way to pathogens. (1 mark)

The implication is of cells not undergoing self-tolerance during maturation. This process is where cells are exposed to different proteins from around the body to ensure they do not initiate an immune response against the host's own cells.

Question 9 A+ Study Notes p.105

a The polio virus has a slow mutation rate and therefore doesn't change its surface antigens. However, four doses are required to increase the memory cells and produce a larger and more rapid response upon subsequent exposures and thus provide long-lasting immunity. (1 mark) The immune response to a previous influenza vaccine would be ineffective against the new antigens of mutated influenza viruses and so a new vaccine is required every year (1 mark).

b

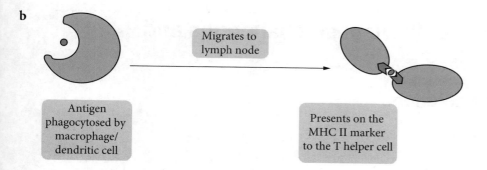

Migrates to lymph node

Antigen phagocytosed by macrophage/ dendritic cell

Presents on the MHC II marker to the T helper cell

c Memory cells remain in the lymph nodes after infection (1 mark) and produce a larger and faster response in subsequent exposure (1 mark).

d

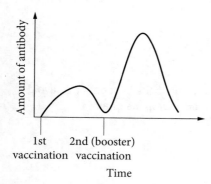

The graph needs to show a steeper and larger second line than the first line. This shows your understanding that the memory cells will proliferate and differentiate quicker producing a larger response on subsequent exposures.

Question 10 A+ Study Notes p.105

a A vaccine contains antigens of EVD
- specific antibodies are produced or antibodies are produced against EVD
- memory cells are produced
- future response is (one of): immediate, faster or greater
- herd immunity.

Many students were able to provide responses that were set out well. It was important that students related their answers to the Ebola virus disease (EVD).

b A humoral response involves B cells and results in the production of antibodies that act against the pathogen.

A cell-mediated response involves T_c cells/cytotoxic T cells (killing infected cells).

Many students' responses were not detailed enough. It is important to note that T cells are in both immune responses. T_c cells are not natural killer cells, and the term 'killer cells' is not appropriate. Students needed to identify which immune response they were referring to. Marks could not be awarded for answers such as, 'One has B cells and the other does not'.

c Implementation of a vaccine program for a whole community means the virus will find it difficult to travel from one person to another, thereby reducing the likelihood of acquiring the virus.

Solutions to Test 9: Disease challenges and strategies U4 Topic 1.1.3

Multiple-choice solutions

Question 1 A+ Study Notes p.106

C Lung disease.

Lung disease is a non-contagious disease that develops over a long period of time. **A** is incorrect because influenza is an infectious virus spread by droplets from infected individuals. **B** is incorrect because smallpox is an infectious virus. **D** is incorrect because measles is a highly infectious disease spread by droplets from infected individuals.

Question 2 A+ Study Notes p.109

A Hand washing

As bacterial gastroenteritis is spread by a bacterial pathogen, hand washing is the best option to avoid spreading the bacteria. **B – D** are incorrect as the pathogen is not spread in droplets or air.

Question 3 A+ Study Notes p.109

D Disrupt the synthesis of new bacterial cell components.

The antibiotic can inhibit the formation of new components, preventing bacterial replication. **A** is incorrect because antibiotics cannot function against viruses. **B** is incorrect because antibiotics target bacteria, which do not cause the fever. The fever is induced by interleukin produced by the body's macrophages. **C** is incorrect because antibiotics are not able to function against viruses.

Question 4 A+ Study Notes p.108

D Sequencing the microbial genome and identifying proteins.

This will allow identification of key proteins that are found in each group of microbial pathogens.
A is incorrect because only cellular pathogens can grow on agar plates. **B** is incorrect because gel electrophoresis can only be used for DNA, RNA and proteins, not whole organisms. **C** is incorrect because all of the antibodies would be similar in structure except for the variable region of the protein.

Question 5 A+ Study Notes pp.110–111

C that people who have negative beliefs about vaccinations often live in the same area.

Herd immunity relies on unvaccinated people not coming into contact with each other. If all these people live in the same area then there is more chance they will interact. **A** and **B** are incorrect as these people may not live in the same area and hence not come into contact with each other. **D** is incorrect because although a correct statement, it is not relevant to this context.

Question 6 A+ Study Notes p.108

B Antibodies that are specific to the virus.

The antibodies are self-proteins produced to protect the body against specific epitopes on the virus. **A** is incorrect because receptors on the surface would be recognised by the body. **C** is incorrect because preventing viruses from leaving an infected cell gives the body time to recognise and kill the infected cell. **D** is incorrect because preventing the viral genome from being inserted ensures that the body can completely clear the virus.

Question 7 A+ Study Notes p.107

C A pandemic spreads across continents, whereas an epidemic is more localised.

Pandemics are multiregional and widespread, whereas an epidemic is a high number of cases in a localised region. **A** and **B** are incorrect because an epidemic is smaller than a pandemic. **D** is incorrect because a pandemic is multiregional and can occur in several continents but does not have to be in all countries, and an epidemic can be within a country.

Question 8 A+ Study Notes p.107

D Type of infective agent.

The symptoms from the pathogen determine how deadly the disease is, not the type of pathogen. **A** is incorrect because the more severe the symptoms, the higher the death rate. **B** is incorrect because the more vulnerable the population, the higher the death rate. **C** is incorrect because the fewer individuals infected, the fewer deaths are associated.

Question 9 A+ Study Notes pp.111–112

B Make the cancer more visible to the immune system.

The antibodies bind to the antigens expressed on the cancer cells to attract other immune cells to the area. **A** is incorrect because cytotoxic T cells are already activated and circulating through the bloodstream. **C** is incorrect because memory B cells are formed during clonal expansion in the lymph nodes. **D** is incorrect because division of cancer cells would increase the disease and cause more harm to the individual.

Question 10 A+ Study Notes p.108

B Identification of a unique DNA or RNA sequence coding for a specific protein.

A is incorrect because only cellular pathogens can grow on an agar plate. **C** is incorrect because only DNA or RNA can be replicated through PCR, not whole pathogens. **D** is incorrect because some pathogens cannot be visualised under a microscope.

Question 11 A+ Study Notes p.106

C Prevent the build-up of resistance in the population.

Not taking the full course of antibiotics can increase the antibiotic resistance of subsequent infections. People with a mild infection and a functioning immune system do not require antibiotics. **A** is incorrect because antibiotics are only prescribed by medical professionals for bacterial infections. **B** is incorrect because the advertisement is about the correct use and administration, not the cost of these drugs. **D** is incorrect because some illnesses are not caused by bacteria.

Question 12 A+ Study Notes pp.109–110

D Getting vaccinated.

The formation of memory cells forms a faster and larger response on subsequent exposure, thereby reducing the symptoms. **A** is incorrect because although probiotics can increase the levels of 'good' bacteria, neither vitamins nor probiotics directly fight viral infections. **B** is incorrect because antibiotics are effective against bacterial infections not viral. **C** is incorrect because antiseptics are only applied to your external surface (skin) and are non-specific.

Question 13 A+ Study Notes pp.197–198

A There are four periods in which the notification rate is greater than six per 100 000.

B is incorrect because September not always the lowest rate. **C** is incorrect because there are peaks and troughs in each year. **D** is incorrect because in 2001, January had the highest peak.

Question 14 `A+ Study Notes` `p.109`

B Giving probiotics before entering the country.

Probiotics are natural bacteria that benefit the body but do not facilitate viral infections. **A** is incorrect because quarantine would prevent asymptomatic individuals from infecting others before symptoms were present. **C** is incorrect because preventing live export would stop hosts for the virus entering the country, thus preventing the virus transmitting to other birds or humans. **D** is incorrect because if one farm becomes infected, it can be isolated to those animals and prevent transmission to all farms.

Question 15 `A+ Study Notes` `pp.111–112`

A A protein on the outside of the cancer cells.

Proteins upregulated on the surface of cancer cells can be bound by antibodies as they face the extracellular fluid. **B** is incorrect because the removal of the virus would not prevent the cancer cell growth. **C** is incorrect because the white blood cells will recognise the antibodies bound and initiate apoptosis of the cancer cells. **D** is incorrect because antibodies are unable to bind to intracellular molecules.

Question 16 `A+ Study Notes` `p.110`

B Outer-membrane proteins pertactin and filamentous haemagglutinin.

These are located on the extracellular surface of these specific bacteria. **A** is incorrect because the operon is within the bacterium and will not enable the immune system to recognise the bacteria as it enters the body. **C** is incorrect because there are many different gram-negative bacteria, some of which are the natural flora of the body. **D** is incorrect because the lactose enzyme is intracellular.

Question 17 `A+ Study Notes` `p.107`

A The viral vector is a mosquito.

The mosquito transmits from one host to another. **B** is incorrect because viruses are obligate parasites. **C** is incorrect because the virus lives in kangaroos and wallabies and is transmitted through the mosquito to different hosts. **D** is incorrect because only infected animals may fall ill and not all would die.

Question 18 `A+ Study Notes` `pp.111–112`

A B plasma cells are only short lived and clear through apoptosis after an infection.

B plasma cells are stimulated through the presence of an infection. **B** is incorrect because the B plasma cell is specific to the cancer cell isolated from the patient, not the myeloma cell. **C** is incorrect because the antibodies produced from a B plasma cell are clones of each other; they are specific to one antigen. **D** is incorrect because the hybridoma cells are not inserted into the patients, only the continual antibodies they produce.

Question 19 `A+ Study Notes` `p.102`

D In females, childhood exposure to *H. pylori* helps to protect against MS.

The exposure to the *H. pylori* was lower (30%) in the MS patients than in the healthy male and females. **A** is incorrect because the antibodies would recognise self-immune cells. **B** is incorrect because the rates of infection by MS were not discussed in the information above, only the *H. pylori* rates. **C** is incorrect because the stomach ulcers are a result of the *H. pylori* bacteria, not the MS autoimmune disorder.

Question 20 `A+ Study Notes` `pp.111–112`

D The constant region of the heavy chain of the monoclonal antibody.

Enable the variable antigen binding sites to be exposed to bind onto the cancerous B cells via CD20. **A** is incorrect because the variable region is where the antibody will bind to the CD20 antigen on the cancer cells. **B** is incorrect because the antigen binding site is specific to the CD20 on the B cells. **C** is incorrect because this is within the body, the antibody facilitates the drug to the cancerous cells.

SOLUTIONS – TEST 9

Short-answer solutions

Question 1 A+ Study Notes pp.111–112

a The antigen is inserted into a mouse for the adaptive immune system to be activated. The B plasma cell is isolated from blood plasma.

(1 mark)

The B plasma cell and myeloma cell are fused to form a hybridoma cell. These cells are cultured to produce antibodies for the specific antigen.

(1 mark)

Antibodies are isolated and inserted into the patient.

(1 mark)

b One of:
- The antibodies are bound with a radioactive or chemotherapy drug, which can initiate apoptosis in these cells.
- The antibodies bind to the cancer cells to attract the immune cells – cytotoxic T cells and natural killer cells.

c There are only a few antigens that are present on cancer cells that are not expressed or present in normal healthy cells.

Due to cancer cells being the individuals own cells, they are very similar to normal functioning cells. Cancer cell are cells which have become immortal and that have accumulated mutations to override apoptotic signals or inhibiting growth signals.

Question 2 A+ Study Notes p.100

	Bacteria	Virus	Prion	Fungus
Contains genetic material	Yes	Yes	No	Yes
Prokaryote, eukaryotic or neither	Prokaryote	Neither	Neither	Eukaryote
Replication via	Binary fission	Host cell machinery	In contact with healthy proteins	Asexual reproduction – spores
Treatment	Antibiotics	Antivirals	None	Antifungals

You would receive 1 mark for each of the correctly identified pathogens.

Question 3 A+ Study Notes pp.109–110

a Virus protein coat, surface proteins, or inactivated or attenuated virus.

b Artificial active immunity.

c

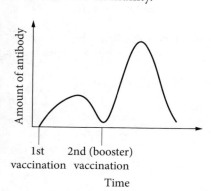

The booster should promote a larger and steeper increase because of the memory cells being present.

d Two of:
- lack of vaccination
- lack of education
- poor sanitation and hygiene
- high population density
- war and hardship preventing access to healthcare.

Question 4 A+ Study Notes pp.93–94; 100; 109

a Non-cellular (no mark) and two of:
- not made of cells
- does not undergo cellular processes
- only reproduced by a host cell.

b Scenario 1 (general viral first line of defence) Intact skin.

Either secretions of mucus or ciliated epithelia.

Scenario 2 (students acknowledged that the mosquito carries the virus)

Two of:
- hair or skin as a physical barrier
- secretions; for example, sweat to deter mosquitoes.

Scenario 3 (students acknowledged that the mosquito breaches first line of defence (intact skin)
Students needed to include a statement that the skin has been breached and give an example of a
defence mechanism in the second line, such as cells releasing interferon.

Many answers were accepted for this question. As stated in the stem of the question, yellow fever is caused
by a virus and is transmitted through the bite of a particular species of mosquito, thus bypassing the first line
of defence mechanisms. Students were required to answer this question with reference to viruses and not
other pathogens, such as bacteria. Note that this is a VCAA question, but interferons are technically second
line of defence.

c Prevent the disease from entering the Australian population and one of:
- quarantine
- treatment
- test for the presence of the virus.

Vaccination alone was not accepted as a solution.

Question 5 A+ Study Notes pp.108–109

a One of:
- place on an agar plate to observe growth
- isolate the genome and proteins from the pathogen from PCR or gel electrophoresis.

b PCR with specific primers for different strains. Only the complementary strands will be
replicated. (1 mark)

Once the DNA had been amplified, a gel can be run with a DNA ladder to determine the
strains. (1 mark)

c Antivirals (1 mark)

One of (1 mark):
- prevent viral release from cells
- prevent viral protein synthesis
- prevent viral attachment to cells.

Question 6 `A+ Study Notes` `pp.106–107`

a A pathogen is a cellular or non-cellular agent that causes disease within the host.

b Pandemic. Multiregional widespread – multiple countries.

c Two of:
- highly contagious
- spread through air droplets and close human contact
- low sanitation
- high population density
- incubation period.

Question 7 `A+ Study Notes` `pp.106–107; 109`

a Lack of immunity due to no exposure to these diseases.

No immunity or education on how the diseases are transmitted or treated.

No treatment for these diseases.

Lack of education about the European pathogens and their transmission/pathogenicity.

b Two of: infected individuals, social distancing between families, increasing sanitation and hygiene.

Protective gear, masks or gloves is not an adequate answer because we cannot guarantee that they existed at the time of the outbreak.

Question 8 `A+ Study Notes` `pp.106; 108`

a Two of:
- bacteria becoming resistant to drugs
- lack of vaccination
- unvaccinated people travelling to high-risk countries
- mutations in *Mycobacterium tuberculosis*.

b Antibiotics

c Place discs with different antibiotics on an agar plate with *Mycobacterium tuberculosis*.   (1 mark)

The larger the zone of inhibition (lack of bacterial growth), the more effective the antibiotic.   (1 mark)

Question 9 `A+ Study Notes` `p.109`

Two of:
- Food. Change of diet from foraging and native foods to processed, high-sugar foods and other Westernised foods could have increased lifestyle diseases.
- Permeant settlements. The build-up of waste without proper sewerage systems can increase communicable diseases.
- Larger communities. The increase in population density can increase host contact with vulnerable individuals. There would be low resistance to new diseases within the populations, increasing morbidity and mortality rates.
- Interaction with foreign individuals. Exposure to new diseases and pathogens.

Question 10 A+ Study Notes pp.93; 100

a   Through a break in skin

b   i   Natural variation in the ability to resist antibiotics exists in bacterial populations. When exposed to antibiotics, resistant bacteria are selected for and survive; they are most fit. Non-resistant bacteria are selected against and do not survive to reproduce. The surviving resistant bacteria reproduce, passing on this resistance to their offspring. The next generation contains a larger proportion of resistant bacteria.

   ii   Hospitals provide an environment where there is extensive use of antibiotics and antiseptics. The wide range of these chemicals means that all but resistant bacteria are selected against, producing populations of resistant bacteria. Hospitals also contain good breeding grounds for bacteria, such as surgical wounds and other injuries.

c   Two of:
   - short breeding time or generation time
   - very large populations and high genetic variation within populations
   - ability to exchange genetic material through exchange of plasmids
   - high mutation rate.

Solutions to Test 10: Genetic changes in a population over time [U4 Topic 2.2.1]

Multiple-choice solutions

Question 1 [A+ Study Notes] [p.132]

D Cell division.

The cell that is dividing will produce no new alleles unless a mutation occurs during the replication of the DNA. **A** is incorrect because migration can introduce new alleles or remove alleles as organisms move into or out of the population. **B** is incorrect because the death of an organism can remove alleles from a population if they are the only organism to have the allele. **C** is incorrect because mutations can increase the number of alleles within a population.

Question 2 [A+ Study Notes] [pp.132–133]

C A mutation that causes a colour change that blends with the surroundings.

A mutation occurs by chance. If the new allele is best suited to the environment, the insects are more likely to reproduce and this allele will remain in the population. A is incorrect because the dark-coloured moths will camouflage better in the environment and be less obvious to predators. **B** is incorrect because the organisms that survive the insecticide are better suited to the environment and will reproduce to increase the number of resistant organisms. **D** is incorrect because the antibiotic will remove some of the bacteria that are not resistant to it.

Question 3 [A+ Study Notes] [p.132]

B Frequency of allele D was 39% in the original population compared to 22% in the recovered population.

7 of the 18 alleles are dominant (D) in the original population, therefore 7/18 × 100 = 39%; whereas 4 of the 18 alleles in the recovered population are dominant (D) therefore 4/18 × 100 = 22%.

Question 4 [A+ Study Notes] [p.132]

B There is extensive genetic diversity in the population.

A large gene pool means that there are a lot of different alleles and variation within the population. **A** is incorrect because there could be a large number with a small amount of genetic variation, e.g. after a bottleneck. **C** is incorrect because if there is variation within the population, particular alleles will be better suited to the environment and/or selection pressures. **D** is incorrect because if there are a large number of alleles, a change in the environment should still have alleles that favour this, altering allele frequency but not resulting in extinction.

Question 5 [A+ Study Notes] [p.147]

C Geographical isolation, natural selection, reproductive isolation.

The two populations become isolated due to geographics, the different selection pressures favour different alleles until speciation occurs – reproductive isolation. **A** is incorrect because at the start of the separation of populations, they would still be able to reproduce and produce viable offspring. **B** is incorrect because geographic isolation would have occurred prior to natural selection in their new environments. **D** is incorrect because genetic drift is a chance event in which only a small unrepresentative population remain.

Question 6 A+ Study Notes p.134

B A bottleneck.

The population is rapidly reduced, thereby reducing the gene pool. **A** is incorrect because the founder effect refers to a situation where a small unrepresentative sample leaves the main population. The whole population has been reduced in this example. **C** is incorrect because a mass extinction involves multiple species and is the complete removal of all organisms. **D** is incorrect because natural selection refers to the survival of the fittest phenotype due to selection pressures. Hunting is not based on nature's fittest phenotype.

Question 7 A+ Study Notes p.149

B The production of sterile hybrids.

The two populations have accumulated so many genetic differences that the gametes can still fuse, but the offspring are not able to reproduce. An example of this is a donkey and horse. **A** and **C** are incorrect because the offspring would still be viable, but the two populations do not encounter each other. **D** is incorrect because the organisms of the same population may prefer different habitats and therefore not interact.

Question 8 A+ Study Notes pp.132–133

A Phenotypes.

Variation must exist within the population. **B** is incorrect because different reproductive seasons would isolate the population into two separate populations. **C** is incorrect because some species would not have a mating call. **D** is incorrect because mutations are by chance.

Question 9 A+ Study Notes pp.132–134

A Genetic drift acts on populations while natural selection acts on individuals.

The reduction in the population due to founder effect or bottleneck affects all organisms, whereas natural selection favours individuals that are the fittest in response to the selection pressure. **B** is incorrect because natural selection favours the fittest phenotype in the population whereas genetic drift is random. **C** is incorrect because genetic drift has a larger effect on small populations. **D** is incorrect because genetic drift occurs by chance and not in response to environmental factors.

Question 10 A+ Study Notes p.139

C Decreases genetic variation.

The crossing of particular plants will limit the gene pool to these few plants. **A** is incorrect because selecting a few favourable plants and continually crossing them will reduce the genetic variation. **B** is incorrect because mutations occur by chance. **D** is incorrect because the farmer is selectively breeding the traits he or she desires. Natural selection is the trait best suited to the environment.

Question 11 A+ Study Notes p.137

B Translocation.

A large section of one chromosome has been broken off and added to another chromosome section (block mutation). **A** is incorrect because a whole section of the chromosome has been changed through block mutation. **C** is incorrect because a point mutation is only a single base change in a gene and cannot be seen on the chromosome. **D** is incorrect because a frameshift mutation is only a single base change in a gene and cannot be seen on the chromosome.

Question 12 A+ Study Notes pp.132–134

C A predator.

The change in coat design camouflages better than the spotty coat. **A** is incorrect because a disease would have favoured resistance not the colour of the beetle's coat. **B** is incorrect because the beetles are the same size for accessing food. **D** More information is needed for climate to be the most likely option.

Question 13 — A+ Study Notes p.135

A Migration.

Immigration and emigration can introduce or remove alleles and change allele frequencies. **B** is incorrect because evolution is a gradual change due to natural selection. **C** is incorrect because the fittest phenotypes survive but no new alleles are introduced. **D** is incorrect because mutations occur by chance and don't always remain in a population.

Question 14 — A+ Study Notes pp.132–133

B Sexually reproducing organisms in a variable environment.

Sexual reproduction produces variation; with changing selection pressures, the fittest phenotype will survive. **A** is incorrect because asexual reproduction produces genetic clones of each other. **C** is incorrect because sexual reproduction produces variation but in a stable environment there are no selection pressures favouring different organisms. **D** is incorrect because asexual reproduction produces genetic clones of each other.

Question 15 — A+ Study Notes p.134

B Low genetic diversity among present-day human populations.

The bottleneck reduced the gene pool of the population, all of which, when increased, will be genetically similar until mutations occur. **A** is incorrect because only a few humans may have been killed and preserved in the ash. **C** is incorrect because high levels of variation would not support the bottleneck. **D** is incorrect because human remains will be present in all deposits. This will not support the bottleneck.

Question 16 — A+ Study Notes p.134

A The founder effect.

An unrepresentative population separated from Sweden with higher than average alleles of the disorder. **B** is incorrect because the individuals who migrated to this area were not selected based on the disorder/trait. **C** is incorrect because the high morbidity and mortality of this disorder would not be the 'fittest' phenotype in the population. **D** is incorrect because the individuals migrating will be mating with others in that isolated population only.

Question 17 — A+ Study Notes p.135

C Both species A and species B are able to produce viable gametes.

Both have the same number of chromosomes, with species B having multiple copies. **A** is incorrect because the plants are located in the same area and flower at the same time. **B** is incorrect because the plants are located in the same area. **D** is incorrect because the species would have had the same number of homologous chromosomes.

Question 18 — A+ Study Notes p.139

B II ⟶ IV ⟶ I ⟶ III

A is incorrect because the desired characteristics need to be chosen first. **C** is incorrect because the trait must be selected before it can be repeated over the generations. **D** is incorrect because the trait needs to be decided before it can be bred over many generations.

Question 19 — A+ Study Notes p.136

A AGA

Adenine is complementary to thymine/uracil. **B** is incorrect because guanine is complementary to cytosine, not uracil. **C** is incorrect because none of the complementary base pairs align. **D** is incorrect because uracil is complementary to adenine, not thymine.

Question 20 A+ Study Notes p.136

B Nucleotide substitution.

C is replaced with A in the DNA. **A** is incorrect because a chromosomal deletion affects multiple genes. **C** is incorrect because only one nucleotide is altered, not all of the codons that follow. **D** is incorrect because the codons that follow are not affected.

Short-answer solutions

Question 1 A+ Study Notes pp.132–135

a The movement of alleles from one population to another due to the migration of individuals or seed dispersal.

b Despite their differences, they are still able to interbreed to produce viable, fertile offspring, or they are not reproductively isolated from each other.

c The tortoises and the prickly pear plants have each operated as the major selection pressure on the evolution of each other. (1 mark)

The plants have evolved growth habits in response to the browsing of the tortoises, and the long-necked tortoises have been selected for as a response to the taller, thickened trunks of the prickly pear. (1 mark)

d **i** Bottleneck effect

   ii Gene pool will be unrepresentative of the original gene pool of prickly pears and some alleles could have been lost from the new population. (1 mark)

   The offspring from the remaining population of prickly pears will be genetically similar. New alleles will only appear through mutations. (1 mark)

Question 2 A+ Study Notes pp.132–133

a Those organisms that are more easily seen by predators may be eaten more often and thus reproduce less. (1 mark)

Camouflaged organisms may be more likely to survive and hence will reproduce in greater numbers and pass this favourable colouring on to their offspring. (1 mark)

b Female birds select mates on the basis of the males' mating displays. Brightly coloured birds may be more likely to be chosen and mate and pass on the genes for bright colours. (1 mark)

These alleles increase in frequency in the population. The alleles of birds that fail to mate are lost from the population. (1 mark)

c Insects with an unpleasant taste only, and no bright colours to act as a deterrent, may be half eaten.

d Selection pressure

e Observation of the predators in the area. If the organisms share a predator, and this predator avoids both species, then this would support the hypothesis that the colouration is a selective advantage.

Question 3 A+ Study Notes pp.132–133

a Through a mutation in an individual that survived and reproduced, passing on the new allele to their offspring.

b By isolating an individual from each population and interbreeding them. (1 mark)

If the two cannot produce viable, fertile offspring, then the two populations have undergone speciation. (1 mark)

They could just be exposed to similar selection pressures – structure is not enough to identify them as the same species.

c Food (access to green seaweed and algae)

d The population would bottleneck and decrease in numbers.   (1 mark)

Only those able to access the limited food (sharper claws or larger to compete for the food) would be able to survive and reproduce.   (1 mark)

Variation could occur if any tortoises are able to digest the brown algae and survive to pass on these alleles onto their offspring.   (1 mark)

Question 4 `A+ Study Notes` `pp.132–133`

a Variation existed between individuals in the ancestral group of sea stars.   (1 mark)

The water temperature changed.   (1 mark)

Different selective pressures act on southern populations compared to the northern populations, such as southern populations becoming exposed to cold-water predators.   (1 mark)

Surviving individuals in each of the populations reproduced and could pass on the advantageous feature.   (1 mark)

> Students would have benefitted from planning their answer prior to writing it. Some students incorrectly treated this as a speciation question, when in fact speciation had already occurred.

b *Cryptasterina pentagona*   (1 mark)

For example, two individuals provide genetic material to the offspring.   (1 mark)

> Many students incorrectly stated that only sexual reproduction occurred in this species, when both species undergo sexual reproduction.

Question 5 `A+ Study Notes` `pp.139–140`

a By inhibiting (two of): cell wall synthesis, protein synthesis, chromosomal replication, cell membrane function or metabolic pathways.

b If variation exists within the population – normally located on a plasmid.   (1 mark)

One of (1 mark):
- Resistant bacteria survive and undergo binary fission increasing the population, which would then be unaffected by antibiotics.
- Horizontal gene transfer occurs, where the bacteria pass antibiotic resistant plasmids to bacteria in the surrounding environment.

c Two of:
- Do not take antibiotics for mild symptoms.
- Take the full course of antibiotics.
- Stay isolated to prevent transmission to others.
- Only take prescribed medication.

d

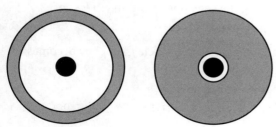

> Zone of inhibition when antibiotic first introduced would be considerably larger than today's. Today's can be minimal or none.

Question 6 A+ Study Notes pp.135; 139

a Non-disjunction in meiosis   (1 mark)

b The favourable traits are selected eg, greater yield; drought tolerance.   (1 mark)

Two plants with the favourable characteristics are crossed, and their offspring with the favourable traits are selected. The offspring with the favourable traits are continuously crossed (1 mark) so that over many generations the favoured trait is present in the farmers crops (1 mark).

c Genetic variation will be reduced.   (1 mark)

The plants with favourable traits will be crossed, reducing the gene pool for all other traits – these traits may not be the best suited to the changed environment and increase loss of crop in changing environments.   (1 mark)

Question 7 A+ Study Notes pp.140–141

a Protein coat with nucleic acid (DNA or RNA).

b Antigenic drift is small genetic changes in the viral DNA that alter the antigens and proteins on the surface of the virus.

c They would require another vaccination because antigens on the surface of the virus may have mutated.   (1 mark)

Because the immune system would only have formed memory cells to the H1N1 and H3N2 strains, new strains would not be recognised.   (1 mark)

d A virus enters a host cell and uses cell organelles to replicate new viral particles.   (1 mark)

New viral particles are assembled, and the virus lyses the cell, releasing the viruses into the extracellular fluid and infecting nearby cells.   (1 mark)

e Two of (1 mark):
- Quarantine and/or isolation when infected with the flu
- Use of masks
- Hygiene and good sanitation.

Question 8 A+ Study Notes pp.134–135; 141

a Variation within the population decreases the likelihood of extinction.   (1 mark)

A large gene pool enables some of the population to survive in changing environmental conditions.   (1 mark)

b Migration (immigration vs emigration) results in immediate change of the gene pool.   (1 mark)

Mutation occurs by chance and would be much slower. The individual with the mutation must survive and reproduce to pass the allele onto the future generations.   (1 mark)

c Bottleneck – sharp reduction in the population, reducing the gene pool, resulting in loss of alleles.   (1 mark)

Founder effect – unrepresentative sample of population with different allele frequency becomes isolated from the main population.   (1 mark)

Question 9 A+ Study Notes pp.133; 135

a Gene flow refers to the movement of alleles between populations by interbreeding. The rufous bristlebird is a weak flyer. The populations are geographically isolated. The rufous bristlebird is unlikely to interbreed.

b A small population size would mean a limited gene pool and reduced variation would reduce the chances of survival.

If there were an environmental change, the chance of a favourable characteristic existing in the population is unlikely because individuals would all be genetically similar.

There is insufficient variation in the population to survive within their current environment.

Inbreeding could result in an increased chance of genetic diseases.

There is a change in allele frequency due to genetic drift.

Many students incorrectly assumed that gene flow is another name for migration. Students could also have answered the question by demonstrating an understanding of natural selection with respect to the bristlebird, such as 'If there were a predator introduced as the population is genetically similar, there may be no suitable variation present. As the bristlebird is ground-dwelling, their eggs may be eaten and the population would be at risk of extinction'.

Question 10 A+ Study Notes p.134

a i Non-representative allele frequencies in a founding population.

ii A small population colonised a new area. The individuals in that small colony had a higher allele frequency than appeared in the parent population's gene pool. The genes were then passed on to subsequent generations.

b Inbreeding makes it more likely that two individuals who are carriers will reproduce and produce affected offspring.

c Genetic drift. Chance events, such as individuals with the allele producing more than the average number of offspring, can cause an increase in gene frequency.

d It is likely that the frequency of the affected gene will increase. More affected individuals survive to adulthood and reproduce, meaning their genes are not lost from the population.

Solutions to Test 11: Changes in species over time U4 Topic 2.2.2

Multiple-choice solutions

Question 1 A+ Study Notes p.148

A Adaptive radiation.

This is rapid evolution in a short period of time from one ancestral species. **B** is incorrect because convergent evolution is where two separate species become more similar. **C** is incorrect because a bottleneck effect is a reduction in the gene pool and population by chance. **D** is incorrect because bottleneck and founder effect are types of genetic drift.

Question 2 A+ Study Notes p.145

D Radiometric dating.

Radiometric dating is an absolute dating technique. **A** is incorrect because index fossils in the same rock strata can give an approximate timescale. **B** is incorrect because the deeper the rock strata, the older the fossil. **C** is incorrect because the older the tree, the more rings it would have.

Question 3 A+ Study Notes p.142

D 4, 2, 1, 3

A is incorrect because multicellular organisms were developed after the engulfment of the mitochondria. **B** is incorrect because eukaryotic cells developed after prokaryotic organisms. **C** is incorrect because species began developing on land before venturing onto land for unoccupied habitats.

Question 4 A+ Study Notes pp.143–144

C Smaller marsupials were better able to survive climatic change and warmer temperatures.

Smaller marsupials had better access to food and habitat change. **A** is incorrect because an increase in food would have enabled megafauna to increase in numbers. **B** is incorrect because the supposed comet striking Earth was millions of years ago. **D** is incorrect because the last ice age was millions of years ago and since then, there has been a gradual rise in sea levels.

Question 5 A+ Study Notes pp.145–146

C Carbon-14.

Carbon-14 has a half-life of 5700 years; it can be used to measure the age of fossil up to 60 000 years old. **A** is incorrect because uranium-235 half-life is 700 million years. **B** is incorrect because potassium-40 half-life is 1.3 billion years. **D** is incorrect because argon-39 breaks down quickly and has a half-life of 269 years.

Question 6 A+ Study Notes p.146

C The molluscs are found in deposits of approximately the same age in each of the locations.

They are found in the same type of rock sediment (same pattern). **A** is incorrect because each mollusc is present in the same type of rock strata (same age), even though they are at different depths in the locations depending on the environment. **B** is incorrect because there is no evidence of when the vertebrates developed. **D** is incorrect because we cannot see all fossils in each of the rock stratum in the diagram.

Question 7 A+ Study Notes p.146

D Location 2 contains the oldest fossils in the rock strata.

When comparing the layers, the oldest layer is present in location 2. **A** is incorrect because, depending on the environment, thickness depends on the amount of sediment compaction. **B** is incorrect because in location 3 there is a rock stratum on top of the bone. **C** is incorrect because, depending on the environment, each may be a different habitat and therefore different species reside in each.

Question 8 A+ Study Notes p.146

B 6700 years.

Approximately 1.25 half-lives of carbon-14 would have passed for 40% of carbon-14 to remain in the sample. 1.25 × 5730 = 7100. **A** is incorrect because it is less than one half-life. **C** is incorrect as it is almost 2 half-lives. **D** is incorrect as it is 3 half-lives.

Question 9 A+ Study Notes pp.148–149

C The altitude of plants from the ocean.

If the plants are at different altitudes, the pollination may occur between plants closer together within that one population. **A** is incorrect because a mountain range is a geographical barrier and would be an example of allopatric speciation. **B** is incorrect because crossing over of flowering season would allow the two populations to have interbred with each other. **D** is incorrect because photosynthetic rate does not affect pollination between two plants.

Question 10 A+ Study Notes pp.148–149

D The new palm is more closely related to *H. forsterianna* than it is to *H. belmorena*.

The gametes of *H. forsterianna* fused and produced an offspring suggesting they are genetically similar and that *H. forsterianna* is the more recent common ancestor. **A** is incorrect because the offspring of the crossed plants are infertile, indicating it is a post-zygotic isolating mechanism. **B** is incorrect because even if the plants were located in the same location, they would be unable to reproduce due to speciation occurring. **C** is incorrect because *H. belmorena* and the new palm are unable to fuse their gametes whereas *H. forsterianna* plants developed, suggesting that the new palm is more closely related to *H. forsterianna*.

Question 11 A+ Study Notes p.147

B The koalas would adapt to their environment and may evolve into separate species over many generations.

A is incorrect because different areas will have different selection pressures such as weather and access to food. **C** is incorrect because Koalas are unable to travel large distances. **D** is incorrect because the koala population are geographically isolated from each other – this is an example of allopatric speciation, not sympatric speciation.

Question 12 A+ Study Notes p.147

D To increase variation within the population to increase survival.

The migration of koalas will increase the gene pool of different populations, reducing the impact of the bottleneck caused by the fires. **A** is incorrect because humans are not breeding the koalas for any specific characteristics for yield or economic benefit. **B** is incorrect because through the exchange of koalas, the artificial migration is increasing the gene pool. **C** is incorrect because the koalas are not being removed from areas.

Question 13 A+ Study Notes p.146

A Was closer to the present-day ground surface than the rock surrounding the *D. pickeringi* fossil.

The deeper the rock strata, the older the rock. Younger layers are deposited onto the already existing layers. **B** is incorrect because carbon dating can only be used to date fossils up to 60 000 years old. These rock strata are millions of years old. **C** is incorrect because 2.3m is the length of the dinosaur. **D** is incorrect because the fossil is in a riverbed and there is no evidence provided it was located near an active volcano.

Question 14 A+ Study Notes p.147

A Antarctica, South America and Africa were joined to Australia in the distant past.

All land masses were once part of a vast continent known as Gondwana 550 million years ago before breaking apart. **B** is incorrect because there is no evidence this species could swim. **C** is incorrect because the small forelimbs could have evolved due to many different selection pressures. **D** is incorrect because fossils need to be buried rapidly and to be undisturbed for long periods to preserve.

Question 15 A+ Study Notes p.146

B Stratigraphy.

Observing layers of rock strata and index fossils is a relative dating technique. **A** is incorrect because although potassium-argon dating can date a fossil at 2 million years, it is not a relative dating technique. **C** is incorrect because radiometric dating is an absolute dating technique. **D** is incorrect because light emitted from surrounding minerals is an absolute dating technique.

Question 16 A+ Study Notes p.146

D An analysis of strata in coal deposits is a more reliable dating technique than carbon dating for Permian fossils; the fossil of *G. major* is younger than the fossil of *G. obliqua*.

Carbon dating is only useful to 60 000 years. After this time, all of the carbon-14 has broken down. **A** is incorrect because they all contain the same genus name. **B** is incorrect because *G. clarkeana* is the youngest of the three fossils because it is the most shallow fossil, so cannot be an ancestor. **C** is incorrect because carbon dating is generally useful to date fossils younger than 60 000 years. These fossils are 290–245 million years old.

Question 17 A+ Study Notes p.145

C Index fossils.

Ammonites are fossils which are widespread and found for a relatively short period of time – in one rock stratum. **A** is incorrect because transitional fossils are species that have characteristics of two or more groups. **B** is incorrect because indirect fossils are impressions or footprints and ammonites are mineralisation of the shells. **D** is incorrect because the ammonite will have mineralised over the 65 million years.

Question 18 A+ Study Notes pp.141–142

D Exoskeletons would not have fossilised.

Hard parts of organisms are more likely to fossilise than organisms with no skeleton. **A** is incorrect because there are still many fossils yet to be discovered that could alter the evolutionary relationships. **B** is incorrect because fossilisation is rare, and rapid burial must occur with a lack of oxygen and scavengers. This must then remain undisturbed for a long period of time. **C** is incorrect because unless rapidly buried, dead organisms are exposed to scavengers and decomposers.

Question 19 `A+ Study Notes` `pp.145–146`

B Measures the ratio of carbon-12 to carbon-14 in organic remains.

Once an organism dies, the carbon-14 decays but the amount of carbon-12 remains constant. The ratio of carbon-12 to carbon-14 can be compared to that in a living organism to determine the age of the fossil. A is incorrect because carbon dating is the only absolute dating technique that directly analyses the organic material in a fossil. Other methods measure materials in the surrounding rock. C is incorrect because carbon-14 has a half-life of about 5700 years, not 30 000 years. D is incorrect because carbon dating is an example of radiometric/absolute dating.

Question 20 `A+ Study Notes` `p.146`

C 1.25 billion years.

50% of the potassium-40 has broken down to argon-40, which is equal to one half-life. **A** is incorrect because 125 million years is less than one half-life so more than 50% of potassium-40 would remain. **B** is incorrect because 310 million years is less than one half-life so more than 50% of potassium-40 would remain. **D** is incorrect because 2.5 billion years is two half-lives so only 25% of potassium-40 would remain.

Solutions to Section B

Question 1 `A+ Study Notes` `pp.145–146`

a It is older than 520 million years and younger than 545 million years.

b Scientists measure the amount of decay of a radioisotope found in the sample to calculate its age.

c Relative dating places fossils in a temporal sequence by noting their positions in layers of rocks, known as strata. As shown in the diagram, trilobite fossils found in lower strata were typically deposited first and are deemed to be older.

d One of:
- The layers may not have been deposited horizontally to begin with.
- The layers may have been overturned by large earth movements.
- Layers of sediment may have been lost to erosion.

e One of:
- The results from the relative dating of the rock stratum need to be compared with other strata from all over the world.
- The samples could be dated by numerical (absolute age) dating.

Question 2 `A+ Study Notes` `pp.145–146`

a

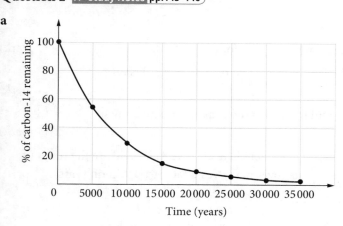

b 13 000 years

Your answer should be greater than 11 000 years and smaller than 14 000 years to receive the mark.

c **i** One of:
- Carbon is a component of all organic chemicals and so is abundant in animal cells.
- Animal remains frequently fall into the time period in which carbon dating is useful.

ii The time a radioactive substance takes to lose half its radioactivity through decay.

d Minerals have replaced the animal tissue, replacing much carbon present. Carbon-14 is not a useful dating tool over long periods because it has a relatively short half-life (approx. 5700 years).

Question 3 A+ Study Notes p.142

a

Order	Choice	Reason
1st	Y	Prokaryotic. Evolved first because they were very simple in structure and function.
2nd	V (all eukaryotic cells have mitochondria)	One of: • Size of organelle is small, like a prokaryote cell • Contains own DNA, which contains circular DNA molecules • Mitochondrion has cell membranes, like a prokaryotic cell • Have ribosomes similar to those of bacteria.
3rd	X (only plant cells have chloroplasts)	One of: • Size of organelle is small, like a prokaryote cell • Contain own DNA, which contains circular DNA molecules • Chloroplast has a cell membrane, like a prokaryotic cell • Have ribosomes similar to those of bacteria.
4th	W	Multicelled eukaryotes were more complex than single-celled eukaryotes and therefore appeared after them in evolution.
5th	Z	Mammals are complex eukaryotes that maintain body temperature, produce milk and prolonged foetal development/care.

b

A	Bacteria or archea
B	Bacteria or archea
C	Animals or fungi
D	Animals or fungi
E	Plants

Question 4 A+ Study Notes pp.147–148

a Allopatric speciation

b Beak shape and size are adaptations to the selection pressure of food, which assist the finches to obtain their food. Finches with beaks better suited to the food source are more likely to survive within the population.

c Large ground finch

d Variation within the population exists. Selection pressure of food is applied. (1 mark)

Small finches with larger beaks are more likely to survive and pass on these alleles to their offspring. (1 mark)

Over time, the allele frequency will change and the size of the beak in the small finch will increase. (1 mark)

e None of the birds is the ancestral species. The birds depicted are all alive today and are the result of evolution from an ancestor (which no longer exists) over many years (at least 3 million years).

Question 5 A+ Study Notes pp.148–149

a Prezygotic isolation prevents organisms of the same species from mating. Examples include flowering season, mating calls and sexual morphology. (1 mark)

Post-zygotic isolation prevents different species from producing viable offspring. Examples include sterile hybrid, gametes being unable to fuse. (1 mark)

b Sympatric speciation

c Flowering occurs at different time periods (prezygotic), preventing the cross-pollination between the palms. (1 mark)

Comparison of *H. forsteriana* and *H. belmoreana* with reference to the data. (1 mark)

d Variation in the population. New selective pressure(s) of pH and altitude. (1 mark)

Better suited phenotypes more likely to survive and reproduce/ no favourable phenotypes and the new group of plants will die out. (1 mark)

Question 6 A+ Study Notes pp.146–147

a Divergent evolution refers to the evolution of more than one distinct population or species from one ancestral population/common ancestor due to different selection pressures.

b Fossil evidence of the existence of an ancestral population displaying the basic structure of the modern populations, and/or biochemical evidence showing a close relationship between the populations. For example, very similar DNA sequences, high affinity in DNA (i.e. DNA hybridisation).

c Gene pool would remain relatively constant. (1 mark)

There would be no selective pressures to favour different alleles and all have equal opportunity to survive and reproduce. (1 mark)

d **i** Sexual selection – behaviour

ii Prezygotic isolation/sympatric speciation

Question 7 A+ Study Notes pp.146–147

a Populations have been isolated because of the lack of water (allopatric speciation). (1 mark)

The populations will have no gene flow and be exposed to different selection pressures. (1 mark)

Variation will exist in the population. Different alleles would be better suited to the selection pressures, meaning the populations would be more likely to survive and reproduce. (1 mark)

Over time, the genetic differences would accumulate, and speciation could have occurred. (1 mark)

b Exposed to different selection pressures and therefore appearing different in their structures and phenotypes. (1 mark)

One of (1 mark):
- Accumulates enough genetic differences to have undergone speciation.
- Have been exposed to similar selection pressures and appear similar in structure still.

c The founder effect refers to when a small, unrepresentative population becomes isolated from the main population. This new population may have a smaller gene pool.

Question 8 A+ Study Notes pp.141–142; 145–146

a Relative age: sedimentary rock strata comparison; the oldest deposits are found at the bottom of rock strata, the newest at the top.

Absolute age: radioisotope dating such as potassium–argon dating.

b **i** There needs to be rapid burial, absence of decay (lack of oxygen and/or low temperature), and the organism must lie undisturbed for a long period.

ii When an animal dies in a terrestrial environment, there is unlikely to be rapid burial. A dead organism exposed to air, in a relatively warm environment, will decay rapidly.

Question 9 A+ Study Notes pp.145–146

a **i** The X should have been placed on the curve at coordinates 2, 0.25.

   ii 12 000 years

b **i** Carbon-dating analysis is not always possible, because the fossil may be older than 50 000 years.

   ii One of:
   - Potassium – argon dating or uranium dating
   - Thermoluminescence/electron spin resonance

c A relative dating technique that could be used is stratigraphy, which is the process of observing the layers of rock (1 mark) and identifying index fossils in the layer or in the rock strata above or below to form a range for the fossil (1 mark).

Question 10 A+ Study Notes pp.145–146

a Continental drift. The countries were once all connected through land bridges, allowing plants and animals to move throughout the land mass.   (1 mark)

   Cynognathus may have only been present in the two regions due to habitats or the amount of distance they were able to travel.   (1 mark)

b Rapid burial with sediment or snow. The fossil must remain undisturbed for a long period.   (1 mark)

   There must be lack of scavengers and oxygen to prevent decay.   (1 mark)

> You must refer to the stem of the question in your answer, if the question mentions the environment that the animal dies in, you must link it with the rapid burial in your answer; for example, if the animal dies in the snow, they are rapidly buried by the snow.

c Indirect/trace fossil

d They are widespread and abundant, and found in a wide range of regions and in large quantities.   (1 mark)

   They are relatively short-lived species and only present in one rock stratum.   (1 mark)

e Transitional fossils include characteristics of two or more groups. An intermediate form – for example, *Archaeopteryx* – has characteristics of bird and reptile, evidence the bird evolved from an ancestor in the reptile family.

Solutions to Test 12: Determining the relatedness of species U4 Topic 2.2.3

Multiple-choice solutions

Question 1 A+ Study Notes p.151

B The degenerate code.

Multiple codons code for the same amino acid. There are 64 possible codons and only 20 amino acids. **A** is incorrect because the sequence of the DNA is different, not the composition of the molecules. **C** is incorrect because conserved genes are genes that have the same function in multiple different species, e.g. cytochrome c. **D** is incorrect because pre-transcriptional modification is not discussed.

Question 2 A+ Study Notes p.150

B Analogous structures.

Analogous structures have similar functions with a different basic structure. **A** is incorrect because they have a similar function but do not have a recent common ancestor; their structures are different. **C** is incorrect because the basic structure of the wings is completely different. **D** is incorrect because there is no recent common ancestor.

Question 3 A+ Study Notes p.152

C B and D

The common ancestor for **B** and **D** is the most recent therefore they are most closely related. **A**, **B** and **D** are incorrect as the common ancestors are not as recent as **B** and **D**.

Question 4 A+ Study Notes p.150

C A vestigial structure.

The pelvis is no longer required for the whale's function and survival. The pelvis will continue to decrease in size until it is eventually lost. **A** is incorrect because a homologous structure is a structure similar to that in another animal but has a different function. **B** is incorrect because analogous structures have similar functions to those of another animal while having a different basic structure. **D** is incorrect because the whale is not transitioning into a new species.

Question 5 A+ Study Notes p.150

B Details of analogous structures found in the two species.

Analogous structures are due to selection pressures being similar in the species' environment; they cannot indicate common ancestry. **A** is incorrect because analysing the DNA sequences would enable identification of the number of mutations between the two species. The fewer mutations, the more recent the divergence. **C** is incorrect because the less difference in amino acids, the more recent the common ancestor – less time for mutation. **D** is incorrect because DNA can be isolated from fossils and analysed.

Question 6 A+ Study Notes p.149

A They are homologous and modified by divergent evolution.

The two limbs have the same basic structure that has evolved over time to suit their functions. **B** is incorrect because the similarity in the structures bones indicates a recent common ancestor. **C** is incorrect because convergent evolution would result in the limbs becoming more similar as a result of similar selection pressures. **D** is incorrect because the bone structures are similar.

Question 7 `A+ Study Notes` `p.150`

A Exploit available niches.

The beak of the bird and mouth of the fish both evolved over time depending on the type of food available for the species in their environment. **B** is incorrect because not all food sources the species eat are large – the finches eat the seeds that grow on the islands and the fish eat the algae off rocks. **C** is incorrect because these structures developed to access food for survival. **D** is incorrect because they are adaptations that have evolved in response to the environment in their habitat.

Question 8 `A+ Study Notes` `p.150`

C Vestigial organs.

Vestigial organs are organs that organisms have retained from their ancestral past but have seemingly lost their function. **A** is incorrect as convergent evolution is when unrelated species evolve similar adaptations in response to a selection pressure. **B** is incorrect as divergent evolution is the basis of speciation. **D** is incorrect as adaptive radiation is when a single species diversifies into many new species.

Question 9 `A+ Study Notes` `pp.143–144`

B Reptiles.

Reptiles were formed as the transition from sea to land began. These were ectotherms and relied on the environment for heat and laid eggs. **A** is incorrect because mammals are endotherms and have therefore developed a complex body system. **C** is incorrect because birds evolved from reptiles as evidenced by the transitional fossil *Archaeopteryx*. **D** is incorrect because although wind-pollinated plants evolved early on, flowering plants required insects for pollination and more complex internal structures and evolved later.

Question 10 `A+ Study Notes` `p.150`

C a shared common ancestor.

No other theory than evolution can explain why these organisms share the same structures. **A** is incorrect as these structures are still used in some form. **B** is incorrect as the features all have the same basic function. **D** is incorrect as they have not separately evolved these features but inherited them from a common ancestor.

Question 11 `A+ Study Notes` `p.146`

D The fish were present in the same time period.

The fossils were both found in the same rock strata. **A** is incorrect because unless the DNA is analysed, it could be assumed that the fish evolved due to similar selection pressures. **B** is incorrect because the fish were determined to be in the same family so had a recent common ancestor. **C** is incorrect because there is no information in the stem of the question to suggest that particular structures were of no use to the fish.

Question 12 `A+ Study Notes` `p.150`

B Analogous structures.

Placed under similar selection pressures, the two frogs have evolved similar defence mechanisms. **A** is incorrect because they evolved independently of each other and do not have a recent common ancestor. **C** is incorrect because the species evolved separately in response to their selection pressures. **D** is incorrect because the two populations are not close enough for the frogs to interact. They are also separate species so offspring would be infertile.

Question 13 A+ Study Notes pp.150–151

A *Drosophila.*

The amino acid sequences are the same between the two species. **B** is incorrect because there are two amino acids different between the two species. **C** is incorrect because there are three amino acids different between the two species. **D** is incorrect because there are four amino acids different between the two species.

Question 14 A+ Study Notes pp.150–151

D Human and yeast.

There are three amino acid differences. **A** is incorrect because there are no amino acids differing between the two. **B** is incorrect because there are two amino acid differences between the two. **C** is incorrect because there is only one amino acid difference between the two.

Question 15 A+ Study Notes pp.150–151

D The numbers in the table account for the minimum number of mutations that have occurred between the species.

Mutations could have occurred before analysis, that have then been reversed by later mutations. **A** is incorrect because a change in the DNA does not always result in a change in the amino acid it codes for. **B** is incorrect because although there are the same amount of amino acid changes, there are a greater number of DNA mutations between the money and the human than the monkey and chimpanzee. **C** is incorrect because some mutations code for the same amino acid, resulting in degenerate code.

Question 16 A+ Study Notes pp.150–151

C More than one DNA triplet can code for each amino acid.

The degenerate code results in multiple codons coding for the same amino acid. **A** is incorrect because a deletion in bases will alter both the amino acids and protein produced. **B** is incorrect because there is inadequate information on the total number of nucleotides and amino acids. A mutation could occur in the nucleotides that then still codes for the same amino acid (degenerate code). **D** is incorrect because a codon will only code for one type of amino acid.

Question 17 A+ Study Notes pp.151–152

A *Pedicinus badii* shares a more recent common ancestor with *Pthirus gorillae* than with *Fahrenholzia pinnata.*

The more recent the line separation, the more recent the common ancestor. **B** is incorrect because *Pediculus humanus* shares a more recent common ancestor with *Pthirus pubis*. **C** is incorrect because each species evolved through divergent evolution. **D** is incorrect because *Pediculus schaeffi* diverged from a common ancestor at the same time as *Pediculus humanus*.

Question 18 A+ Study Notes pp.151–152

B The minimum number of differences in the nucleotide sequence of the *Pediculus humanus* and *Pthirus gorillae* is 5.

It is the minimum because there could have been other mutations that still coded for the same amino acids. **A** is incorrect because more than 11 nucleotides can be altered and code for the same amino acid (silent mutation). **C** is incorrect because *Fahrenhoizia* diverged from the other species first, indicating that it has had the longest time to accumulate mutations. **D** is incorrect because *Pthirus gorillae* and *Pthirus pubis* have the most recent common ancestor.

Question 19 `A+ Study Notes` `pp.150–151`

C There would be more amino acid differences between *Aptenodytes patagonicus* and *Spheniscus demersus* than the two in the *Spheniscus* genus.

The divergence occurred further back and therefore there was more time for mutations to accumulate. **A** is incorrect because *Aptenodytes* diverged from each other 10 million years ago, compared to 5.3 million years ago. **B** is incorrect because there is insufficient evidence. **D** is incorrect because *Aptenodytes* diverged from each other 10 million years ago, compared to *Spheniscus* which diverged 5.3 million years ago.

Question 20 `A+ Study Notes` `pp.146–147`

C Differences in environmental temperatures and habitat.

The change in selection pressures may not have favoured one of the sizes. For example, warmer conditions would favour smaller penguins (reducing the amount of fat and increase surface area to volume). **A** is incorrect because all penguins spend a large amount of time in the ocean for food; this is a similar selection pressure. **B** is incorrect because this could result in speciation but does not link to the size difference. **D** is incorrect because all penguins produce offspring through eggs.

Short-answer solutions

Question 1 `A+ Study Notes` `pp.132–133; 149`

a Natural selection

b Homologous structures

c Each mammal has the same basic bone structure, even though these have been modified for different lifestyles. Bones no longer used are still present.

d Darwin would have explained this using natural selection. Ancestral bats naturally vary with respect to the length of bones in their wings. Those ancestral bats with longer bones could glide more effectively and hence evade predators or gain a better food supply. They survived longer and hence produced more offspring, who inherited the favourable trait.

e Structural genes determine what protein is produced. The amount of the protein produced is under the control of master genes, which determine where and at what stages these genes are transcribed and hence expressed.   (1 mark)

Changes in these master genes can cause changes in the amount of protein product, the timing of the production and the cells in which it's produced. These three factors can affect the phenotype, in this case bone length.   (1 mark)

Question 2 `A+ Study Notes` `p.152`

a

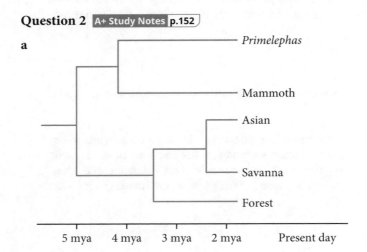

You would receive 1 mark for the time scale and 1 mark for every two correct species. The African savanna and forest elephants can split any time between 5 and 2 million years ago.

b Physical characteristics could be observed through comparative anatomy of fossils.

OR

DNA/amino acid analysis to determine the number of differences between the species. The more differences, the further back the common ancestor.

c Founder effect

d One of the following methods of genetic analysis:
- conserved gene analysis
- DNA or amino acid comparison
- DNA hybridisation.

You cannot interbreed to observe viable offspring because the mammoth is extinct.

Question 3   A+ Study Notes pp.146–147; 150–151

a i The evolution of more than one distinct population or species from the one ancestral population.

   ii Populations of a species of animal or plant may become isolated from other populations, creating a geographical barrier. No gene flow and variation within the population.   (1 mark)

Different selection pressures (there may be variation in environmental conditions, climate or food supply). Different environments cause different features to be selected for, causing increased variation between the populations.   (1 mark)

Over time, these variations and genetic differences accumulate enough to prevent breeding between populations – speciation occurs.   (1 mark)

b DNA or amino acid sequences of a conserved gene between the two species.   (1 mark)

Genetic analysis would be more accurate as the further back the two trees diverged, the more time for mutations to have accumulated.   (1 mark)

Question 4   A+ Study Notes pp.147; 152

a Lack of successful reproduction between populations; either they do not mate at all or viable, fertile offspring are not produced.

b The separation of breeding populations by some physical barrier, usually water or mountains.

c A barrier is introduced. Variation in the isolated populations.   (1 mark)

Conditions in the separated regions may vary so that different characteristics are selected for in the new environments. This leads to increasing genetic differences.   (1 mark)

One of (1 mark):
- Behavioural differences may evolve also, particularly variations in mating calls or display behaviours, making interbreeding impossible or unlikely.
- If placed together the difference in genetics results in infertile offspring – speciation has occurred.

d Either diagram:

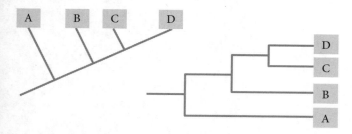

You would receive 1 mark for every two correctly identified species.

Question 5 A+ Study Notes p.151

a i Gorilla

   ii There is only one difference in the amino acid sequences of the beta haemoglobins. Since mutations occur over time, the organism with the least number of mutations would have had a common ancestor most recently.

b The DNA code would have undergone 1 mutation/ would differ in at least one location forming the different amino acid. (1 mark)

   Because of the redundant nature of the DNA code, the DNA codes may vary without producing a change in the amino acid sequence of the haemoglobins – multiple codons for each amino acid. (1 mark)

c $8 \times 3.5 = 28$ million years ago

d Two of:
   • DNA hybridisation
   • DNA sequencing
   • comparative embryology
   • mitochondrial DNA sequencing
   • homologies
   • fossils
   • amino acid sequencing of related proteins.

Question 6 A+ Study Notes pp.147–148; 151–152

a Variation existed within the population. The ancestral species became isolated on the islands – no gene flow. (Geographical barrier because islands too far from each other) (1 mark)

   Different environments – different selection pressures favoured different alleles on each of the 14 islands. (1 mark)

   Over many generations, these genetic differences would have accumulated until speciation occurred and the 14 species would be unable to produce viable offspring with each other. (1 mark)

> This question refers to the steps involved in allopatric speciation/adaptive radiation.

b DNA analysis has become more accurate. DNA can identify the specific number of genetic mutations between a conserved gene. The further back the common ancestor, the more mutations will have occurred. (1 mark)

   Comparing the species anatomy can be misleading due to adaptations to similar selection pressures. (1 mark)

c Similar selection pressures resulting in convergent evolution of the organisms. (1 mark)

   Similar phenotypes were favoured. More likely to survive and pass on these alleles onto their offspring. (1 mark)

d *Geospiza fortis* is the most recent common ancestor.

   Diverged.

Question 7 A+ Study Notes pp.151–152

a

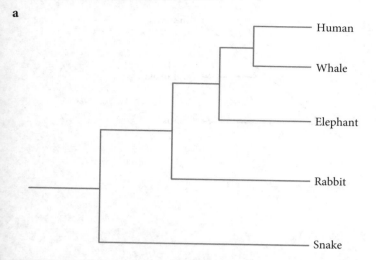

- Human
- Whale
- Elephant
- Rabbit
- Snake

You would receive 2 marks for all correct species, 1 mark for three of the five correct species. 0 marks would be received for two or less correctly identified species.

b The chimpanzee would be more closely related to the human than the whale, due to more recent divergence, because they are both primates and mammals. (1 mark)

Yeast would break off before the common ancestor of all the animals because it is more simplistic than the multicellular organisms. (1 mark)

Question 8 A+ Study Notes pp.146; 151–152

a Divergent evolution

Convergent was a common incorrect answer.

b Comparative anatomy

Students may have mistaken this question for a natural selection, homologous/ analogous question or thought of vestigial organs with the link of fossils.

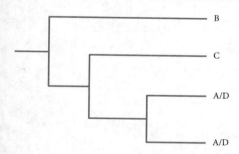

- B
- C
- A/D
- A/D

c i *I. major* and whales share a recent common ancestor and pigs have two differences from *I. major*. (1 mark)

ii Hippopotami have one difference. (1 mark)

Question 9 A+ Study Notes p.142

a i Approximately 45 million years ago

ii Approximately 30 million years ago

b Lesser panda because it has the most recent common ancestor at 40 million years ago.

c Their similarities would be due to convergent evolution. Similar environments and food sources have led to the selection of similar features in the two species.

Question 10 A+ Study Notes pp.141–142; 151–152

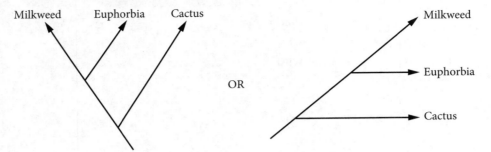

OR

b Natural selection

c One of:
- lack of sedimentation
- lack of hard parts.

For students to gain the mark, it was important that they named a necessary condition for fossilisation or identified that the lack of a specific condition would have prevented fossilisation. Vague statements such as 'incorrect conditions for fossilisation' did not score a mark.

d Cross-pollinate the two plants/interbreed the two plants (1 mark)

If the offspring is viable, then the two cacti can be classified as the same species. If the offspring is infertile, then the two cacti can be classified as separate species. (1 mark)

Solutions to Test 13: Human change over time U4 Topic 2.2.4

Multiple-choice solutions

Question 1 A+ Study Notes pp.159–160

A The Australian Aboriginal population has a closer genetic link to the *Homo sapiens* in Papua New Guinea than the founding African population.

Small populations of *Homo sapiens* migrated away from Africa, reducing the genetic variation. **B** is incorrect because migration of *Homo sapiens* occurred 70 000–80 000 years ago. **C** is incorrect because *Homo sapiens* interbred with the Neanderthals. **D** is incorrect because the land masses were connected millions of years ago.

Question 2 A+ Study Notes pp.153–155

B An increase in brain size and cranium.

The increase in brain size followed the bipedal locomotion and occurred as a result of in movement of the foramen magnum and increased protein in diet. **A** is incorrect because a change in the habitat would have required the species to detect and respond to predators. **C** is incorrect because standing upright exposed less direct sunlight to the body. **D** is incorrect because there would be less energy expenditure to walk on two limbs than all four, enabling them to move further distances.

Question 3 A+ Study Notes pp.152–155

C Bipedal locomotion. Stereoscopic colour vision.

All hominins had some form of bipedal locomotion and all primates have 3D vision. **A** is incorrect because hominin only consists of the humans and bipedal ancestors. **B** is incorrect because the foramen magnum is only centred in species with bipedal locomotion. The foramen magnum of apes is still at the rear of the head. **D** is incorrect because both have a large brain relative to their body size and the S-shaped spine developed due to locomotion.

Question 4 A+ Study Notes p.157

B *H. denisovans*

Homo neanderthals and *H. denisovans* had a common ancestor and both interbred with *H. sapiens*. **A** is incorrect because *H. erectus* evolved before the evolution of *H. sapiens*. **C** is incorrect because *Australopithecus* were present prior to the *Homo* genus. **D** is incorrect because *Australopithecus* were present prior to the *Homo* genus.

Question 5 A+ Study Notes p.160

C Is useful for researchers to trace the maternal line back in time.

The egg is the only gamete that retains its mitochondria and passes it down to all offspring. **A** is incorrect because mitochondrial DNA replicates via binary fission in the gametes. **B** is incorrect because the mitochondria in the sperm are lost when it fuses with the egg. **D** is incorrect because there is no crossing over or recombination of the mitochondrial DNA as in the gamete formation in meiosis.

Question 6 A+ Study Notes p.160

A The Neanderthal sequence branches before the divergence of the various human mitochondrial DNA lineages, but after the split from chimpanzees.

The three human sequences are more similar to each other than to the Neanderthal or chimpanzee sequences. **B** is incorrect because mutations have accumulated within the sequences since the divergence of each species. **C** is incorrect because, although they can have a common ancestor, there is insufficient evidence to suggest humans evolved from Neanderthals (they were also present at the same period). **D** is incorrect because the chimpanzees break away from the other species much earlier, with Neanderthals being closer to the modern human sequences.

Question 7 `A+ Study Notes` `p.155`

C   A short bowl-shaped pelvis relative to body length.

This feature suggests a change to bipedalism to account for the added weight of the trunk. **A** is incorrect because all primates have hands and feet with flat nails. **B** is incorrect because all primates have an opposable toe or thumb. **D** is incorrect because all primates have the ability to rotate their shoulders for movement in the canopy.

Question 8 `A+ Study Notes` `p.153`

C   Increased protein in the diet from hunting and running bipedally is linked with increased brain capacity.

An increase in protein in the diet and the movement of the foramen magnum to be more centralised assisted the development of the cranial capacity. **A** is incorrect because *H. floresiensis* has a smaller brain size. **B** is incorrect because bipedalism occurred before the increase in cranial capacity. **D** is incorrect because you cannot determine the behaviours through the skeletal fossils.

Question 9 `A+ Study Notes` `p.160`

A   Modern humans bred with Neanderthals after modern humans left Africa but before they spread to Asia and Europe.

For the DNA to be present in these populations, the archaic *Homo sapiens* would have had to live in similar areas and have interbred before dispersing out further. **B** is incorrect because the Neanderthals would need to be present in this location for the % of DNA to be present in today's populations. **C** is incorrect because there is no evidence of Neanderthals in Africa (no DNA present). **D** is incorrect because the DNA is present throughout all continents except Africa.

Question 10 `A+ Study Notes` `p.153`

B   The older the fossil, the smaller the braincase that surrounds the cerebral cortex.

The brain size developed through hominin evolution with the exception of *Homo floresiensis*. **A** is incorrect because the older the fossil, the more posteriorly located the foramen magnum. **C** is incorrect because the pelvis shortened and became more bowl shaped throughout hominin development with the added weight of the trunk with bipedalism. **D** is incorrect because with the use of tools and fire, the muscles and teeth decreased in size.

Question 11 `A+ Study Notes` `p.160`

C   A man who has the condition must have inherited it from his mother.

The mitochondria are always passed down the maternal lineage. **A** is incorrect because all individuals have mitochondria down the maternal pathway. **B** is incorrect because the male mitochondria is not passed down. The mitochondria are located in the tail of the sperm, which breaks off with fusion with the egg. **D** is incorrect because if a woman in the family is affected, it will be passed onto all the offspring.

Question 12 `A+ Study Notes` `p.159; 161`

C   Denisovans were more closely related to *H. neanderthalensis* than to *H. sapiens*.

Denisovans and *Homo neanderthalensis* have a more recent common ancestor. **A** is incorrect because there is no fossil evidence of Denisovans existing in Europe and no DNA evidence in modern humans. **B** is incorrect because Homo erectus became extinct before Denisovans evolved. **D** is incorrect because the modern Papuan people are Homo sapiens and they diverged earlier than Neanderthals and Denisovans.

Question 13   A+ Study Notes   p.156

C   Show early evidence that hominins were bipedal.

The *Australopithecus* demonstrated facultative bipedalism with bipedal characteristics but still a small ape-like brain. **A** is incorrect because *Australopithecus* were present prior to *Homo erectus*. **B** is incorrect because primates have existed for 50 million years, and include monkey lemurs, apes and humans. **D** is incorrect because the term should be 'hominin', not 'hominoid'. This was a common mistake made by students.

Question 14   A+ Study Notes   p.160

D   Greater variation in the mitochondrial DNA in Africa than all other continents

The founding population existed for longer (allowing mutations to accumulate) so would have a large population – increased variation compared to the small populations that migrated away. **A** is incorrect because the evolution of *Homo sapiens* independently with continuous gene flow supports the multiregional hypothesis. **B** is incorrect because there is no conclusive evidence to whether the *Homo sapiens* were genetically the same or evolved independently from viewing fossils. **C** is incorrect because *Homo erectus* is a different species from *Homo sapiens*.

Question 15   A+ Study Notes   p.157

C   They had relatively large brow ridges and noses.

Their muscle attachments were more prominent than in modern *Homo sapiens*. **A** is incorrect because the Neanderthal brain was larger in mass than that of *Homo sapiens*. **B** is incorrect because with their larger brain size, the pelvis was required to be more bowl shaped for birth and the weight of their trunk. **D** is incorrect because through the development from *Australopithecus*, arms had decreased in size proportional to their legs.

Question 16   A+ Study Notes   pp.141–142

C   Anaerobic conditions.

The lack of oxygen can slow or prevent the decay. **A** is incorrect because the stem of the question identifies the location as an old riverbed, suggesting rapid burial by sediment. **B** is incorrect because scavengers damage and increase the decomposition of the fossil. **D** is incorrect because for maximum fossilisation to occur, the specimen needs to remain undisturbed for a long period of time.

Question 17   A+ Study Notes   p.156

D   More prominent brow ridges and muscle attachments.

The skull of *Homo erectus* is thicker with more prominent muscle attachments due to the diet. **A** is incorrect because the arm proportions shortened to similar to *Homo sapiens*. The bones were more robust. **B** is incorrect because these were the first erect posture, fully bipedal ancestors with a foramen magnum moving more centrally. **C** is incorrect because *Homo erectus* was fully bipedal and the opposable toe is transitioning to a shorter toe.

Question 18   A+ Study Notes   p.153

B   Milk-producing mammary glands and fur or hair over body surface.

This is a characteristic of all mammals. **A** is incorrect because all primates have opposable toes and thumbs. Humans have lost the opposable toe for bipedalism. **C** is incorrect because primates have long gestation times and reduced number of young to enable their development before birth. **D** is incorrect because all primates have depth perception and colour vision.

Question 19 `A+ Study Notes` `p.159`

B *H. floresiensis* was a species that showed considerable social cooperation and methods for passing on knowledge.

The communal fireplaces and large carcasses indicate group hunting techniques. **A** is incorrect because with a small population, there will have been a limited gene pool. **C** is incorrect because there is no presence of these *Homo* species in Australia. **D** is incorrect because *H. neanderthalensis* had larger brains than *H. sapiens*.

Question 20 `A+ Study Notes` `p.154`

C Was an important anatomical development that assisted toolmaking in hominins.

The precision of the thumb allowed for more complex tools. **A** is incorrect because all primates have opposable thumbs and all but the *Homo* genus have opposable toes. **B** is incorrect because the loss of the opposable toe is important for bipedalism. **D** is incorrect because the change in arm to leg ratio developed following bipedalism.

Short-answer solutions

Question 1 `A+ Study Notes` `p.153`

a Approximately 7 million years ago

b The ancestors of the apes (earlier primates) appeared after the separation of Australia from the other continents.

c **i** About 30 million years

    ii Human and chimpanzee have a more recent common ancestor with 1.6% difference compared with human and rhesus monkey at 5.5%. (1 mark)

    The longer the divergence the longer the time for mutations to accumulate. (1 mark)

d **i** 48

    ii Fusion of two chromosomes into one.

Question 2 `A+ Study Notes` `p.153`

a Skull 3

b Skull 3 has the smallest brain case of the three species shown. (1 mark)

    Skull 3 has the most prominent mouth; whereas the flatter faces of skulls 1 and 2 appeared later in human evolution. (1 mark)

To achieve full marks, your answer needs to include a comparative term, such as 'whereas' or 'compared to'.

c Skull 3 was found in Africa. It belongs to an *Australopithecus*, a genus not found outside Africa.

You should be aware that the *Australopithecus* genus had characteristics of an ape skull – a small cranial capacity, low sloping forehead, prominent brow ridges.

d The changes are very unlikely.

    The suggested changes are based on use and disuse of organs and tissues. This does not cause evolutionary change, as acquired traits cannot be inherited. (1 mark)

    For evolution to occur, a trait must have a selective advantage. None of the suggested changes would afford humans a greater chance of survival and successful reproduction. (1 mark)

Question 3   A+ Study Notes   pp.153–155

a Two of:

- shorter
- more bowl-shaped pelvis for increased weight of the trunk
- S-shaped spine
- more centralised foramen magnum
- angled knees for stability
- loss of opposable toes.

b One of:

- freeing hands for tools and hunting
- holding young more securely
- able to see predators
- walking further distances.

c Cranium and brain size increased following bipedalism.   (1 mark)

Increased protein in the diet from hunting increased the brain size in addition to the movement of the foramen magnum to become more centralised, allowing an increase in behaviours such as tool making.   (1 mark)

d Two of:

- teeth decreased in size as individuals used tools and cooked their food
- decreased cheek bones for muscle attachments
- more parabolic jaw and loss of diastema.

Your answer must link the changes in skull structure to diet of *Homo* genus.

Question 4   A+ Study Notes   pp.145–146; 161

a The two Indigenous populations are more genetically similar because they became separated into the distinct population more recently, suggesting that interacting and interbreeding occurred between the populations.

b Larger, more robust frames would have been better suited to the environment for hunting and harsher environments.   (1 mark)

Selection pressures for modern *Homo sapiens* have been removed with the assistance of medicine, technology, infrastructure.   (1 mark)

c The sites provide understanding of how the species interacted and the correlation with the brain development and increased complexity of tools. Fossils of humans may not have been in optimal conditions for fossilisation.

d Radiometric dating – carbon dating. The known rate of carbon-14 breakdown into nitrogen-14 (half-life 5700 years) to compare the ratio of carbon 14 to carbon-12.   (1 mark)

$5700 \times 7 = 40\,000$ (7 half-lives)   (1 mark)

Your answer needs to demonstrate that you can work out how many half-lives through observing the age of the fossil and working backwards to observe how many times the half-life goes into the age of fossil.

Question 5 `A+ Study Notes` `pp.153–155`

a

Classification	Species included	Key characteristic
Primate	Order that includes all monkeys, lemurs, apes, humans and their ancestors	One of: 3D vision, flat nails with five digits, opposable toe or thumb, flexible skeleton, long gestation time and extensive young brain development
Hominin	Extinct *Homo* species and all immediate ancestors (*Homo, Australopithecus,* para and ardi)	Erect posture/bipedal locomotion, specialised tool use
Hominoid	All modern apes and *Homo* species and their extinct ancestors	Relatively large brains, no tail, shortened spine, broader pelvis and wide chest

b One of:

- shorter more bowl-shaped pelvis
- arm to leg ratio
- evidence of tool use and hunting
- facial structure – less protruding eyebrows, more vertical face and smaller muscle attachment, non-opposable toe.

c Prior to DNA analysis, scientists studied fossil evidence, identifying common structures and behaviours between various populations. (1 mark)

With the development in mtDNA and nuclear DNA analysis isolated from the fossil remains, the genetic similarities and differences (Neanderthal DNA) could inform scientists about the divergence and similarities between populations. (1 mark)

Question 6 `A+ Study Notes` `pp.141–142; 159`

a

Species	Environment	Condition
H. erectus georgicus	Near the banks of the Black Sea	• Low oxygen • Burial in sedimentation • Lack of scavengers/decomposers • Highly mineralised water
H. naledi	Cave in South Africa	• Constant humidity • Constant temperature • No scavengers, decomposers or predators • No wind, water or sunlight

b Two of:

- forehead slopes mor
- face is less flat
- brow ridges prominent
- more prominent zygomatic arches
- teeth types are different sizes
- skull is less rounded
- U-shaped jaw
- less parabolic.

c One of:

- change in pelvis shape to support body in more upright position
- S-shaped spine
- shorter arm-to-leg ratio
- large heel bone.

A significant number of students answered this question incorrectly. Some students were confused when using ratios. To avoid confusion, in this case, students could have stated that the arms are shorter than the legs. The question specifically stated 'other than skull structure'; therefore, the position of the foramen magnum was not a suitable answer.

d Two of:

- art
- artefacts
- stone tools
- positioning of the bodies to indicate burial
- fashioned shell or bone objects.

Question 7 A+ Study Notes p.156

a Similarity: Both have the *Homo* genus originating from Africa.   (1 mark)

Difference: Gene flow continually occurred in the multiregional theory, resulting in the development of multiple groups of *Homo sapiens* at the same time, whereas the out of Africa theory supports all other species becoming extinct and the archaic *Homo sapiens* migrating out.   (1 mark)

b DNA analysis to determine the degree of similarity between the two.   (1 mark)

Location of fossils to determine if there was interaction between the two groups.   (1 mark)

Interbreeding the two and determining whether the offspring are viable is not possible because they are now extinct.

Question 8 A+ Study Notes p.160

a **i** One of:

- Interbreeding occurred between *H. neanderthalensis* and ancestors of (present-day European, East Asian and Australian Aboriginal) *H. sapiens*.
- DNA was passed from one generation to the next.

Many students correctly identified that gene flow occurred. However, others incorrectly identified this as genetic drift.

ii They are not separate species.

Students should be aware that if two populations of organisms are classified as separate species, they are unable to reproduce and form viable/fertile offspring. As this was not the case for the *H. sapiens* and *H. neanderthalensis*, it cannot be assumed they were different species. The key to this question is the last part of the sentence – 'common definition of species'.

b Out of Africa theory. Humans first evolved in Africa where there were no Neanderthal populations because there is no Neanderthal DNA in populations in Africa.   (1 mark)

Populations of humans moved out of Africa and encountered other *Homo* species.   (1 mark)

Supported by the presence of *Homo neanderthalensis* DNA in all modern humans except African populations.   (1 mark)

c Migration occurred sometime less than 80 000 years ago.

Question 9 A+ Study Notes pp.152–155

a

Characteristic	Chimp	*Homo sapiens*
Pelvis structure	Long and narrow	Short and bowl shaped
Foramen magnum	Posteriorly located for quadrupedal locomotion	Centralised to assist in bipedalism
Arm to leg ratio	Longer arms than legs	Shorter arms than legs
Hands/feet	Opposable toe and thumb	Loss of opposable toe for bipedal motion

b *Australopithecus* had characteristics from its ancestors and evolved into *Homo sapiens*. (1 mark)

Evidence of two of (1 mark):
- chimp-like structures – prominent eyebrow ridges and protruding face, small brain
- *Homo sapiens* structures – short bowl-shaped pelvis, bipedal.

Question 10 A+ Study Notes pp.153; 159

a One of:
- the position of the foramen magnum at the base of the skull, rather than in the back
- the arrangement of the teeth (a more parabolic shape in hominins, rather than the rectangular arrangement of the apes)
- reduced gap between front and back teeth, compared to apes
- even-sized teeth with no prominent canines; no zygomatic arch.

b i Later

  ii Two of:
- larger cranial capacity
- smaller teeth
- flatter face.

c i Two of:
- prominent brow ridges
- no chin
- smaller cranial capacity
- more protruding face.

  ii The presence of handcrafted stone or bone tools in the same deposits in which the *Homo floresiensis* remains were discovered. (1 mark)

The presence of bones of butchered animals is evidence of cutting. (1 mark)

d mya

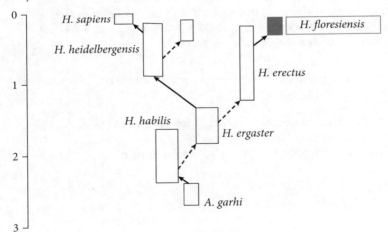

Solutions to Test 14: Experimental design Topic 5.1

Multiple-choice solutions

Question 1 A+ Study Notes p.197

C

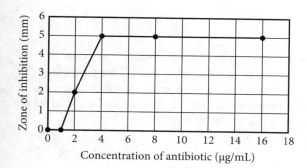

The dependent variable is located on the *y*-axis and the independent variable is located on *x*-axis. **A** is incorrect because the two are not different trials. **B** is incorrect because the data is not categorical. **D** is incorrect because the zone of inhibition is the dependent and should be located on the *y*-axis, and the concentration is the independent variable and should be located on the *x*-axis.

Question 2 A+ Study Notes pp.197–198

A The reaction time of the subject improved as the trials progressed.

B is incorrect because the distance decreased as the trials increased. **C** is incorrect because the decrease in time is not linear. **D** is incorrect because the ruler remains at 1 m for every trial (controlled variable).

Question 3 A+ Study Notes p.195

A High precision and low accuracy.

The collected data is close together (precise) but not close to the true result (not accurate). **B** is incorrect because the collected data would be next to/over the true result if it were high accuracy. **C** is incorrect because low precision would result in the collected data being widespread. **D** is incorrect because the data would be spread around the true result.

Question 4 A+ Study Notes p.197

B Between points B and C.

At point B, the solution is exposed to oxygen and there is a rapid decrease in the amount of alcohol present as it is metabolised to acetic acid. **A** is incorrect because between A and B, the yeast is undergoing anaerobic respiration and producing alcohol. **C** is incorrect because although the alcohol is still being metabolised, it is occurring at a slower rate. **D** is incorrect because there is no data to know what is occurring after point D.

Question 5 A+ Study Notes p.198

D Metabolism of alcohol by *Acetobacter*.

The presence of oxygen allowed for the breakdown of the alcohol. **A** is incorrect because there is no information to support statement that the yeast cells died. In some cases, a high level of alcohol can be poisonous to yeast, but this is not the information provided in the question. **B** is incorrect because there is no data stating glucose is a limiting factor. **C** is incorrect because yeast can undergo aerobic or anerobic respiration.

Question 6 A+ Study Notes p.197

A

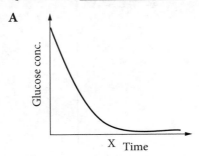

The yeast underwent anerobic respiration with glucose as a substrate, therefore decreasing the glucose levels. **B** is incorrect because there would be a decrease in glucose not an increase as the time progressed. Glucose is a substrate, not a product. **C** is incorrect because there would only be a decrease in the glucose. **D** is incorrect because with a high level of glucose, the enzyme–substrate collisions would increase and slow over time as it became less concentrated so there would be no plateau at the start.

Question 7 A+ Study Notes p.197

B The enzyme sample had been contaminated and denatured.

None of the trials resulted in any reaction, indicating there was an issue with the amylase used in the experiment. This could have been denatured before the experiment due to a change in pH or temperature during storage/isolation. **A** is incorrect because lowering the temperature would decrease the rate of reaction but would not inhibit the reaction completely. **C** is incorrect because a lower amount of starch would react quicker and reduce the colour of the iodine. **D** is incorrect because a random error is one that only occurs in some trials and would result in all other trials working.

Question 8 A+ Study Notes p.195

A Can be reproduced.

When the experiment is repeated, the data will be similar to each other with low to no variation. **B** is incorrect because this would result in low precision. **C** is incorrect because a high standard deviation would mean a wide variation in the data collected, resulting in low precision. **D** is incorrect because high numbers of random errors and a lack of controlled variables would result in a varied set of results and a wide range of data points, resulting in low precision.

Question 9 A+ Study Notes p.193

A The amount of oxygen produced was 30 mL min^{-1}.

The quantity of oxygen produced is given in numerical form hence the data is quantitative. **B** is incorrect as it describes a colour change. **C** is incorrect as it describes the production of a gas. **D** is incorrect as it describes the production of bubbles.

Question 10 A+ Study Notes p.196

D Addition of the wrong amounts of substrates.

Carelessness can be controlled by the experimenter. **A** is incorrect because if a measuring apparatus has a limited scale, approximation is required, leading to variation that cannot be controlled. **B** is incorrect because miscalibration is a systematic error that cannot be controlled directly by the experimenter within the experiment. **C** is incorrect because natural variations can be due to a slight variation in conditions that cannot be controlled, e.g. scales.

Question 11 A+ Study Notes p.195

A Measuring the zone of inhibition around each of the discs.

The zone of inhibition will indicate how effective the antibiotic is depending on the concentration. The larger the zone, the more effective the concentration of antibiotic. **B** is incorrect because the density of the bacteria could differ across the plate and will not indicate which antibiotic concentration is most effective. **C** is incorrect because the mass of the plate will not indicate where the bacterial growth is found. **D** is incorrect because the independent variable is the concentration of antibiotics on one type of bacterium.

Question 12 A+ Study Notes p.195

A Including a control disc with no antibiotic.

A control group will act as a baseline to compare the concentration of antibiotic to and ensure that it is the change in concentration which is inhibiting the bacterial growth and not another factor. **A** is incorrect because this is already included in the experimental set up above. **C** is incorrect because another type of bacteria will respond differently to the antibiotic. **D** is incorrect because heating the plate to 70°C will kill the bacteria.

Question 13 A+ Study Notes p.195

D Allow the pH to be measured accurately.

The probe is more accurate than a pH indicator **A** is incorrect because miscalibration of a probe is a systematic error, not a random error. **B** is incorrect because other conditions would need to be controlled, such as the number of reactants, temperature for a controlled experiment. **C** is incorrect because if the experiment does not have controlled variables, the pH will vary regardless of the calibrated probe.

Question 14 A+ Study Notes p.195

A Repeat the experiment several times to find out if they would obtain the same data.

Repeating the experiment increases reliability if the results are close together. **B** is incorrect because altering the style of the data will not increase the reliability of the results. **C** is incorrect because changing the independent variable will alter the experiment and the data obtained. **D** is incorrect because the experiment will be altered and therefore the experimental results will differ.

Question 15 A+ Study Notes p.194

A Act as a baseline to compare results to.

The control group lacks the independent variable to compare the effect to. **A** is incorrect because controlled variables ensure only one variable is being altered in an experiment. **B** is incorrect because random selection is part of the experimental set-up. **D** is incorrect because the control group sets the baseline to observe the effect of the independent variable.

Question 16 A+ Study Notes p.195

C Accurate but not precise.

The data was close to the true value (0.1 ppm) but spread out (100 ppm). **A** is incorrect because the data was close to the true value and not close to each other. **B** is incorrect because the variance between the data was large at 100 ppm. **D** is incorrect because the data was close to the true value (0.1 ppm).

Question 17 A+ Study Notes p.195

C Group 3's measurements are precise but not accurate.

The results varied by 0.5 (precision) but were not close to the true value. **A** is incorrect because the results varied by 0.5 from each other but from the thermometer by 3°C. **B** is incorrect because are precise (close to each other within 0.5) but not accurate. **D** is incorrect because the results varied between each other and the true value.

Question 18 A+ Study Notes p.194

B Glucose concentration.

> The experiment was investigating the effect of glucose concentration on yeast. **A** is incorrect because temperature was what was being measured. It was dependent on changes of glucose concentration. **C** is incorrect because the experiment was investigating the change of glucose on yeast, therefore the yeast amount be affected by changes to glucose concentration. **D** is incorrect because the ethanol production is what would be produced and was not measured in the experiment.

Question 19 A+ Study Notes p.196

C The amount of yeast added into each chamber.

> A spatula of yeast can vary between each trial and experimental group, leading to random error and increasing the variance between each result. **A** is incorrect because the calibration is a systematic error and would affect the accuracy of the results but not precision. **B** is incorrect because the yeast species would remain the same throughout the experiment. **D** is incorrect because a control group acts as a comparison that lacks the independent variable to compare the experimental groups to.

Question 20 A+ Study Notes pp.197–198

A Optimum temperature at which the enzyme operates is higher in the bacteria than in the river buffalo.

> The peak of the temperature in bacteria is higher (60°C) than in the buffalo (35°C). **B** is incorrect because enzymes only denature above their tolerance range. Lower temperatures slow the collision rate. **C** is incorrect because the buffalo have an optimal pH of 7.4 but the bacterium have an optimal pH of 5.8. **D** is incorrect because 40°C is above the optimal temperature of the buffalo's enzymes. This would result in the enzymes starting to denature.

Short-answer solutions

Question 1 A+ Study Notes pp.194–197

a Control group. (1 mark)

Acts as a baseline for the experimental groups to compare to. Lacks the independent variable. (1 mark)

b The greater the temperature (or the lower the temperature), the more acidic the solution (the higher the rate of fermentation).

> You must refer to the independent variable and dependent variable in your answer.

c Qualitative data – subjective data. (1 mark)

The colour of the paper is determined by the individual observing the paper, so is open to bias/potential random errors. (1 mark)

d One of:
- repeating the experiment to form an average, which would reduce the effect of outliers
- changing the data from qualitative to quantitative by using a pH probe instead of litmus paper to give a specific pH.

Question 2 A+ Study Notes pp.194–195

a Two of:
- CO_2 concentration
- water availability
- time in the sunlight
- temperature.

b **i** Independent variable: different light intensities – low, medium, high (control group no light).

(1 mark)

Dependent variable: Rate of oxygen produced (oxygen probe).

(1 mark)

Controlled variables: Same species of plant, same amount of water, same CO_2 concentration, same temperature, same surface area of leaves.

(1 mark)

Measurement: One of: concentration of oxygen by an oxygen probe over 12 hours/large sample size/ repeated to form an average.

(1 mark)

ii The greater the light intensity (high), the greater the rate of oxygen produced in the 12 hours; therefore, the greater rate of photosynthesis.

Question 3 A+ Study Notes pp.194–195

a The Petri dish acts as a control group that lacks the independent variable (treatment), and acts as a baseline to compare the experimental groups to.

b Two of:
- same mass of cancer and normal cells
- same type of cancer cells
- same temperature
- same time period.

c Type of treatment

The independent variable is the variable that is purposely being altered in an experiment.

d Petri dish 3: Antibodies normally function against extracellular pathogens, through the binding of the antibodies to the cells they are unable to perform any action in isolation. These antibodies may be recognised by the innate immune system but with a small response.

(1 mark)

Petri dish 4: Antibodies bound onto the cells will activate their chemotherapy drug, which will initiate apoptosis in the cancer infected cells reducing the cell growth by 80%.

(1 mark)

The antigens are only present on the cancer cells and therefore no antibodies will bind to the normal cell surface.

e Plate 4.

(1 mark)

The cancer cells are 80% reduced in growth without any impact on the natural cells.

(1 mark)

Plate 6 has the highest cell reduction rate but also reduces the normal cell growth by 60%, so there are high side effects. In your answer, you must refer to the data of both the normal and cancer cells.

f Cancer cells are self-cells and produce proteins that the immune system recognises as self. (1 mark)

The cancer cells are only recognised as foreign if they present faulty proteins that the immune system recognises as foreign.

(1 mark)

Question 4 A+ Study Notes p.194

Independent variable: Temperature or pH (use a range) and how this will be changed, either with water baths or buffer solution.

Dependent variable: O_2 concentration (measured by an O_2 probe).

Controlled variables: Amount of hydrogen peroxide, mass (g) of liver, time of experiment.

Can choose pH or temperature – whichever is not your independent variable.

Experimental groups: temperature or pH control group. Each without the enzyme.

Large sample size (multiple flasks in each at any time).

Repeat the experiment.

> You can never be 100% sure on where the 4 marks will come from when your exam is marked. To ensure you receive full marks, you needed to include all of the following in your response: IV, DV, CV, large sample size, how it will be measured, units, repeat the experiment, experimental groups and control group. Label all clearly.

Question 5 A+ Study Notes pp.194; 200

a Two of:
- removal of agar plates and equipment in biohazard bins
- sterilisation of equipment
- sealed agar plates
- process in an incubator
- disinfectant in areas used for the experiment.

b Zone of inhibition (mm)

c One of:
- The experiment was not repeated.
- Lids were not placed over the agar plates.
- The bacteria were not kept in an incubator to keep them at a constant temperature.
- Concentration of antibiotic on the discs.

d Erythromycin and tetracycline at 5 mm zone of inhibition.

e *Mycobacterium tuberculosis* due to the 12 mm zone of inhibition.

Question 6 A+ Study Notes pp.194; 197

a The rate of reaction increases for enzymes in baths 1, 2 and 3 over the course of 10 minutes as the oxygen concentration increases.   (1 mark)

20°C was the optimal temperature in the experiment, producing 22 AU of oxygen at 10 minutes.   (1 mark)

Rate of reaction at 60°C remained minimal due to the enzymes denaturing because the bath was hotter than the optimal temperature.   (1 mark)

> Always refer to data points when answering a graph referring to data.

b One of:
- Repeat the experiment.
- Have a control group.

c The hypothesis is not supported.   (1 mark)

The enzyme has denatured – irreversible conformational shape change due to the temperature being above optimal.   (1 mark)

> You should notice that in the stem of the question, there was no mention of forgetting to add any solutions or adding any of the test tubes into the baths. Therefore, it must be assumed that the experiment was started. This is further evidenced with a slight increase in the oxygen level before it plateaus due to the denaturation.

d One of:

- The oxygen probe being miscalibrated. This would need to be recalibrated to give a more accurate value.
- The measuring beaker being misprinted so recording volume incorrectly, which would be corrected by replacing the beaker with a more precise piece of equipment.

Repeating an experiment will only reduce the effect of random errors, not systematic errors; therefore, it is not an acceptable answer.

Question 7 A+ Study Notes p.194

a That a deficiency of a particular nutrient will inhibit the growth of the plants.

b If the growth of any of the plants is reduced or halted by the deficiency, then the hypothesis is supported.

c **i** A part of a controlled experiment in which all variables are kept constant. It allows comparison with the experimental groups in which one variable is altered.

  ii Set up a fifth flask with a quantity of the plant and include all four of the nutrients.

Question 8 A+ Study Notes pp.194; 197

a Either aerobic or cellular respiration; or anaerobic and aerobic.

b Dependent variables: CO_2 and O_2 levels. Independent variable temperature of the chamber/environmental temperature

'Heat lamp' and 'temperature' were not awarded any marks.

c To establish a baseline for the experiment.

d For example:

Make sure that the experiment is conducted at the same time of day because the cockroaches are more active at different times; therefore, could increase aerobic respiration.

Feed the cockroach the same food each day to ensure the same initial glucose levels, which could affect cellular respiration.

Ensure that the environment the cockroach is kept in between experiments is the same so that other factors such as external temperature do not affect the cellular respiration rate before the experiment.

1 mark for stating the control measure and 1 mark for explaining.

e **i** When the temperature is constant, the levels of CO_2 increase sharply and slowly rise with decreasing temperature. (1 mark)

  The levels of O_2 decrease sharply and then slowly decrease with decreasing temperature. (1 mark)

  ii Conclusion: Aerobic respiration is occurring, or the rate of cellular respiration is dependent on the temperature of the chamber.

  Evidence: Oxygen is an input and is therefore decreasing. Carbon dioxide is an output and is therefore increasing. Low temperatures lower the rate of reaction.

You should refer to each of the following in your response: the cellular process named in part a; the variables identified in part b; the evidence collected during Matthew's experiments. This question gave students the opportunity to put the information together, drawing on the data given and then providing explanations and conclusions.

Question 9 A+ Study Notes p.194

a   To provide energy or ATP

b   Ethanol/alcohol, carbon dioxide and ATP

c   Experiment set up with same amount of glucose.   (1 mark)

   Experiment set up with same amount of alcohol dehydrogenase.   (1 mark)

   Dependent variable: measure the amount of product produced; for example, carbon dioxide.   (1 mark)

   Independent variable: presence or absence of furfural.   (1 mark)

> Students who were able to understand the information and identify the key components of experimental design scored at the highest level. Many students were confused by the terms and their responses lacked clarity.

d   Furfural has a shape complementary to the active site of alcohol dehydrogenase that prevents the substrate from attaching to the enzyme.

Question 10 A+ Study Notes pp.194–195

a   All plants grew taller with the hormone than with the water treatment.   (1 mark)

   Referring to the data. For example, the rose grew 0.9 cm taller with the hormone.   (1 mark)

> Always refer to the data and units.

b   One of:
   - The student wanted to ensure that any result was due to the effect of the hormone, rather than characteristics of the plant.
   - To determine whether the same hormone stimulates root development in different species of plants.

c   i   Rosemary: the difference between the hormone and water is 2.5 cm.

  ii   Repeat the experiment with each plant several times.

d   i   To ensure that results were due to the presence or absence of the hormone and not to another factor.

  ii   Two of:
   - time in the hormone or water
   - light intensity
   - temperature
   - amount of water use during the experiment.

e   Control group. Acts as a baseline to compare the effect of the hormone to, lacks the independent variable and increases the validity of the experiment.

Practice exam 1

Multiple-choice solutions

Question 1 A+ Study Notes p.21

D Equal to the genome of any genetic material in the prokaryote.

Prokaryotes are more simplistic. **A** is incorrect because prokaryotic organisms can only undergo transcription and translation, which often occur simultaneously. **B** is incorrect because prokaryotic organisms can possess plasmids with additional genetic material that codes for proteins. **C** is incorrect because the proteome is the total number of proteins that can be produced within the cell; not all proteins need to be expressed at once.

Question 2 A+ Study Notes p.6

D Golgi apparatus. Modification and folding of synthesised polypeptides.

The Golgi apparatus folds the polypeptide chain to form a functional protein, then modifies by adding molecules and packages into a vesicle for export out of the cell. **A** is incorrect because the ribosome is involved in translation of mRNA into a polypeptide chain. **B** is incorrect because, although the function and organelle are correct, the chloroplast is not involved in the production of a protein. **C** is incorrect because mitochondria are responsible for production of ATP for the synthesis and exportation of the protein.

Question 3 A+ Study Notes pp.13–14

B RNA processing with exon retention and intron removal.

153 amino acids require 153 codons, composed of 459 bases from DNA. **A** is incorrect because, although multiple codons code for one amino acid, there are only three bases for each amino acid (153 × 3). **C** is incorrect because, although each codon has three bases, not all bases in the DNA code for amino acids in the polypeptide chain. **D** is incorrect because transcriptional factors bind to the promoter upstream at the same point of all genes.

Question 4 A+ Study Notes p.18–20

C Quaternary.

The single tertiary structures of the polypeptide chain do not function in the catalase enzyme until combined. **A** is incorrect because the amino acid sequence (polypeptide chain) is not functional. **B** is incorrect because the alpha helices and beta pleated sheets do not form a functional protein. **D** is incorrect because the tertiary structure is composed of a single polypeptide chain, which in some cases is the final functional stage. The tertiary structures of the catalyse enzyme cannot function in isolation. The functionality of the protein occurs in this protein at the quaternary level when the four tertiary polypeptides are bonded together.

Question 5 A+ Study Notes pp.16–18

C The presence of tryptophan in the bacterial surroundings.

Tryptophan binds to the repressor protein, causing a conformational shape change and enabling it to bind to the operator. **A** is incorrect because the absence of tryptophan will induce the synthesis of the operon. The *trp* operon is repressible and therefore is always active unless tryptophan is present in the environment. **B** is incorrect because the repressor protein binds to the operator region of the operon. The RNA polymerase binds to the promoter region. **D** is incorrect because when the operon is repressed, synthesis of the structural genes is unable to occur.

Question 6   A+ Study Notes   pp.31–33

B   IV ⟶ III ⟶ V ⟶ I ⟶ II

The plasmid and gene of interest are both cut with a restriction enzyme to form sticky ends, which are then reformed to make a recombinant plasmid. The bacteria become transformed through the uptake of these plasmids and are selected to be used for isolation of proteins or a gene of interest. **A** is incorrect because the DNA ligase needs to re-form the bonds between complementary sticky ends to form the recombinant plasmids. **C** is incorrect because the plasmid and DNA of interest must be cut with the same endonuclease (restriction enzyme) (IV), before having the complementary sticky ends reformed with DNA ligase (III). **D** is incorrect because the plasmid needs the gene of interest incorporated before the bacteria take up the plasmid and become transformed.

Question 7   A+ Study Notes   pp.26–27

B   Homologous with foreign DNA sequences from plasmids in bacteriophages.

Sections of viral DNA are retained in the CRISPR system to prevent against future infection. **A** is incorrect because the Cas9 (endonuclease) occurs upstream of the spacers and non-repetitive sequences. **C** is incorrect because RNA polymerase binds to the promoter region of genes. **D** is incorrect because the spacers are sections of proteins from a virus to protect from the same strain of virus in future exposures.

Question 8   A+ Study Notes   pp.29–30

D   Lane 2 contains more genetic material than lane 5.

Lane 2 contains four fragments of DNA compared with lane 5, which has three. The strands in lane 2 are also longer than those in lane 5. **A** is incorrect because no DNA is present in lane 4. This could be due to a random error of DNA not being mixed with the buffer, or the sample not being properly inserted into the wells. **B** is incorrect because the less movement of the DNA, the larger the fragment. **C** is incorrect because lane 3 may not have been exposed to an endonuclease, or it may have no recognition sites. There is no DNA present in lane 4.

Question 9   A+ Study Notes   p.34

C   Knock out of genes that alter the fatty acids in soybeans. Rice modified with daffodil genes to increase beta carotene.

A transgenic organism has had DNA inserted from another species. In this case, the rice has had a gene artificially inserted from a daffodil. **A** and **B** are incorrect because altering the expression of a gene through transcriptional factors is genetic modification but is not transgenic, because the gene is not from another species. **D** is incorrect because both are examples of genetic modification but are not transgenic.

Question 10   A+ Study Notes   pp.64; 68

B CO_2, O_2; O_2, H_2O

CO_2 is used in the light-independent stage of photosynthesis, and O_2 is an output of the light-dependent reaction when water is split. O_2 is an input for the electron transport chain in cellular respiration and combines with the hydrogen ions and electrons to form H_2O. **A** is incorrect because photosynthesis releases O_2 in the light-dependent reaction and uses CO_2 as an input in the light-independent stage. **C** is incorrect because glucose is an output of photosynthesis and CO_2 is an input. H_2O and CO_2 are outputs for cellular respiration. **D** is incorrect because there is no net ATP for photosynthesis and NADPH is an electron carrier involved in photosynthesis.

Question 11   A+ Study Notes   p.68

A   Oxygen concentration.

Oxygen is an input for the electron transport chain of cellular respiration. The lower the oxygen concentration, the fewer electrons can be released into the membrane, slowing the process of ATP synthesis. **B** is incorrect because carbon dioxide is an output of the Krebs cycle. **C** is incorrect because water is an output of the reaction. Water is formed when oxygen combines with the electrons and the hydrogen ions from the membrane. **D** is incorrect because light is involved in photosynthesis, not cellular respiration.

Question 12 A+ Study Notes pp.61–62

D

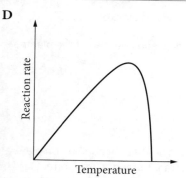

The rate of reaction would increase because the particles have more kinetic energy and collisions occur more frequently up to the optimal temperature. Above the optimal temperature, the enzymes undergo an irreversible conformational shape change and are unable to bind to the substrates. This rapidly decreases the rate of reaction. **A** is incorrect because the curve is equal on both sides of the graph. This indicates the response of an enzyme to pH. **B** is incorrect because as the temperature rises above the optimal for the enzyme, the enzyme will undergo an irreversible conformational shape change, and the rate of reaction will decrease. This indicates a response to unlimited substrate. **C** is incorrect because this graph represents enzyme saturation levels. The rate of reaction would decrease as the temperature rises above the optimal for the enzyme.

Question 13 A+ Study Notes pp.195–196

D Group 2. Group 3.

Group 2 had an outlier in trial 3, which could have been due to a random error, and group 3 had the smallest variance of 0.1. **A** is incorrect because group 3 has the most precise results, whereas group 2 has a random error in trial 3. **B** is incorrect because group 1's results are all close together, and group 4's have a large variation, making it the least precise. **C** is incorrect because group 2 has the largest outlier of 3.

Question 14 A+ Study Notes pp.61–62

C The pH was out of the tolerance range for the enzymes involved in respiration.

The enzymes involved in fermentation had denatured and undergone an irreversible conformational shape change, preventing them from binding to the substrates. **A** is incorrect because fermentation produces a very small amount of energy. For the functioning cells to survive, the yeast would need to continually undergo respiration. **B** is incorrect because the yeast cells do not perform fermentation. The enzymes within the yeast cells facilitate the process. **D** is incorrect because the miscalibration of the probe would have produced inaccurate results in all the experiments.

Question 15 A+ Study Notes p.65

A Photosynthesis decreases and photorespiration increases.

When CO_2 levels are low, oxygen binds to the Rubisco, resulting in a compound that cannot be used for other reactions. **B** is incorrect because increasing CO_2 levels would facilitate photosynthesis, not cellular respiration. **C** is incorrect because stomata closing would slow the gas exchange. **D** is incorrect because plants that form a compound that can utilise at night are CAM plants. C3 plants do not have that ability.

Question 16 A+ Study Notes p.68

C A depletion of citrate in the mitochondrial matrix.

A build-up of the oxaloacetate and the acetyl CoA would occur, resulting in a decrease in the product citrate because the enzyme is unable to synthesise this from the substrates. **A** is incorrect because pyruvate is the output of glycolysis, not the input. **B** is incorrect because for the ATP synthase to function, the electron transport chain requires the NADH and $FADH_2$ from the Krebs cycle. **D** is incorrect because only NADH is formed in glycolysis.

Question 17 `A+ Study Notes` `p.24`

B Produced; Modified by enzymes and packaged into secretory vessels; Released by exocytosis.

Proteins are produced in the endoplasmic reticulum, modified in the Golgi bodies and released at the plasma membrane. **A**, **C** & **D** are all incorrect.

Question 18 `A+ Study Notes` `pp.31–32`

B Plate 6 only.

The exposure to the ampicillin and the presence of some bacterial growth indicate that the bacteria on plate 6 have taken up the plasmid with the antibiotic-resistant gene and have been transformed. **A** is incorrect because all bacteria can grow on a nutrient agar plate. There is no selection pressure for the bacteria. **C** is incorrect because there is no selection pressure for the bacteria on plate 3. **D** is incorrect because plate 6 demonstrates sensitive bacteria, which have taken up the plasmid for resistance.

Question 19 `A+ Study Notes` `pp.31–32`

D None of the plates.

A is incorrect because plate 1 has sensitive bacteria and does not contain any plasmids. All bacteria will grow on nutrient agar; therefore, you cannot identify which antibiotic they are resistant to. **B** is incorrect because plate 6 and potentially plate 3 have resistant bacteria. **C** is incorrect because plate 3 has been exposed to the plasmid with the bacteria, but because there is no antibiotic present, all bacteria will grow. Plate 6 has growth in the antibiotic; therefore, the plasmid has been taken up and is resistant to both Bactrim and Ampicillin.

Question 20 `A+ Study Notes` `pp.31–32`

B Some will be resistant to both Bactrim and Ampicillin.

There is no selection for the transformed bacteria against the antibiotic on the plate. Some bacteria would have no plasmids, and some would have taken up the plasmids from the surroundings. You cannot tell which bacteria on this plate are transformed. **A** is incorrect because the bacteria are exposed to the plasmids so they could have taken up the plasmids from their surroundings. **C** and **D** are incorrect because the plasmids contain resistance to both Bactrim and Ampicillin.

Question 21 `A+ Study Notes` `p.97`

B Interleukin.

Interleukin is released from macrophages to the hypothalamus, increasing the body temperature to slow bacterial growth. **A** is incorrect because interferon is involved in the upregulation of antiviral genes in cells surrounding infected cells. **C** is incorrect because complement proteins become activated and bind to foreign material to lyse, agglutinate or neutralise pathogens. **D** is incorrect because neutrophils migrate to the area of infection, engulf the foreign material and apoptose, leading to the formation of pus.

Question 22 `A+ Study Notes` `p.108`

A Polymerase chain reaction.

Through the binding of specific primers, only the viral strain with a complementary sequence will amplify. **B** is incorrect because viruses cannot grow on agar plates. These pathogens are obligate parasites and require a host cell to replicate. **C** is incorrect because the virus is too small to view under the microscope, so it would not be possible to identify the type of virus present. **D** is incorrect because viruses have the same basic structure of a protein coat. The virus differs through protein receptors on the surface.

Question 23 A+ Study Notes p.108

C H1N2

Both the haemagglutinin and the neuraminidase proteins were amplified due to the primers. **A** is incorrect because only the haemagglutinin 1 could amplify. **B** is incorrect because neither the haemagglutinin nor the neuraminidase DNA were amplified. **D** is incorrect because there is no primer for H2 or N1.

Question 24 A+ Study Notes p.105

C Artificial and passive immunity.

The injection is administered to the individual and is composed of antibodies; therefore, no memory cells will be formed from the injection. **A** and **B** are incorrect because injection provides artificial immunity. No memory cells are formed so the type of immunity is passive. **D** is incorrect because no memory cells are formed so the type of immunity is passive.

Question 25 A+ Study Notes pp.102–103

B The humoral response was initiated due to memory being formed on first exposure.

The exposure to the injection would have initiated clonal selection and expansion into B memory cells and B plasma cells. **A** is incorrect because the individual's immune response contains antibodies and memory cells and therefore does not require the injections he received after the first exposure. **C** is incorrect because memory cells would have been formed during the vaccine injection administered after the first exposure. **D** is incorrect because antiseptics are applied to the surface of your skin and should not be ingested.

Question 26 A+ Study Notes pp.111–112

D Bind to cytotoxic T cell receptors, inhibiting recognition of myelin proteins on MHC I markers.

This would inhibit the destruction of the T cells, which recognise the myelin cells as foreign. Decreasing the damage and inflammation would result in less chemotaxis to the area. **A** is incorrect because binding a radioactive drug to the antibodies would speed up progression of the disease. **B** is incorrect because the naïve B cells that reside in the lymph node are not the cause of the disease and will have no impact on its progression. **C** is incorrect because increasing leukocyte migration increases inflammation and would increase the damage to the myelin cells.

Question 27 A+ Study Notes pp.94–95

B Chemotaxis of innate immune cells to site of infection from cytokine release.

The continued damage to cells in the area would release cytokines, attracting more leukocytes to the area. These cells would become infected by lysing of the intracellular pathogen, and the cycle would continue. **A** is incorrect because B and T cells are lymphocytes, and proliferation in the lymph nodes would not affect the lungs. **C** is incorrect because it is the migration of the macrophages and infection of these that progresses the infection. **D** is incorrect because pus formation is due to the migration of neutrophils that engulf the foreign material before undergoing apoptosis.

Question 28 A+ Study Notes pp.98–99

C Dendritic cells presenting the antigen of the bacteria on the MHC II markers to T helper cells and MHC I markers on naïve T cells.

The naïve T cells require direct activation from dendritic cells in addition to interleukin from T helper cells for clonal expansion. **A** is incorrect because neutrophils are not antigen-presenting cells. **B** is incorrect because antigen-presenting cells present their antigen on the MHC II marker to T helper cells. **D** is incorrect because naïve B cells require the interleukin from activated T helper cells before proliferation can occur and macrophages do not lyse to release the antigens.

Question 29 `A+ Study Notes` `pp.100–101`

C Are composed of lymphoid tissue with clusters of lymphocytes residing within the lymphatic system.

The lymph nodes filter the fluid of any foreign pathogens exposed to the naïve and memory B and T cells to activate the adaptive immune response. **A** is incorrect because the lymphatic system does not have a pump. It uses a series of valves and muscle movement to push the lymph throughout the body. **B** is incorrect because the valves create a unidirectional flow of lymph. **D** is incorrect because the return of the fluid back to the circulatory system occurs at the ducts, not at the nodes throughout the body.

Question 30 `A+ Study Notes` `p.148`

B Sympatric speciation.

There is no geographical barrier mentioned in the stem of the question, but different flowering times and different soil conditions can result in the populations becoming separated and only pollinating the flowers in their area. **A** is incorrect because the two species only recently evolved and became less similar (divergent evolution). **C** is incorrect because there were not many species that developed from a single ancestral species. An example of adaptive radiation is the Galapagos finches. **D** is incorrect because there is no geographical barrier separating the two populations.

Question 31 `A+ Study Notes` `p.145`

B Index fossil.

These fossils are only located in one rock stratum and are found worldwide, enabling other fossils in the same stratum to be given an approximate age. This is known as relative dating. **A** is incorrect because a transitional fossil contains characteristics of two or more groups. **C** and **D** are incorrect because there is no information about the type of remains that are present in the fossil.

Question 32 `A+ Study Notes` `pp.145–146`

C 17 000 years old.

12.5% of the fossil remaining is 3 half-lives. 1 half-life = 50%, 2 half-lives = 25%, 3 half-lives = 12.5% **A** is incorrect because it cannot be assumed that the percentage of carbon remaining is the age of the fossil. **B** is incorrect because the percentage remaining is not a quarter of the 60 000. **D** is incorrect because the age of the fossil is not calculated by working out how many times 12.5 goes into 100 (8 times) and multiplying this by the half-life.

Question 33 `A+ Study Notes` `p.146`

D Rock stratum F is older than strata A and E but younger than stratum B.

All rock layers are above B because this is the oldest stratum in the diagram. **A** is incorrect because the older the rock stratum, the deeper it is located. Stratum A is the youngest rock stratum in the image. **B** is incorrect because, although C is the deepest rock stratum at location 2, there is the same rock stratum in location 1 with a stratum underneath it. **C** is incorrect because stratum D has different rock properties and is younger than stratum B. The same stratum is above stratum C in location 2 and is therefore younger than stratum B.

Question 34 `A+ Study Notes` `p.152`

B Harbour seal and Antarctic fur seal.

These two species had a common ancestor around 25 mya; therefore, there was more time for differences to accumulate. **A** is incorrect because these two species diverged around 17 mya, which would not allow enough time for differences to accumulate. **C** is incorrect because these two species diverged around 5 mya, which would not allow enough time for differences to accumulate. **D** is incorrect because these two species diverged around 15 mya, which would not allow enough time for differences to accumulate.

Question 35 `A+ Study Notes` `p.145`

A Speciation due to different selection pressures favouring different alleles.

Over many generations, different alleles may be favoured due to the selection pressure of predation. As these alleles accumulate, speciation may occur. **B** is incorrect because there is no indication of the size of the populations or the variation within each. **C** is incorrect because sympatric speciation occurs in the same population, not in isolated populations. **D** is incorrect because there is no migration so genetic drift will not occur.

Question 36 `A+ Study Notes` `pp.150–151`

C The yeast and chicken are more closely related than the monkey and the snake.

There are only 27 amino acid differences between the chicken and yeast, compared with 37 between the monkey and the snake. **A** is incorrect because there are more amino acid differences between the monkey and the yeast than between the monkey and snake. **B** is incorrect because the code is degenerate. Multiple codons code for the same amino acid; therefore, a silent mutation may have occurred and there are more differences in the DNA. **D** is incorrect because the human only has 3 differences with the monkey, whereas it has 23 with the whale.

Question 37 `A+ Study Notes` `pp.153–154`

B Skull set 2 represents a *Homo erectus* skull; skull set 1 represents a *Homo sapiens* skull.

Homo erectus had a sloping face, more prominent muscle attachment, smaller brain case and a less parabolic jaw shape. **A** is incorrect because the skull is vertical, the foremen magnum is central, and the muscle attachments are small. **C** is incorrect because the foramen magnum is located at different points and there is a large muscle attachment variation between the two. **D** is incorrect because *Homo sapiens* have reduced eyebrow ridges and smaller muscle attachments.

Question 38 `A+ Study Notes` `pp.152–153`

B Primates have five digits with flat nail beds, whereas mammals have hooves or claws on their digits.

All primates have an opposable toe or finger and flat nails. This facilitates precision grip. **A** is incorrect because all mammals have some form of hair or fur covering. **C** is incorrect because all mammals are endotherms (warm blooded), and maintain a relatively stable internal environment despite the temperature of their surroundings. **D** is incorrect because all mammals give birth to live young and all are able to produce milk in some form.

Question 39 `A+ Study Notes` `p.152`

D *H. oceania* contains more DNA from Neanderthals and Denisovans than European *Homo sapiens*.

The European *Homo sapiens* only came into contact with the Neanderthals, not the Denisovans. **A** is incorrect because the African *Homo sapiens* does not contain any Neanderthal or Denisovan DNA. **B** is incorrect because the Neanderthals interbred with all of the *Homo sapiens*, except for African. **C** is incorrect because *Homo erectus* is present at the same time as *Homo sapiens*, Neanderthals and Denisovans, and so cannot be a common ancestor.

Question 40 `A+ Study Notes` `pp.156–157`

B The *Homo* genus had a larger cranial capacity and reduced prognathism compared to the *Australopithecus* genus.

The brain capacity was significantly larger for the *Homo* genus, enabling more complex skills such as tool use and development of settlements. **A** is incorrect because some in the *Australopithecus* genus, such as *A. aferensis*, were facultatively bipedal. **C** is incorrect because the first of *Homo* species to leave Africa was *Homo erectus*; before this, all these remained in Africa. **D** is incorrect because sexual dimorphism was present in all *Homo* species. *Homo sapiens* have the least. Arm length was reducing in the *Australopithecus* genus.

Short-answer solutions

Question 1 `A+ Study Notes` `pp.18–20`

a

Structure of protein (levels)	Bonds and interactions
1 Primary	Peptide bonds between adjacent amino acids
2 Secondary	Hydrogen bonds between distant amino acids Beta pleated sheets, alpha helices and random coils
3 Tertiary	Hydrogen bonds, ionic bonds and disulfide bonds Interactions between variable groups (R groups)
4 Quaternary	Hydrogen bonds, ionic bonds Two or more polypeptides bonded together

b   Most proteins only contain a single polypeptide chain and function at the tertiary level (1 mark), whereas quaternary structure requires two or more polypeptide chains bonded together before the protein is functional (1 mark).

> Do not refer to the subunits of the quaternary structure as proteins because they are not functional without bonding together. You must also use a comparative statement because the question is asking for the difference between two things.

c   • Ribosomal complex forms around the start codon on mRNA with the first tRNA.
   • tRNA with anticodons complementary to the codons on the mRNA bring a specific amino acid to the ribosome.
   • Peptide bond is formed between amino acids through a condensation polymerisation.
   • Stop codon is reached and a release factor causes the ribosomal complex to dissociate and the polypeptide is formed.

> Students commonly get mixed up between transcription and translation. A question will not always provide a diagram or specifically ask for the steps in either. It is important that you know the inputs and outputs of each process to enable correct interpretation of the question.

Question 2 `A+ Study Notes` `pp.28–29`

a   Denaturing: DNA is heated to 94°C to break the hydrogen bonds between the two strands.   (1 mark)

Annealing: DNA is cooled to 50–55°C, enabling the annealing of primers to either end of the DNA.   (1 mark)

Extension: DNA is heated to 72°C. Taq polymerase catalyses the synthesis of the newly formed complementary strands.   (1 mark)

b   The most likely type of pathogen is a virus, and the best form of treatment is antivirals.

Question 3 `A+ Study Notes` `pp.61; 65–66`

a   Photosynthesis is catalysed by enzymes. These enzymes have an optimal temperature at which the rate of reaction is highest.   (1 mark)

One of (1 mark):
   • Temperatures higher than optimal denature the enzymes causing a conformational shape change, and prevent the reactions from occurring.
   • Low temperatures reduce kinetic energy of the molecules, slowing the collision rate between the substrate and enzymes.

b No, the hypothesis is not accurate. Green wavelengths are reflected, giving the plant its green colour.

(1 mark)

One of (1 mark):

- Red and blue wavelengths are absorbed at a greater rate for photosynthesis.
- White light is absorbed at a greater rate than green for photosynthesis, because it contains all wavelengths.

You would not receive any marks for a 50/50 response. You must clearly state that the hypothesis is not supported.

c Repeat the experiment and/or increase the sample size.

d C3 plants take in carbon dioxide from surroundings. C3 plants fix carbon dioxide directly into a 3-carbon compound that can be used in photosynthesis (1 mark), whereas CAM plants intake the CO_2 during the night and convert it into a 4-carbon molecule to be used during the day (1 mark).

To achieve full marks, your answer needs to include a comparative term, such as 'whereas' or 'compared to'.

e The plant would display C3 properties in a low-stress environment and CAM properties in a high-stress environment.

(1 mark)

A low-stress C3 pathway enables the plant's stomata to be open during the day for photosynthesis. For C3 plants, O_2 is released during the day in this scenario.

(1 mark)

A high-stress CAM pathway enables the stomata to remain closed during the day, meaning less water is lost. O_2 would be released at night.

(1 mark)

Question 4 A+ Study Notes pp.93–99; 106

a Two of:

- intact skin
- respiratory tract
- gastrointestinal tract
- mucosal membranes.

b Natural killer cell (1 mark) recognises a lack of MHC marker on virally infected cells and releases perforins and granzyme to lyse (1 mark).

c Dendritic cells would phagocytose the bacteria and present on their MHC II marker to migrate to the lymph node and present the cholera antigen to the T helper cell.

(1 mark)

T helper cell releases interleukins to naïve B cells, which have encountered the free antigen specific and complementary to the receptor in a process known as clonal selection.

(1 mark)

The naïve B cells proliferate and differentiate (clonal expansion) into B plasma cells, which produce antibodies that bind onto the bacteria (to opsonise, neutralise, agglutinate), and B memory cells, which reside in the lymph nodes, prepared for subsequent exposures.

(1 mark)

You should always try to refer specifically to the question in your response, in this case by using the name of the protein and pathogen. Remember that viruses are both intracellular and extracellular, and therefore both the humoral and cell-mediated responses would play a role in the removal of this pathogen. In most cases (unless otherwise noted in the question), bacteria are extracellular pathogens and activate the humoral response.

d Once activated, complement proteins bind to the foreign pathogen and material to agglutinate (clump) and/or coat to neutralise (prevent toxin release or entry into cells) and/or opsonise (attract other leukocytes).

e The mosquito acts as a vector to transmit the virus from one host to another.

Question 5 A+ Study Notes pp.31–33

a   Insulin is made up of two chains – chain A and chain B which need to be produced separately.   (1 mark)

b   Bacterial DNA does not contain introns, only exons. (1 mark) Leaving introns in the DNA will cause the genes not to be expressed.   (1 mark)

c   Detection of successful gene insertion   (1 mark)

d   The bacteria on the agar plate with the amp antibiotic are the desired bacteria because they can survive with or without the gene of interest, but have taken up the plasmid.   (1 mark)

The surviving bacteria on the tcl plate have taken up the plasmid without the insulin gene incorporated. Therefore, the resistance gene is still functioning, and the bacteria survived.   (1 mark)

Question 6 A+ Study Notes pp.146; 152; 160

a   Clade B

b   mtDNA mutates at a faster rate than nuclear DNA. No recombination through meiosis. Passed down through the maternal line. High copy number (multiple copies within a cell; therefore, easier to access).

c   Variation within the different isolated populations would have enabled those with suitable beaks to obtain food, depending on the different selection pressures.   (1 mark)

Over many generations without gene flow, these populations could have undergone speciation due to the accumulating genetic differences.   (1 mark)

> There are two possible answers to this question: speciation (as in the above answer) and that the birds would become extinct.

c   Radiometric dating/carbon dating.   (1 mark)

The known rate of decay of carbon-14 into nitrogen-14 compared to carbon-12. Half-life of 5700 years.   (1 mark)

> 'Absolute dating' is not specific enough, and is stated in the stem of the question. You must be able to demonstrate a clear understanding of this process to receive the second mark.

Question 7 A+ Study Notes pp.152; 156–159

a   The Denisovans were a small population and only interacted to interbreed with *Homo sapiens* in the areas in which they inhabited.

b

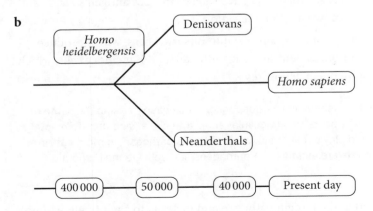

> To receive full marks you should include a timeline, with only the *Homo sapiens* line extending to the present day. Note that the tree could also be drawn with a 'fork' shape.

c   The definition of speciation is the inability to produce viable fertile offspring with other populations of different species. The Denisovans, Neanderthals and *Homo sapiens* were still all able to interbreed and produce viable offspring because they were all same species.

Question 8   A+ Study Notes   pp.96–97; 99

a   Following first exposure, the body initiates a humoral response where IgE antibodies are formed. These bind to the mast cells in the body via the heavy chain, resulting in the individual becoming sensitised.   (1 mark)

Upon subsequent exposure, the allergen binds to the IgE receptors cross-linking on the mast cell. This results in the degranulation of histamine and an inflammatory response.   (1 mark)

> Always refer to the specific antibodies and how they bind when discussing an allergic response.

b   Two of: redness, heat, swelling, pain, loss of function in the area.

c   Two of:
- Cytokines: Enable the movement of leukocytes via chemotaxis (macrophages, neutrophils, dendritic cells) to the area of infection and inflammation.
- Histamine: Results in increased vascular permeability and vasodilation.
- Interleukin: Released from macrophages; travels to the hypothalamus to increase the core body temperature, causing fever to develop.

d   The lymph nodes are where the specific B plasma cells that produce the IgE antibodies are formed, and where the memory cells reside for subsequent exposure.

Question 9   A+ Study Notes   pp.141–142; 145–147

a   Two of:
- Rapid burial by sediment to prevent oxygen decay
- Fossils must remain undisturbed in the sediment rock for a long period of time
- Lack of scavengers

> Refer back to the stem of the question and mention where the fossil is buried in your response.

b   Original evidence, including the use of comparative anatomy and the fossil record would have been compared to new evidence (1 mark) in the form of the use of molecular homology of a conserved gene, either amino acids or DNA (1 mark).

c   Convergent evolution

d   Evolution occurred through variation within the populations.   (1 mark)

Similar selection pressure such as food for both echidna and anteater.   (1 mark)

The animals better suited to the environment have more advantageous traits, making them more likely to survive and pass on the similar alleles onto their offspring.   (1 mark)

Question 10   A+ Study Notes   pp.16–18

a   One of:
- Multiple genes with similar function are transcribed at the same time, resulting in the production of the same amount of proteins.
- An operon increases the number of genes on the chromosome due to less promoters for RNA polymerase attachment.

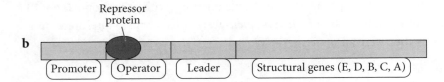

Repressor protein

b

Promoter | Operator | Leader | Structural genes (E, D, B, C, A)

You would receive 1 mark for the promoter, operator, leader and structural genes. The E, D, B, C and A are not required but it is beneficial to include these if you are confident. The second mark is allocated to the repressor being bound onto the operator.

c Repression involves the *trp* repressor protein binding to the operator sequence and blocking the initiation of transcription. (1 mark)

Attenuation does not require the repressor protein and prevents the completion of transcription (1 mark)

Question 11 A+ Study Notes pp.69; 71–72

a Ethanol

b 2

c To ensure no oxygen enters the chamber so that anaerobic respiration can occur. (1 mark)

To maximise the production of ethanol, temperature, pH and access to substrates for the bacterium will influence the rate of reaction enzymes involved. (1 mark)

If you explain the impact that a variable such as temperature had on enzymes you would receive 2 marks.

d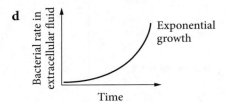

Exponential growth

Bacterial rate in extracellular fluid

Time

To receive full marks, you must draw the line of the graph and label both axes.

Question 12 A+ Study Notes pp.111–112

a The antigen is inserted into a mouse, where it is recognised as foreign and a humoral response is activated. (1 mark)

The B plasma cells are isolated from the blood of the mouse. (1 mark)

b The cell produced by the process is hybridoma. It is formed because B plasma cells are short lived, whereas myeloma cells are immortal. When combined, it produces antibodies continually.

c Artificial and passive immunity. (1 mark)

The anti-venom provides antibodies for the venom but there are no memory cells formed. (1 mark)

d To ensure that the results are reliable.

Practice exam 2

Multiple-choice answers

Question 1

C Collagen is an example of a protein with a secondary structure.

Collagen has no folding of its alpha-helix. **A** is incorrect because a tertiary structure protein involves folding of the alpha-helix or beta-pleated sheet. Collagen shows no folding of its secondary structure. **B** is incorrect because a fibrous protein is a secondary structure protein. Myoglobin clearly shows folding. **D** is incorrect because a quaternary structure protein has more than one polypeptide chain bonded together. The picture only shows one chain.

Question 2

C Amino acids

Amino acids make up proteins. **A** is incorrect because glucose is a saccharide monomer that makes up some complex carbohydrates. **B** is incorrect because carbohydrates are a group of molecules consisting of monomers and many larger complex carbohydrates. **D** is incorrect because glycerol is a type of carbohydrate.

Question 3

A Proteins that can be expressed by the genome of an individual.

B is incorrect because this is the definition of a genome. **C** and **D** are incorrect because they are not relevant to the question.

Question 4

A DNA polymerase.

This is the enzyme found within the human nucleus that is used to join the nucleotides A, T, C and G together to makes new strands of DNA. **B** is incorrect because RNA polymerase is used to synthesise strands of RNA, not DNA. **C** is incorrect because Taq polymerase is found in thermophilic bacteria (heat-loving) and used in the process of PCR. **D** is incorrect because DNA ligase is to used bond nucleotides together during genetic modification processes.

Question 5

C Enzymes provide activation energy for reactions.

The question which asks which of the statements is incorrect. Enzymes do not provide activation energy for a reaction. They lower the activation energy required. **A** is incorrect because enzymes are proteins that act as biological catalysts that speed up the rate of a reaction; therefore, this statement cannot be incorrect. **B** is incorrect because enzymes have a specific active site that only recognises molecules that have a complementary shape; therefore, this statement cannot be incorrect. **D** is incorrect because enzymes are not used in a reaction and they can be used many times; therefore, this statement is correct.

Question 6

D Synthesise glucose in the Calvin cycle. Synthesise glucose in the Calvin cycle plus an additional C4 cycle.

C4 plants have an additional set of reactions that occur during the 'dark' in order to synthesise glucose. **A** is incorrect because both types of plant use both the light and 'dark' reactions to synthesise glucose. **B** is incorrect because both plants need light energy to synthesise glucose. **C** is incorrect because both plants synthesise glucose primarily but are able to convert some of these sugars to proteins by using nitrates from the soil.

Question 7

A Modifying proteins and packaging them for export from the cell.

Organelle X is the Golgi apparatus, which is responsible for modifying and packaging proteins in preparation for secretion from the cell. **B** is incorrect because the rough endoplasmic reticulum is responsible for synthesising proteins ready for movement around the cell. **C** is incorrect because the plasma membrane is responsible for controlling what enters and leaves the cells. **D** is incorrect because mitochondria are responsible for generating energy in the form of ATP.

Question 8

B T cytotoxic cells.

T_C cells are responsible for attacking and destroying infected body cells. If a T_C cell encounters a body cell displaying unusual antigens, it will attack it. Cells of a transplanted organ display different antigens from the host's normal body cells, so T_C cells are responsible for attacking and destroying the organ. **A** is incorrect because T helper cells are responsible for stimulating both B and T_C cells during the adaptive response. While it may be useful to suppress these cells so that they don't stimulate the T_C cells, it would have no direct impact on the success of the organ transplant. **C** is incorrect because B cells are responsible for producing antibodies that eliminate bacteria and other free-floating pathogens. **D** is incorrect because mast cells are responsible for producing histamine to aid in the inflammatory response.

Question 9

D Prions already found in the brain tissue being consumed.

Kuru is caused by a prion, which is taken up by brain cells. It is a deformed protein and cannot be destroyed by disinfectants. **A** is incorrect because a viral infection would infect the brain cells; however, viruses can be destroyed by disinfectant. **B** is incorrect because a protist is a single-celled organism and therefore would not invade the neurons. It could also be destroyed by chemical means. **C** is incorrect because bacteria do not infect individual cells because they are generally too big. They can also be destroyed by chemical means.

Question 10

D All of the above.

All of options A–C are correct so D is the most appropriate answer.

Question 11

D Endonuclease

Endonucleases cut DNA at known sequences. **A** is incorrect as protease is a digestive enzyme that digests protein. **B** is incorrect as polymerase is an enzyme that synthesises DNA. **C** is incorrect as ligase is an enzyme that joins DNA.

Question 12

A Introns must be removed, and exons spliced together.

When a strand of DNA is transcribed, the non-coding sections are called introns and the coding sections are called exons. The introns are removed and the exons are spliced together to form a functional piece of mRNA. **B** is incorrect because translation occurs after transcription has occurred to produce the protein. **C** is incorrect because exons are the coding parts of the gene. Therefore, they must be kept and spliced together. **D** is incorrect because it does not make sense to remove the coding parts of the DNA, as this would produce a strand of mRNA that does not produce a protein.

Question 13

C Restriction enzymes that recognise the same sequences on the plasmid and the foreign DNA.

Restriction enzymes cut DNA and must be used to cut both the plasmid and the foreign gene to ensure that the sticky end bases match for base pairing. **A** is incorrect because restriction enzymes cut the DNA; however, the same enzyme must cut both the plasmid and the foreign DNA that is to be inserted into the plasmid. This is to ensure the sticky end bases match and can bond by complementary base pairing. **B** is incorrect because reverse transcriptase is an enzyme that synthesises a strand of DNA from its corresponding RNA. **D** is incorrect because DNA ligase is the enzyme that is used to bond the bases together on the plasmid and the foreign DNA.

Question 14

C Calcium chloride and heat shocked to 42°C.

The bacteria would be more likely to take up the plasmid if the bacteria are bathed in calcium chloride and heat shocked. **A** is incorrect because the bacteria must be heat shocked in order to take up the plasmid. Culturing at 37°C would not allow them to take up the plasmid. **B** and **D** are incorrect because the bacteria require calcium chloride, not sodium chloride, in order to take up the plasmid.

Question 15

C 1 fragment, 4515 bp.

This is the resulting size of both the plasmid and the inserted gene. There is only one fragment. **A** is incorrect because if there is only one cutting site on the strand, one fragment of DNA would be made. **B** and **D** are incorrect because to determine the size of the fragment, you need to add the size of the plasmid (3678 bp) and the size of the gene (837 bp). The resulting fragment is 4515 bp.

Question 16

B Smaller, because there would be less variants of different genes introduced into the community.

The gene pool is defined as the sum total of all genes and alleles in a population. The Amish community would have a smaller gene pool because there is less gene migration due to in breeding; therefore, fewer alleles are introduced into the population. **A** is incorrect because the gene pool is defined as the sum total of all genes and alleles in a population. The Amish community would have a smaller gene pool because they have migrated away from the original population and over time fewer variants would have been introduced through breeding. Mutations would have occurred, but this would not increase the gene pool significantly in a small population. **C** is incorrect because the gene pool would not stay the same as the original population. The isolated Amish community would not have new variants or alleles introduced into it in the way the original population would through breeding with others from different populations. **D** is incorrect because the gene pool would be smaller than the original population because a fewer new variants of allele have been introduced.

Question 17

A Their alleles become more and more different due to random mutations.

If the birds were separated and unable to physically mate, over time each population would accumulate their own random mutations. They would also be subject to different selection pressures, and potentially become different enough to prevent breeding. This leads to the formation of a new species. **B** is incorrect because although each population would have adapted to their new habitat over time, this is due to the random mutations that would have occurred, and is not something that can be manipulated consciously by the owls. **C** is incorrect because the birds may have chosen to breed to with similar birds in their new habitat, but this would not necessarily have led to the formation of a new species. **D** is incorrect because if the birds were unable to find their mates, this would not have led to the formation of a new species.

Question 18

C 4

The top layer of the site is always the youngest, and site 4 has a segment that none of the other sites have; therefore, it is the youngest. **A** is incorrect because the top layer of site 3 is also found in sites 1 and 4 with newer layers on top. **B** is incorrect because the top layer in site 2 is also found in sites 1, 3 and 4 with new layers on top.

Question 19

B Determining the age of a fossilised organism.

Radiometric dating can determine the absolute age of a fossil by measuring the amount of a radioisotope found in a specimen. **A** is incorrect because radiometric dating is the method of determining half-lives of radioisotopes found in a fossil specimen. This is not useful for determining relationships between fossils. **C** is incorrect because this dating technique does not look at the DNA within a specimen. **D** is incorrect because this dating technique cannot be used to determine the structure of the fossil.

Question 20

D The organisms from the early species lacked bony parts to fossilise.

Single-celled organisms do not have any bony parts and therefore do not form fossils. **A** is incorrect because single-celled organisms cannot be terrestrial animals. **B** is incorrect because single-celled organisms were never fossilised in the first place because they do not have any bony parts. **C** is incorrect because this statement is incorrect. Land formation is not relevant for fossil formation.

Question 21

B Substitution mutation.

C has been replaced by T, so it is a substitution mutation. **A** is incorrect because a base has not been deleted in mutation 1; it has simply been replaced by another base. **C** is incorrect because silent mutation does not affect the resultant amino acid. In this case, the substituted base has caused Lys rather than Glu to be placed into the amino acid sequence. **D** is incorrect because a nonsense mutation causes a stop signal to be placed into the genetic sequence. From the amino acid sequence, it is clear that this has not happened because Lys has replaced Glu.

Question 22

C A frameshift mutation that causes the remaining triplet codes to change and therefore all amino acids in the sequence will also change, creating a different or non-functional protein.

Mutation 2 is an insertion of the base C on the DNA strand. This has caused every triplet code from that point on to shift forwards, changing the codons on the mRNA, and the sequence of the resultant amino acid sequence. Mutation 3 is a deletion of the base A, which shifts all the of triplet codes on the DNA, the codons on the mRNA and the resultant amino acid sequence. **A** is incorrect because mutations 2 and 3 show that individual bases have either been deleted or inserted. This does not mean a whole gene has been deleted. **B** is incorrect because if there was a frameshift mutation due to an insertion or deletion of a base, this would change the sequence of codons, which would change the sequence of amino acids and alter the resultant protein structure and function. **D** is incorrect because RNA polymerase would bind to the DNA and transcribe the DNA regardless of the frameshift mutation changes.

Question 23

D Chicken and iguana are the most closely related species.

From the information in the table, there is only one difference between the chicken and iguana, so this statement is the most correct. **A** is incorrect because it is not clear which organism is the most closely related to the bullfrog, as both the bullfrog and iguana have two differences in amino acid sequence, but the bullfrog and shark also have two differences. **B** is incorrect because is not possible to tell from the table if the gorilla has the most distant common ancestor. **C** is incorrect because it is not possible to tell from the table if the shark and iguana have a recent common ancestor.

Question 24

B A member of the family of organisms that includes humans, gibbons and great apes.

A is incorrect because hominins are a subdivision of the hominoids, which include modern humans, extinct human species and all of our immediate ancestors. **C** is incorrect because a hominoid is classified from many structural and genetic features and not just the size and shape of their skulls. **D** is incorrect because this is not the definition of a hominoid.

Question 25

B Artificial selection.

Artificial selection is the mechanism of specifically choosing a trait and only breeding those individuals with that particular trait, so that all offspring are born with the desired trait. In this case, cows with larger muscle mass would have been selected for breeding, and, over time, offspring would end up with larger muscle mass. **A** is incorrect because natural selection is the process by which nature selects the genetically fittest individuals. In this case, the Belgian blue would not be selected by nature as the genetically fittest organism. **C** is incorrect because this generally refers to the modification of the gene in a lab. This is not the case with the Belgian blue. **D** is incorrect because 'survival of the fittest' refers to natural selection. Belgian blues are not selected naturally, as stated above.

Question 26

A It can be caused by damaged DNA.

Damaged DNA can induce apoptosis because it signals to a cell that the cell needs to be destroyed. **B** is incorrect because apoptosis is highly regulated and controlled by the cell. Uncontrolled cell death is dangerous for the organism. **C** is incorrect because apoptosis is highly regulated and therefore does not allow toxic material into the extracellular environment. **D** is incorrect because phagocytosis does not trigger apoptosis. When a cell has undergone apoptosis, the apoptotic bodies are phagocytosed.

Question 27

D Production of ATP from the breakdown of glucose.

The purpose of aerobic respiration is the production of ATP. **A** is incorrect because conversion of light energy to chemical energy occurs in photosynthesis. **B** is incorrect because aerobic respiration does not produce lactic acid. This is the product of anaerobic respiration. **C** is incorrect because aerobic respiration does not store energy as starch. Energy is stored as ATP.

Question 28

D Well V Sample 2, Well W Sample 4, Well X Sample 1, Well Y Sample 3, Well Z Sample 5.

The sizes of the fragments in the samples match the bands on the gel plate.
All other options are incorrect because the sizes of the fragments in the samples do not match the gel plate.

Question 29

C Sympatric speciation.

The two species are thought to have diverged through sympatric speciation, which is the isolation of the two species due to barriers in their reproductive strategies. It is thought that the flowering times of the two palms differed, so eventually the two species were unable to breed. **A** is incorrect because allopatric speciation is the formation of new species due to geographic isolation. These palm trees have developed in close proximity on the same island so they cannot have developed from allopatric speciation. **B** is incorrect because artificial selection refers to the selection of traits in organisms and the breeding of individuals with those traits to increase the frequency of the traits in the population. This is not the case in this example. **D** is incorrect because genetic modification tends to refer to the artificial changes made to a genome in a lab. This is not the case with the palm trees of Lord Howe Island.

Question 30

D Group A is more accurate and less precise than group B.

Group A is more accurate because its measurements are closer to the true value, but it is less precise than group B because there is greater deviation around the mean. **A** is incorrect because group A is less precise than group B because there is more deviation from the mean, and it is more accurate because it is closer to the true value. **B** is incorrect because groups A and B have the same number of repeated trials and group B has greater deviation so is therefore less reliable. **C** is incorrect because group B is more precise than group A because it has a smaller deviation from the mean.

Question 31

A Cyanide is a non-competitive inhibitor of cytochrome oxidase.

Because cyanide binds to an allosteric site on the enzyme and exposure to it leads to death, it is more likely that it is a non-competitive inhibitor that changes the active site of the enzyme and therefore prevents it from catalysing its reaction. Cytochrome oxidase is present in the electron transport chain, which is the final pathway in the formation of ATP required for cellular activity. If it cannot function, this would lead to death. **B** is incorrect because competitive inhibitors bind to the active site but can be displaced by the substrate. This type of inhibitor does not prevent the reaction because it tends to be reversible. This would not lead to death of the individual. **C** is incorrect because the information in the question states that cyanide affects the enzyme cytochrome oxidase. There is nothing to indicate that it kills the mitochondria, which are organelles, not cells. **D** is incorrect because cyanide does not directly initiate apoptosis.

Question 32

B Transgenic and genetically modified.

Bt cotton has had its genome modified; therefore, it is a genetically modified organism. It has had DNA from another organism inserted into it; therefore, it is also considered transgenic. **A** is incorrect because Bt cotton is transgenic but it is also genetically modified because it has had its genome modified. **C** is incorrect because Bt cotton is genetically modified because it has had its genome modified, but it is also transgenic because it has had foreign DNA inserted into its genome. **D** is incorrect because Bt cotton is both transgenic and genetically modified.

Question 33

D Thylakoid: Light-dependent reaction. Stroma: Light-independent reaction

The light-dependent reaction occurs on the thylakoid membranes, and the light-independent reactions occur in the stroma. **A** is incorrect because the light-independent reaction occurs in the stroma, and the light-dependent reaction occurs on the thylakoid membranes. **B** is incorrect because the Calvin cycle is another name for the light-independent reaction, which occurs in the stroma. Glycolysis is a reaction of respiration. **C** is incorrect because glycolysis and the Krebs cycle are reactions of respiration, not photosynthesis.

Question 34

A Injecting a mouse with an antigen.

The first step when synthesising monoclonal antibodies is to inject a mouse with a specific antigen of a particular pathogen. The mouse will produce B cells and antibodies in response to the antigen. **B** is incorrect because injecting a mouse with antibodies would not stimulate its immune system. **C** is incorrect because removing the spleen would prevent the mouse's immune system working effectively. It is not involved in producing monoclonal antibodies. **D** is incorrect because antibodies are not fused with myeloma cells. Antibodies are proteins, not cells.

Question 35

B It incorporates carbon dioxide into the Calvin cycle to enable the formation of glucose.

Rubisco is found in large amounts in chloroplasts and its role is to incorporate carbon dioxide into the reaction.
All other options are incorrect because Rubisco is not involved in any of these reactions.

Question 36

C A pandemic.

The stem of the question states that the disease crossed borders and affected more than one country. It is therefore considered a pandemic. **A** is incorrect because an epidemic refers to outbreaks that only affect one country; whereas the bubonic plague crossed borders and affected more than one country. **B** is incorrect because this answer is nonsensical. **D** is incorrect because endemic means that a species is only found in one place or country.

Question 37

D Non-self antigen of cellular origin.

A bacterium is a non-self antigen of cellular origin. This means that it contains protein or carbohydrate markers that your body would recognise as non-self. It is also a cellular organism. **A** is incorrect because the bacterium is non-self, not self. A self-antigen would be recognised by your body as belonging to it. This is not the case for any invading pathogen. **B** and **C** are incorrect because a bacterium is a cellular organism.

Question 38

A Intact skin.

Intact skin is a first line of defence against pathogens and prevents the entry of pathogenic material. **B** is incorrect because antibody production is part of the third line or adaptive immune response. **C** is incorrect because phagocytosis is part of the second line or innate immune response. **D** is incorrect because the production of cytokines is part of the second line or innate response.

Question 39

D 690.

If 310 bases are cytosine, then 310 bases must also be guanine, because of complementary base pairing. $310 + 310 = 620$. The remaining bases in the gene $(2000 - 620 = 1380)$ must be allocated to thymine and adenine in equal number due to complementary base pairing. Therefore $\frac{1380}{2} = 690$.

All other options are incorrect. See explanation above.

Question 40

D Natural passive immunity.

Because the baby does not make the antibodies, this type of immunity is passive, and natural because there is not artificial creation of the proteins. **A** is incorrect because breast milk contains antibodies that give the infant some immunity to pathogens. **B** is incorrect because the immunity is not artificial. **C** is incorrect because the immunity is not active.

Short-answer solutions

Question 1

a mRNA – produced during the process of transcription and provides a template of the DNA code, which can then be translated in the ribosome to form a protein. (1 mark)

tRNA – used to bring specific amino acids to the ribosomes to form a polypeptide chain during translation. (1 mark)

rRNA – used in the formation of ribosomes. (1 mark)

b
- In the presence of tryptophan, the tryptophan binds to the *trp* repressor protein and changes its structure. (1 mark)
- This allows the repressor to bind to the operator region which prevents RNA polymerase from binding. (1 mark)
- In the presence of tryptophan and the repressor protein is not bound to the operator the ribosome does not pause at the two trp codons. (1 mark)
- This causes a different hairpin loop to form which terminates transcription. (1 mark)

Question 2

a One of:
- T cytotoxic cell – destroys virally infected cell by recognising foreign antigens presented on MHC class I molecules of somatic (body) cell and releasing chemicals (such as perforin), which causes lysis of cell membrane OR releases death ligand which promotes apoptosis of cell.
- T helper cell – releases chemicals that stimulate other immune cells (T cells and B cells) to proliferate and activate.
- Dendritic cell – engulfs free-floating viruses and presents antigens on MHC class II molecules to stimulate T_h cells.

b Three of:
- It is not a useful indicator as antibodies will begin to decrease in blood after the pathogen has been removed from the body.
- A more useful indicator would be to measure memory (B or T_c/T_h) cells, which are produced during the primary adaptive immune response.
- This is because T_c cells are directly involved in destroying infected somatic (body) cells and so memory T_c cells are a good indicator of long-term immunity to the virus.
- Memory T_h cells are involved in stimulating and proliferating the T_c cells and B cells involved in the response so these would be a good indicator of long-term immunity.
- B memory cells are produced during the primary immune response and would provide long-term immunity. Antibodies are produced to destroy free-floating viruses (Note: the idea that B cells/ antibodies destroy a virally infected cell is incorrect. They are only involved in destroying any viruses circulating in body fluids).

c Two of:
- Wearing face masks to reduce spread of virus through coughing, sneezing and talking
- Social distancing to prevent contamination between individuals
- Quarantine of individuals to prevent those infected from coming into contact with non-infected individuals
- Lockdown groups/communities to prevent transmission
- Sanitise surfaces and hands to prevent contamination.

d • The viral mRNA injected into an individual is taken up by the cells and translated using the cellular biochemical pathway (i.e. cellular ribosomes etc.).

• Viral proteins are produced by the cell AND displayed by the cell OR the proteins are secreted into the blood, plasma or tissue fluid.

• The immune system is stimulated into responding to foreign antigens or viral proteins as it recognises them as non-self.

• $(T_c/T_h/B)$ memory cells are produced, and the individual acquires long-term immunity against the virus.

e Active artificial (1 mark). The individual is stimulated to produce their own memory cells via artificial means (i.e. not through acquiring the disease naturally).

Question 3

a Dependent variable: number of leaf discs floating to the surface.   (1 mark)

Independent variable: light intensity/distance of the lamp from the beaker.   (1 mark)

b If the lamp is closer to the beaker or light intensity is greater, more leaf discs will float to the surface (or vice versa).

c Two of:
• number of leaf discs
• concentration of sodium bicarbonate
• volume of sodium bicarbonate (Note: amount of sodium bicarbonate is not acceptable)
• type/age of plant the leaf discs are taken from
• same light bulb in the lamp
• distance measured from the beaker measured from the same point.

d As light intensity increases, or distance from the lamp decreases, the number of discs floating to the surface increases over time (1 mark). For example, at 30 minutes, the number of discs floating to the surface when the lamp is only 5 cm away is 30, compared with only three discs floating at 25 cm away from the beaker (1 mark).

Use of data to demonstrate description will gain the second mark.

e Photosynthesis is occurring in the leaf discs and an output of this reaction is oxygen gas.   (1 mark)

Oxygen gas collects in the cells of the leaf disc/on the surface of the discs.   (1 mark)

Causes the discs to float to the surface.   (1 mark)

f **i** The results of 30 cm from the beaker do not seem to match the trend as the number of discs floating to the surface is higher than at 25 cm.

ii Personal error was involved, such as one of (1 mark):
• incorrect number of discs placed in beaker
• incorrect concentration of sodium bicarbonate
• contamination of solution/incorrect measuring of discs floating to the surface
• incorrect placement of lamp from the beaker (i.e. it was closer due to faulty measuring)
• random error such as environmental temperature was not controlled and therefore affected results.

Question 4

- There would have been variation in the rat population due to random mutation where some rats would have warfarin resistance. (1 mark)

- The rats with the mutation for warfarin resistance would have a selective advantage. (1 mark)

- They would be more likely to survive and reproduce. (1 mark)

- The offspring produced would have the warfarin-resistance gene and these rats would continue to reproduce and the percentage of the population with the warfarin-resistant allele would increase. (1 mark)

Question 5

a   i   Cytosol of cell

   ii   Matrix of the mitochondria

b

Stage	Coenzyme(s) required	Purpose of coenzyme(s)
Glycolysis	One of: • ATP • ADP • NAD$^+$	One of: ATP –provides energy for the reaction ADP – receives energy from the reaction and is converted to ATP NAD$^+$ – stores and shuttles protons and electrons from the reaction
Krebs cycle	One of: • ATP • ADP • NAD$^+$ • FAD	One of: ATP - provides energy for the reaction ADP - receives energy from the reaction and is converted to ATP NAD$^+$ – stores and shuttles protons and electrons from the reaction FAD – stores and shuttles protons and electrons
Electron transport chain	One of: • ADP • FADH$_2$ • NADH	One of: ADP – receives energy from the reaction and is converted to ATP FADH$_2$ – provides protons and electrons for the reaction NADH – provides protons and electrons for the reaction

1 mark for each correct row = 3 marks in total

c   A coenzyme is a substance that is required for an enzyme to function correctly.

d   Without the B vitamins, the reactions of aerobic respiration would not proceed effectively/there would be no reactions/reactions would be prevented from completing. (1 mark)
This would result in less ATP being formed during the reaction. (1 mark)
Therefore, the person would become fatigued and have less energy. (1 mark)

e   26 or 28ATP

f   Description: As temperature increases, rate of respiration increases. (1 mark)

Explanation: This is because cellular respiration is an enzyme-controlled reaction; as temperature increases, the kinetic energy of the molecules increases and therefore there are more collisions between enzymes and substrates (1 mark) and the rate of reaction increases. (1 mark)

g   Lactose resulted in the lowest rate of respiration. (1 mark)

One of (1 mark):
- The enzymes involved in cellular respiration in the yeast were unable to break down lactose.
- Active site of enzymes did not have a complementary shape to the lactose.
- Yeast do not contain lactase to break down lactose before glycolysis.

h One of:
- Yeast may have stored glycogen/sugar and be utilising this for respiration, as an initial substrate.
- Control may have been contaminated with a sugar.
- Incorrect measurement of rate of respiration.

Stating that the yeast may have undergone anaerobic respiration is incorrect because this reaction still requires a substrate such as sugar for the reaction to occur.

Question 6

a **i** The bacterial plasmids are used as a vector to transfer the herbicide-resistant gene into the plant.

ii Endonucleases are used to cut the plasmid and the gene.

iii Ligases are used to bond the nucleotides together.

b The same endonuclease enzyme must be used on both the plasmid and the target gene because the same enzyme will produce complementary sticky ends. (1 mark)

This allows for the sticky ends to bond together via complementary base pairing, which makes it easier for the gene to be inserted into the plasmid. (1 mark)

c Any advantage of (1 mark):
- Higher yield of crops/canola
- Farmer doesn't have to use as much herbicide to grow the crop, which is better for other organisms within the ecosystem
- Less chance of weeds surrounding the canola developing herbicide resistance

Any disadvantage of (1 mark):
- Concern that GM crops could pollinate with other non-GM crops
- Effect on ecosystems could be detrimental
- Effect on human microbiome/digestive system could be negative
- Ethical concerns about 'playing God'.

Question 7

a Substrate: beta-carotene
Product: vitamin A
Enzyme: beta-carotene oxygenase 1

b When a protein becomes denatured, the high heat energy causes the bonds between amino acids to break. (1 mark) This causes the tertiary structure of the enzyme to unfold (1 mark). The active site changes shape and it can no longer catalyse the reaction (1 mark).

Question 8

a

Skull	Species	Justification
A	*H. neanderthalensis*	More prominent eyebrow ridges than *Homo sapiens* so cannot be this species, less sloping forehead than species B so unlikely to be *heidelbergensis*
B	*H. heidelbergensis*	More sloping forehead and prominent eyebrow ridges than species A so likely to be *heidelbergensis*

1 mark for each row = 2 marks

b Neanderthal DNA has been discovered in the modern *Homo sapiens* genome. (1 mark) This indicates that Neanderthals and *H. sapiens* must have interbred. (1 mark) Therefore, Neanderthals cannot have been an ancestor of *H. sapiens* but must have existed at the same time and been closely related/same species (1 mark).

Question 9

a Single guide RNA that is complementary to the targeted DNA sequence is introduced into the cell. Attached to the single guide RNA is the CRISPR-Cas9 protein. (1 mark)

The single guide RNA detects a DNA sequence in the genome and the Cas protein cuts it. (1 mark)

A newly synthesised strand of modified/repaired DNA may be incorporated into the genome or the faulty gene may be switched off/silenced. (1 mark)

b Universal (1 mark)

All organisms contain the same genetic code, which means genes from one organism can be used in another. (1 mark)

c 90–92°C (1 mark)

The original DNA strands would fail to separate due to failure to break the hydrogen bonds. (1 mark)

This would mean there would be no template strands for the free-floating nucleotides to use and therefore no new fragments/genes would be amplified. (1 mark)